O	In chi square, the observed frequency.
r_s	A correlation coefficient for ranked data; developed by Spearman.
r	Pearson product-moment correlation coefficient.
r^2	The coefficient of determination; the proportion of variance in variable Y that is associated with variance in variable X.
S	The standard deviation of a sample; used to describe the sample.
s	The standard deviation of a sample; used to estimate the standard deviation of a population.
s^2	Variance of a sample; used to estimate the variance of a population.
s_D	Standard deviation of the distribution of differences between correlated scores (direct-difference method).
$s_{\bar{D}}$	Standard error of the difference between correlated means (direct-difference method).
$s_{\bar{X}_1 - \bar{X}_2}$	Standard error of the difference between means.
$s_{\bar{X}}$	Standard error of the mean for a sample.
SS	Sum of squares; the sum of the squared deviations from the mean.
T	A nonparametric statistic used to test the difference between two correlated samples (Wilcoxon matched-pairs signed-ranks test).
t	A ratio of the difference between means to the standard error of the difference between means; a sampling distribution of such ratios.
U	Nonparametric statistic used to determine whether two sets of data based on two independent samples came from the same population (Mann-Whitney U test).
UL	Upper limit of a confidence interval.
X	A score.
x	A deviation score.
$\bar{X}$	The mean of a sample.
$\bar{\bar{X}}$	The mean of a set of means.
X_H	The upper limit of the highest score in a distribution.
X_L	The lower limit of the lowest score in a distribution.
Y'	The Y value predicted from some X value.
$\bar{Y}$	The mean of the Y variable.
z	A standard score; a score expressed in standard-deviation units.
z_X	A z value for a score on variable X.
z_Y	A z value for a score on variable Y.

BASIC STATISTICS

TALES OF DISTRIBUTIONS

3RD EDITION

BASIC STATISTICS

TALES OF DISTRIBUTIONS

3RD EDITION

CHRIS SPATZ
Hendrix College

JAMES O. JOHNSTON
Oglala Lakota College

BROOKS/COLE PUBLISHING COMPANY
Monterey, California

Consulting Editor: Roger E. Kirk, Baylor University

Brooks/Cole Publishing Company
A Division of Wadsworth, Inc.

Printed in the United States of America
10 9 8 7 6 5 4 3 2

Library of Congress Cataloging in Publication Data

Spatz, Chris
  Basic statistics.

  Bibliography: p.
  Includes index.
  1. Statistics. I. Johnston, James O. II. Title.
QA276.12.S66 1984   519.5   83-26317

ISBN 0-534-03209-5

Sponsoring Editor: **C. Deborah Laughton**
Production and Manuscript Editor: **John A. Servideo Editorial/Production**
Permissions Editor: **Carline Haga**
Interior and Cover Design: **Charles Carter Design**
Interior Illustration: **Phil Carver and Friends**
Typesetting: **J. W. Arrowsmith Ltd.**
Cover Printing, Printing, and Binding:
  R. R. Donnelley & Sons Co., Crawfordsville, Indiana

*T*o Carolyn and Thea

*E*ven if our statistical appetite is far from keen, we all of us should like to know enough to understand, or to withstand, the statistics that are constantly being thrown at us in print or conversation—much of it pretty bad statistics. The only cure for bad statistics is apparently more and better statistics. All in all, it certainly appears that the rudiments of sound statistical sense are coming to be an essential of a liberal education.

—R. S. Woodworth

*P*REFACE

*B*asic Statistics: Tales of Distributions (third edition) is designed for a one-term, introductory course in statistics. In writing and revising this book, we have tried to be comprehensible and complete for the student who takes only one course, and comprehensible and preparatory for the student who will take additional courses.

Detailed directions and examples are given for each statistical procedure, but we also have concentrated heavily on conceptualization and on interpretation of statistical results. In addition we included discussion and examples of experimental design.

Our specific goals have been to teach and help students to

1. solve statistical problems.
2. develop insight and intuitive understanding of statistical reasoning.
3. interpret the results of a statistical analysis.
4. choose the proper statistical technique for a particular experimental design.

If we accomplish these goals, students who complete this book should be able to understand the statistical concepts in many journal articles and to design, analyze, and interpret a simple experiment.

We think you will like this book. We believe students will find it a pleasant book to read and study since it is written in a conversational rather than a formal style. Most of the examples and problems deal with situations that students readily understand and relate to.

Besides the writing style, this book has a number of features that are designed to make statistics easier to learn. For example, the problems are an integral part of the text. The answers, complete with all necessary steps or explanations, are at the back of the book. Concepts that will be important in later chapters are identified as "Clues to the Future" and set off in boxes. The "Error Detection" boxes call attention to ways of catching and correcting mistakes. At the beginning of each chapter, a list of objectives serves to orient the reader. This same list also serves as an effective summary of the chapter.

There are three glossaries: those of words and formulas are at the back of the book; the symbol glossary is conveniently located on the inside covers of the book.

Finally, the major divisions of the book are identified with separate "Transition Pages" that summarize the material already covered and describe in what ways the following material will be different.

A superficial examination of this third edition will show that the basic outline of the first two editions remains. There are still thirteen chapters and a summary, which is designated Chapter 14. The chapter titles and material covered are practically the same. A closer examination, however, shows many revisions. The emphasis on calculations from grouped data has been omitted. This affected Chapters 3 and 4. Chapter 3 (Central Values and the Organization of Data) is completely reorganized. It is now shorter, but it has more material on graphs. Chapter 4 (Variability) is also revised with a greater emphasis on s, the estimate of σ, which now comes first in the chapter. A more complete table of the normal curve is introduced in Chapter 6, giving areas on both sides of a z score. Chapter 12 (The Chi Square Distribution) has been rewritten to reflect recent recommendations on the use of Yates' Correction. In addition to these major changes many, many sections have been updated, and refined.

It is a pleasure to acknowledge that this book has benefited from the work of students and colleagues. Our students used this material and pointed out its strengths and weaknesses. Joyce McKenzie did the typing for the third edition and the workbook. Roger E. Kirk of Baylor University made many important suggestions and saved us from making several errors. C. Deborah Laughton and Todd Lueders of Brooks/Cole and their reviewers contributed a number of helpful criticisms and suggestions. We would like to thank the reviewers of this edition: Mary Attig of Pennsylvania State University, Kenneth J. Berry of Colorado State University, W. J. Bridgeman of Miami University, Elizabeth Cole of Gonzaga University, Roger E. Kirk of Baylor University, Kathleen Kowal of the University of North Carolina, Maureen Powers of Vanderbilt University, Michael J. Reich of Lewis University, David H. Robinson of Henderson State University, Kathryn Schwartz of Scottsdale Community College, and Paul E. Thetford of Texas Women's University. We are grateful to the Literary Executor of the late Sir Ronald A. Fisher, F. R. S., and to Dr. Frank Yates, F. R. S., and to Longman Group Ltd., London, for permission to reprint Tables III and IV from their book *Statistical Tables for Biological, Agricultural and Medical Research* (6th edition, 1974).

Our final and most important acknowledgment, however, goes to our families, who helped us in many ways while we were writing and revising the book.

Chris Spatz
James O. Johnston

Contents

1—INTRODUCTION—1

What do you mean, "statistics"?   2
What's in it for me?   3
Some terminology   5
Problems and answers   8
Scales of measurement   9
Statistics and experimental design   11
Statistics: then and now   13
How to use this book   14

2—REVIEW AND MORE INTRODUCTION—18

Part 1: Pre-test   20
Part 2: Review of fundamentals   21
Part 3: Rules, symbols, and shortcuts   30

3—CENTRAL VALUES AND THE ORGANIZATION OF DATA—35

Arranging scores in descending order   36
Simple frequency distributions   37
Grouped frequency distributions   39
Graphic presentation of data   43
Describing distributions   47
Measures of central value   51
Finding central values of simple frequency distributions   54
Finding central values of grouped frequency distributions   57
The mean, median, and mode compared   59
Determining skewness from the mean and median   61
The mean of a set of means $(\overline{\overline{X}})$ 61

4—VARIABILITY—65

The range   67
The standard deviation   68
The standard deviation as a descriptive index of variability   69
s as an estimate of σ 75
The variance   78
z scores   79
Transition page   83

5—CORRELATION AND REGRESSION—84

The concept of correlation   85
The correlation coefficient   91
The meaning of r 96
Strong relationships but low correlations   100
Other kinds of correlation coefficients   102
Correlation and regression   103
The regression equation   104
Transition page   110

6—THEORETICAL DISTRIBUTIONS INCLUDING THE NORMAL DISTRIBUTION—111

A rectangular distribution   112
A binomial distribution   114
Comparison of theoretical and empirical distributions   115
The normal distribution   116
Comparison of theoretical and empirical answers   127

7—SAMPLES AND SAMPLING DISTRIBUTIONS—129

Representative and nonrepresentative samples   132
Sampling distributions   137
The sampling distribution of the mean   137
Calculating a sampling distribution when σ is not available   142
Confidence intervals   146
Other sampling distributions   150
A taste of reality   151

8—DIFFERENCES BETWEEN MEANS—153

A short lesson on how to design an experiment   154
The logic of inferential statistics   156
Sampling distribution of a difference between means   159
A problem and its accepted solution   161
How to construct a sampling distribution of differences between means   162
A shortcut—tails of distributions   164
An analysis of potential mistakes   168
One-tailed and two-tailed tests   171
Significant results and important results   173
How to reject the null hypothesis—the topic of power   174
Transition page   177

9—THE t DISTRIBUTION AND THE t TEST—178

The t distribution   180
Degrees of freedom   184
Independent-samples and correlated-samples designs   185
Using the t distribution for independent samples   189
Using the t distribution for correlated samples   192
Using the t distribution to establish a confidence interval about a
  mean difference   197
Assumptions for using the t distribution   200
Using the t distribution to test the significance of a correlation
  coefficient   201
Transition page   204

10—ANALYSIS OF VARIANCE: ONE-WAY CLASSIFICATION—205

Rationale of ANOVA   207
More new terms   211
Sums of squares   212
Mean squares and degrees of freedom   216
Calculation and interpretation of F values using the F distribution   217
A learning experiment   219
Comparisons among means   220
Assumptions of the analysis of variance   228

11—ANALYSIS OF VARIANCE: FACTORIAL DESIGN—231

Factorial design and interaction   232
Main effects and interaction   234
Restrictions and limitations   238
A simple example of a factorial design   240
Analysis of a 3×3 design   248
Comparing levels within a factor   254
Transition page   259

12—THE CHI SQUARE DISTRIBUTION—260

The chi square distribution   262
Chi square as a test for goodness of fit   262
Chi square as a test of independence   265
Shortcut for any 2×2 table   267
Chi square with more than one degree of freedom   269
Small expected frequencies   272
When you may use chi square   276

13—NONPARAMETRIC STATISTICS—279

The rationale of nonparametric tests   280
Comparison of nonparametric and parametric tests   281
The Mann-Whitney U test   282
The Wilcoxon matched-pairs signed-ranks test   288
The Wilcoxon and Wilcox multiple-comparisons test   293
Correlation of ranked data   296
Our final word   300

14—TALES OF DISTRIBUTIONS—PAST, PRESENT, AND FUTURE—301

Tales of distributions—past   302
Tales of distributions—future   302
Tales of distributions—present   303

APPENDIXES—309

References   310
Tables   314
Glossary of Words   333
Glossary of Formulas   338
Answers to Problems   344
Index   401

BASIC STATISTICS

TALES OF DISTRIBUTIONS

3RD EDITION

1 INTRODUCTION

Objectives for Chapter 1: After studying the text and working the problems in this chapter, you should be able to:

1. distinguish between descriptive and inferential statistics,
2. define the words *population, sample, parameter, statistics,* and *variable* as these terms are used in statistics,
3. identify the lower and upper limits of a number obtained by measuring a quantitative variable,
4. identify four scales of measurement and give examples of each,
5. distinguish between statistics and experimental design, and
6. define some experimental-design terms—*independent variable, dependent variable,* and *extraneous variable*—and identify these variables when you are given the description of an experiment.

This is a book about statistics written by people who are not statisticians for people who do not plan to be statisticians. The authors are psychologists who, for several years, have used the branch of mathematics called statistics for the practical purpose of analyzing data generated by experiments. During this time, we have taught courses on how to use statistical tools. Our students, we have found, can learn in one term to identify the questions an experiment is asking, determine the statistical procedures that will answer those questions, carry out the procedures, and then interpret the answers in plain English. The best way for *you* to learn all this is to take an active approach to learning. Specialists in the psychology of learning have known for many years that active learning is superior to passive learning, and this book is based on that active-learning approach—so get ready from the start to *do* some statistics and to actively *think* about some statistics.

WHAT DO YOU MEAN, "STATISTICS?"

Before we can say much more about statistics, we need to discuss the term itself. The word *statistics* has two meanings. A **descriptive statistic**[1] is an index number that summarizes or describes a set of data. You are already familiar with at least one of these index numbers, the arithmetic average, called the **mean**. You have probably known how to compute a mean since elementary school—just add up the numbers and divide by the number of numbers. As you already know, the mean gives you one number that is somewhat typical of all the numbers. In this sense, the mean is a descriptive statistic. You will learn about other descriptive statistics starting with Chapter 3.

The basic idea is simple: a descriptive statistic summarizes a characteristic of a set of data with one number. Historically, "statistics" referred to "state numbers"—numbers that deal with population, taxes, and land. State numbers are descriptive statistics.

More recently (within the last 100 years), another procedure has been developed that is referred to as **inferential statistics**. This procedure is a method of drawing inferences about a population when you have only a sample from that population.

Very often, you will want to do more than just describe the data you have collected. You will want to generalize your findings to the larger population of organisms and situations that are similar to the ones you measured. Inferential statistics are used for this purpose. Thus, you use inferential statistics to decide whether or not the relationships observed in the samples should be generalized to the larger populations. The conclusions of almost every experiment are based on inferential statistics. Examples of such inferences are common. On the basis of a small percentage of election returns, it is possible to tell, with little chance for error, who will win. On the basis of experiments with animals, the U.S. Food and Drug Administration determines whether substances are safe for human consumption. On the basis of market trends over the past three years, businesses

---

[1]Words and phrases printed in boldface type are defined in the glossary of words, which begins on page 333.

make plans for the coming year. In all these examples, however, the conclusion must be preceded by a determination of whether or not the observed relationships are the result of chance fluctuations.

Perhaps an example will best explain how inferential statistics are used. Today, there is a lot of evidence that uncompleted tasks are remembered better than completed tasks. This is known as the *Zeigarnik Effect*. Let's go back to a time when this phenomenon was about to be discovered by Bliuma Zeigarnik.[2] Zeigarnik asked the participants in her experiment to do about 20 tasks such as to construct a box from cardboard, work a puzzle, and make a clay figure. For each participant, half the tasks were interrupted before completion. Later, when participants were asked to recall the tasks they had worked on, they listed more of the interrupted tasks (about seven) than the completed tasks (about four). However, as you well know, scores on tests are subject to a certain amount of chance or luck. So, the question arises: was the difference in recall due to the interruption or was it due to chance? One way to answer the question is to do the experiment again and again and observe whether the first results continue to occur. However, you can answer the question after only *one* experiment by using inferential statistics. The way inferential statistics works is that it tells you the probability that the "chance is responsible" answer is correct. If the probability is very low, the "chance is responsible" explanation can be discarded. An answer based on inferential statistics would be a statement like, "The difference between the means is due to interruption and not due to chance because chance would produce results like these only one time in a hundred." Probability, then, is at the heart of statistics.

Statistical methods (both descriptive and inferential) are ways to accomplish certain goals. In this book, we will make every attempt to specify as precisely as possible the goals for which each statistical method is suited. In general, statistical methods are used in experiments to discover relationships among the factors studied.

Clue to the Future

The first major part of this book is devoted to descriptive statistics (Chapters 3 to 5) and the second major part to inferential statistics (Chapters 6 to 13). The two parts are not separate, though, because to understand inferential statistics you must first understand descriptive statistics.

WHAT'S IN IT FOR ME?

That is a reasonable question. We have several answers; pick the ones that apply to you and add any others that are true in your own case. One thing that is in it for you is that, when you have successfully completed this course, you

[2]A summary of this study can be found in Ellis (1938). The complete reference, and all others in the text, can be found in the list of references starting on page 310.

will understand how statistical analyses are used to *make decisions.* You will understand both the elegant beauty and the ugly warts of the process. H. G. Wells, a novelist who wrote about the future, said "Statistical thinking will one day be as necessary for efficient citizenship as the ability to read and write." So chalk this course up under the heading "general education about decision making, to be used throughout life."

Another result of the successful completion of this course is that you will know how to use a tool that is basic to many disciplines, such as psychology, education, medicine, biology, sociology, forestry, and economics. Each of these disciplines has its own interests, but when it comes to deciding whether there is any difference between two methods of therapy, two methods of teaching, two rates of administering hormones, two kinds of advertising, and so forth, they all rely on statistics. The ideas of statistics are multidisciplinary; because they are used by all, this course opens several doors for you at once.

More and more disciplines are finding that statistical methods are useful in expanding knowledge. In American History, for example, the authorship of 12 of the Federalist Papers was disputed for a number of years. (The Federalist Papers, a group of 85 short essays, were written under the pseudonym, "Publius," and published in New York City newspapers in 1787 and 1788. Written by James Madison, Alexander Hamilton, and John Jay, they were designed to persuade the people of the state of New York to ratify the Constitution of the United States.) On each of the 12 disputed papers, a quantitative value analysis was performed in which the importance of values such as national security, a comfortable life, justice, and equality was assessed. This value analysis was compared with value analyses of papers known to have been written by Madison and Hamilton. The authors of the study concluded that James Madison wrote all 12 papers (Rokeach et al., 1970). This conclusion is in accordance with the conclusion reached by historians who used more traditional methods. A knowledge of statistics, then, may be essential if you are to be an intelligent consumer of your own professional literature.

If you are an independent sort—a person who distrusts authority and prefers to find out things for yourself, statistics offers you a way to analyze the data that come from your own observations and experiments. We happen to think that experiments are lots of fun, and we recommend them to you. The best ones are those you think up yourself. Once you have designed an experiment and gathered the data, a statistical analysis is often the first step in the process that ends with you telling exactly what your experiment has shown.

We hope that this course will encourage you or even teach you to ask questions. Consider the questions, one good and one better, asked of a restaurateur in Paris who served delicious rabbit pie. He served rabbit pie even when other restaurateurs could not obtain rabbits. A suspicious customer asked this restaurateur if he were stretching the rabbit supply by adding horse meat.

"Yes, a little," he replied.

"How much horse meat?"

"Oh, it's about 50–50," the restaurateur replied.

The suspicious customer, satisfied, began eating his pie. Another customer, one who had learned to be more thorough in asking questions about statistics, said:

"50–50? What does *that* mean?"

"One rabbit and one horse," was the reply.

Perhaps an overall answer to the question "What's in it for me?" is that a knowledge of statistics enables you to communicate with others who use statistics (a diverse group, as we have shown). Scientists and others find that communication with their colleagues is greatly improved by the use of statistics (which means that they are more apt to believe you if your results are presented statistically). In short, we think there will be lots of lasting value for you in this course.

SOME TERMINOLOGY

Like most courses, statistics introduces you to a number of new words. In statistics, most of the terms keep being used, time and time again. Your best move, when introduced to a new term, is to stop, read the definition carefully, and memorize it. The more you use the terms, the more comfortable you will be with them.

Populations, Samples, and Subsamples

A **population** consists of all members of some specified group.[3] A **sample** is a subset of a population. A **subsample** is a subset of a sample. A population is arbitrarily defined by the investigator and includes all relevant cases. A population might be:

- all fifth-grade pupils in the United States in 1984,
- all female freshmen presently enrolled at Purdue University,
- the children of families in which one parent has been institutionalized as schizophrenic,
- all Scottish terriers, or
- all students, teachers and children.

Investigators are always interested in some population. However, you can see from these examples that populations are often so large that not all the members can be measured. The investigator must often resort to measuring a sample that is small enough to be manageable which is still representative of the population. A sample taken from the first population listed above might include only 200 fifth graders. From the last population on the list, Zeigarnik used a sample of 164. Ways of assuring that samples are representative of their populations will be discussed at various points in this book.

Going a step further, samples are often divided into subsamples and relationships among the subsamples are determined. The sample of fifth graders, for example, might be subdivided on the basis of sex, socioeconomic status, race, or arithmetic textbook used. The investigator would then look for similarities or differences among the subsamples.

---

[3]Actually, in statistics, a population consists of the *measurements* on the members and not the members themselves. Thus, for a group of fifth graders, there might be a population of achievement-test scores, a population of IQ scores, and a population of scores for the 50-meter dash.

Resorting to the use of samples and subsamples introduces uncertainty into the conclusions because different samples from the same population nearly always differ from one another in some respects. Inferential statistics are used to determine whether or not such differences should be attributed to chance. You will meet this problem head-on in Chapter 7.

Authors of research articles often fail to identify the population to which their results can be generalized. You have to figure out the population for yourself, but the task isn't difficult. For example, the results of a study on 14 overweight adults might be generalized to all overweight adults. The conclusions about 25 albino rats might be generalized to all albino rats, or all rats, or all rodents, or perhaps to all mammals.

Parameters and Statistics

A **parameter** is some numerical characteristic of a population. An example is the mean IQ score of *all* fifth-grade pupils in the United States. A **statistic** is some numerical characteristic of a sample or subsample. Examples of statistics could be the mean IQ score of 200 fifth graders (a sample) and the mean IQ score of 100 male fifth graders (a subsample). A parameter is constant; it does not change unless the population itself changes. There is only one number that is the mean of the population; however, it often cannot be computed because the population is too large to be measured. Statistics are used as estimates of parameters, although, as we suggested before, a statistic tends to differ from one sample to another. If you have five samples from the same population, you will probably have five different sample means. Remember that parameters are constant; statistics are variable.

Variables

A **variable** is something that exists in more than one amount or in more than one form. Memory is a variable. The Wechsler Memory Scale is used to measure people's memory ability, and variation is found among the memory scores of any group of people. The essence of *measurement* is the assignment of numbers on the basis of variation.

Height and sex are both variables. The variable height is measured using a standard scale of feet and inches. The notation 5'7" is a numerical way to identify a group of persons who are similar in height. Of course, there are many other height groups, each with an identifying number. Sex is variable. With sex, there are (usually) only two groups of people. Again, we may identify each group by assigning a number to it, usually 0 for females and 1 for males. All zeros are similar in sex. We will often refer to numbers like 5'7" and 0 as *scores* or *test scores*. A score is simply the result of a measurement. So we may refer to an "age score" or "height score" or "anxiety score." Score just means number.

Quantitative Variables

Most of the variables you will work with in this text can be classified as **quantitative variables**. When a quantitative variable is measured, the scores tell you something about the amount or degree of the variable. At the very least, a larger score indicates more of the variable than a smaller score does.

Take a closer look at an example of a quantitative variable, IQ. IQ scores come only in whole numbers like 94 and 109. However, it is reasonable to think that of two persons who scored 103, one just barely made it and the other almost scored 104. Picture the variable IQ as the straight line in Figure 1.1.

Figure 1.1 shows that a score of 103 is used for a range of possible scores: the range from 102.5 to 103.5. The numbers 102.5 and 103.5 are called the **lower limit** and the **upper limit** of the score. The idea is that IQ can take any fractional value between 102.5 and 103.5, but all scores in that range are rounded off to 103.

In a similar way, a score of 12 seconds is used to express all scores between 11.5 and 12.5 seconds; 11.5 seconds is the lower limit, and 12.5 is the upper limit of a score of 12 seconds.

Sometimes scores are measured and expressed in tenths, hundredths, or thousandths. With a good stopwatch, for example, time can be measured to the nearest tenth of a second. Scores like 3.8, 12.1, and 6.0 result. These scores also have upper and lower limits, and the same reasoning holds: each score represents a range of possible scores. The score 3.8 is the middle of a range of scores running from 3.75 to 3.85. Upper and lower limits of the scores 12.1 and 6.0 are 12.05 and 12.15, and 5.95 and 6.05.

To find the upper and lower limits of any score, then, subtract half a unit from and add half a unit to the last digit on the right of the score. Thus, for the score 6.13 seconds, the lower limit is $6.13 - .005 = 6.125$ seconds and the upper limit is $6.13 + .005 = 6.135$ seconds.

The basic idea is that the limits of a score have one more decimal place than the score itself, and this last decimal place is 5. A good way to be sure you understand this is to draw a few number lines like the one in Figure 1.1, put scores like 3.0, 3.8, or 5.95 on them, and find the lower and upper limits.

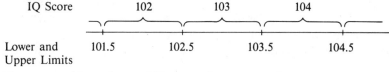

Figure 1.1   The upper and lower limits of IQ scores of 102, 103, and 104.

Clue to the Future   The upper and lower limits of a score on a quantitative variable will be important in Chapter 3, "Central Values and the Organization of Data."

Qualitative Variables

In addition to quantitative variables, sometimes you will deal with **qualitative variables**. With such variables, the scores (numbers) are simply used as names; they do not have quantitative meaning. For example, political affiliation is a qualitative variable. Terms like "Democrat," "Republican," "Independent," and

"Other" might be assigned the numbers 1, 2, 3, and 4, but these numbers clearly have no quantitative meaning. The same thing is true for the 0 and 1 used to identify sex. The numbers are simply used as substitutes for names.

PROBLEMS AND ANSWERS

At the beginning of this chapter, we urged you to take up the active-learning approach. Did you? For instance, did you draw those lines and label them, as we suggested two paragraphs back? If not, try it now for the score 3.05. If you didn't get 3.045 as the lower limit, you need to reread that material. If you still have trouble, get some help from your instructor or from another student who understands it. Be active. Make it your policy throughout this course to never let anything slip by you.

Part of active learning is active reading. Read this text actively. You will know you are doing this if you reread a paragraph that you did not understand the first time, or wrinkle your brow in concentration, or say to yourself "Hmm, I really don't understand that; I'll ask my instructor about it tomorrow," or, best of all, nod to yourself, "Oh yes."

From time to time, we will interrupt you with a set of problems so that you can practice what you have just been reading about. Working these problems correctly is also evidence of active reading. You will find the answers in the Appendix at the end of the book. Here are some ways to utilize these problems and answers in the most efficient way.

1. Buy yourself a spiral notebook for statistics. Work all the problems for this course in it, since, for several problems, we refer you back to a problem you worked previously. When you make an error, don't erase it—circle it and work the problem correctly below. Seeing your error later serves as a reminder of what not to do on a test. If you find that *we* have made an error, write to us as a reminder of what not to do in the next edition.
2. Never, *never* look at an answer before you have worked the problem (or at least tried twice to work the problem).
3. When you come to a set of problems, work the first one and then immediately check your answer against the answer that we give. If you make an error, find out why you made it—faulty understanding, arithmetical error, or whatever.
4. Don't be satisfied with just getting the answer. Be sure you understand what the answer means and the process by which you got it.
5. When you finish a chapter, go back over the problems immediately, reminding yourself of the various techniques you have learned.

O.K., here is an opportunity to see how actively you have been reading.

Problems   1. Mark each of the following as D (descriptive) or I (inferential) statistics.
    —— a. Results of the poll of 1000 citizens indicate that Buglemouth is favored by 60% of the voters.

—— b. Final enrollment figures at Crabgrass A & M for the past three years lead to a projection of decreased enrollment for next year.

—— c. The mean score for this class was 68% on the midterm exam.

—— d. The legislature appropriated a 12% increase in salaries for state employees.

—— e. The superior performance of Group 2 subjects supports the hypothesis that moderate anxiety improves the learning of a simple task by college students.

2. Define *population, sample, statistic,* and *parameter.*

3. The variables below are all quantitative ones. What are the lower and upper limits of each measurement?

a. 67 inches

b. 30.5 seconds

c. 50.75 miles per hour

d. 2 tablespoons of flour

e. 3.0 pounds

f. IQ = 140

g. 2.013 liters

SCALES OF MEASUREMENT

Numbers mean different things in different situations. Numbers are assigned to objects according to rules. You need to distinguish clearly between the *thing* you are interested in and the *number* that symbolizes or stands for the thing. For example, you have had lots of experience with the numbers 2 and 4. You can state immediately that 4 is twice as much as 2. That statement is correct if you are dealing with numbers themselves, but it may or may not be true when those numbers are symbols for things. The statement is true if the numbers refer to apples; four apples are twice *as many* as two apples. However, the statement is not true if the numbers refer to the order that runners finish in a race. Fourth place is not twice anything in relation to second place—not twice *as slow* or twice *as far behind* the first-place runner. The point is that the numbers 2 and 4 are used to refer to both apples and finish places in a race, but the numbers mean different things in those two situations.

S. S. Stevens (1946) identified four different measurement scales that help distinguish different kinds of situations in which numbers are assigned to objects. The four scales are: nominal, ordinal, interval, and ratio. The first of these, the nominal scale, is not a scale at all in the usual sense. In the **nominal scale**, numbers are used simply as names and have no real quantitative value. It is the scale used for qualitative variables. Numerals on sports uniforms are an example; here, 45 is *different from* 32, but that is about all we can say. The person represented by 45 is not "more than" the person represented by 32, and certainly it would be meaningless to try to *add* 45 and 32. Designating different colors, different sexes, or different political parties by numbers will produce nominal scales. With a nominal scale, you can even reassign the numbers and still maintain the original meaning, which is only that things with different numbers are different. Of course, all things that are alike have the same number.

A second kind of scale, the **ordinal scale**, has the characteristic of the nominal scale (different numbers mean different things) plus the characteristic of indicating "greater than" or "less than." In the ordinal scale, the object with the number 3 has less or more of something than the object with the number 5. Finish places in a race are an example of an ordinal scale. The runners finish in rank order, with "1" assigned to the winner, "2" to the runner-up, and so on. Here, 1 means less time than 2. Other examples of ordinal scales are house numbers, Government Service ranks like GS-5 and GS-7, and statements like "She is a better mathematician than he is."

The third kind of scale is the **interval scale**, which has properties of both the ordinal and nominal scales, plus the additional property that intervals between the numbers are equal. "Equal interval" means that the distance between the things represented by "2" and "3" is the same as the distance between the things represented by "3" and "4." The centigrade thermometer is based on an interval scale. The difference in temperature between 10° and 20° is the same as the difference between 40° and 50°. The centigrade thermometer, like all interval scales, has an arbitrary zero point. On the centigrade thermometer, this zero point is the freezing point of water at sea level. Zero degrees on this scale does not mean the complete absence of heat; it is simply a convenient starting point. With interval data, we have one restriction: we may not make simple ratio statements. We may not say that 100° is twice as hot as 50° or that a person with an IQ of 60 is half as intelligent as a person with an IQ of 120.[4]

The fourth kind of scale, the **ratio scale**, has all the characteristics of the nominal, ordinal, and interval scales, plus one: it has a true zero point, which indicates a complete absence of the thing measured. On a ratio scale, zero means "none." Height, weight, and time are measured with ratio scales. Zero height, zero weight, and zero time mean that no amount of these variables is present. With a true zero point, you can make ratio statements like "16 kilograms is four times heavier than 4 kilograms."[5]

Having illustrated with examples the distinctions among these four scales—a practice common to most elementary statistics textbooks—we must tell you that it is sometimes difficult to classify the variables used in the social and behavioral sciences. Very often they appear to fall between the ordinal and interval scales. It may happen that a score provides more information than simply rank, but equal intervals cannot be proved. Intelligence test scores are an example. In such cases, researchers generally treat the data as if they were based on an interval scale.

The main reason why this section on scales of measurement is important is that the kind of descriptive statistics you can compute on your numbers depends to some extent upon the kind of scale of measurement the numbers represent. For example, it is not meaningful to compute a mean on nominal data such as the numbers on football players' jerseys. If the quarterback's number is 12 and a running back's number is 23, the mean of the two numbers (17.5) has no meaning at all.

---

[4]Convert 100° C and 50° C to Fahrenheit, and suddenly the "twice as much" relationship disappears.

[5]Convert 16 and 4 kg to pounds (1 kg = 2.2 lbs), and the "four times heavier" relationship is maintained.

STATISTICS AND EXPERIMENTAL DESIGN

Here is a story that will help distinguish between statistics (applying straight logic) and experimental design (finding out what actually happens). This is an excerpt from a delightful book by E. B. White, *The Trumpet of the Swan* (1970).

The Fifth-graders were having a lesson in arithmetic, and their teacher, Miss Annie Snug, greeted Sam with a question.

"Sam, if a man can walk three miles in one hour, how many miles can he walk in four hours?"

"It would depend on how tired he got after the first hour," replied Sam.

The other pupils roared. Miss Snug rapped for order.

"Sam is quite right," she said. "I never looked at the problem that way before. I always supposed that man could walk twelve miles in four hours, but Sam may be right: that man may not feel so spunky after the first hour. He may drag his feet. He may slow up."

Albert Bigelow raised his hand. "My father knew a man who tried to walk twelve miles, and he died of heart failure," said Albert.

"Goodness!" said the teacher. "I suppose *that* could happen, too."

"Anything can happen in four hours," said Sam. "A man might develop a blister on his heel. Or he might find some berries growing along the road and stop to pick them. That would slow him up even if he wasn't tired or didn't have a blister."

"It would indeed," agreed the teacher. "Well, children, I think we have all learned a great deal about arithmetic this morning, thanks to Sam Beaver."

Everyone had learned how careful you have to be when dealing with figures.

Statistics involves the manipulation of numbers and the conclusions based on those manipulations (Miss Snug). Experimental design deals with all the things that influence the numbers you get (Sam and Albert). This could have been a "pure" statistics book, from which you would learn to analyze numbers without knowing where they came from or what they referred to. You would learn about statistics, but such a book would be dull, dull, dull. On the other hand, to describe procedures for collecting numbers is to describe experimental design—and this book is for a statistics course. Our solution to this conflict is to generally side with Miss Snug but to include some aspects of experimental design throughout the book. We hope that the information on design that you pick up from this book will encourage you to try your hand at your own data gathering. We'll start our discussion of experimental design now.

In the design of a typical simple experiment, the experimenter is interested in the effect that one variable (called the **independent variable**) has on some other variable (called the **dependent variable**). Much research is designed to discover cause-and-effect relationships. In such research, differences in the independent variable are the presumed cause for differences in the dependent variable. The experimenter chooses values for the independent variable, administers a different value of the independent variable to each group of subjects, and then measures the dependent variable for each subject. If the scores on the dependent variable differ as a result of differences in the independent variable, the experimenter may be able to conclude that there is a cause-and-effect relationship.

An example might help to clarify this important concept. Assume for the moment that you are a budding gourmet cook attempting to improve your roast beef gravy. You want to know what effect adding a little Worcestershire sauce will have on the taste perceptions of those who eat it. So, you give two dinner parties that feature roast beef. At one party, the gravy contains Worcestershire sauce. At the other party, it does not. At both parties, you count the number of favorable comments about the gravy. The independent variable in this example is Worcestershire sauce. The dependent variable is the guests' perceptions of it as measured by favorable comments. The assumption is that differences in the number of favorable comments about the gravy are caused by the presence or absence of Worcestershire sauce.

Table 1.1 shows some independent-dependent variable cause-and-effect relationships that are often accepted as true.

Table 1.1   **Some Cause-and-Effect Statements and their Independent and Dependent Variables**

Independent Variable	Dependent Variable	Cause-and-Effect Statement
1. Amount of practice	Degree of success	Practice makes perfect
2. Adequacy of wardrobe	Degree of success	Clothes make the man
3. Amount of time passed	Amount of healing	Time heals all wounds
4. Number of cooks	Quality of broth	Too many cooks spoil the broth
5. Age of the broom	Effectiveness of the broom	A new broom sweeps clean
6. Motion of a stone	Amount of moss gathered	A rolling stone gathers no moss
7. Dietary habits	Health	An apple a day keeps the doctor away

One of the problems with drawing cause-and-effect conclusions is that you must be sure that changes in the scores on the dependent variable are the result of changes in the independent variable and not the result of changes in some other variable(s). Variables other than the independent variable that can cause changes in the dependent variable are called **extraneous variables**. Many extraneous variables could be operating in the roast beef gravy example—so many, in fact, that it would be a mistake to draw a cause-and-effect conclusion. Some extraneous variables in that situation might be differences in the quality of beef on the two occasions, differences in the "party moods" of the two sets of guests, and differences in the quality and exact proportions of gravy ingredients other than Worcestershire sauce on the two occasions. These and other variables could affect the number of favorable comments about the gravy.

It is important, then, that experimenters be aware of and control extraneous variables that might influence their results. The simplest way to control an extraneous variable is to be sure that all subjects are equal on that variable. For example, if only well-rested subjects are used and the experiment is always conducted in the same room by the same experimenter, the extraneous variables of fatigue, environment, and experimenter are controlled.

Independent variables are often referred to as *treatments* because the experimenter frequently asks "If I treat this group of subjects this way and treat another group another way, will there be a difference in their behavior?" The ways that the subjects are treated constitute the **levels** of the independent variable being studied, and experiments typically have two or more levels.

The experiment by Zeigarnik described on page 3 can be used to illustrate these experimental-design terms. In her experiment, the independent variable was whether or not a task was interrupted. There were two levels of this variable: yes and no. The dependent variable was whether or not a task was recalled.

The design of the experiment used by Zeigarnik is called a *within-subject design* because each subject is tested under all levels of the independent variable. One of the nice characteristics of a within-subject design is that it controls many extraneous variables so well. Thus, variables like the age of the subjects, how good their memories are, and their motivation are the same for both levels of the independent variable.

The general question that experiments try to answer is "What effect does the independent variable have on the dependent variable?" At various places in the following chapters, we will explain experiments and statistical analyses of their results in terms of independent, dependent, and extraneous variables. These explanations will usually assume that all extraneous variables were well-controlled. That is, you may assume that the experimenter knew how to design the experiment so that changes in the dependent variable can be correctly attributed to changes in the independent variable. However, you will be presented with a few investigations (like the roast beef gravy example) that we hope you recognize as being so poorly designed that conclusions about the relationship between the independent and dependent variable cannot be drawn.

STATISTICS: THEN AND NOW

According to Walker (1929), the first course in statistics given in a university in the United States was probably "Social Science and Statistics," taught at Columbia University in 1880. The professor was a political scientist and the course was offered in the Economics Department. In 1887 at the University of Pennsylvania, separate courses in statistics were offered by the Departments of Psychology and of Economics. By 1891, Clark University, The University of Michigan, and Yale had been added to the list of schools and Anthropology had been added to the list of departments. Biology was added in 1899 (Harvard) and Education in 1900 (Columbia).

You might be interested to know when statistics was first taught at your school and in what department. College catalogues are probably the most accessible source for this information.

Kirk (1978) divides present-day statisticians into four categories: (1) those who are able to understand material that is presented statistically; (2) those who are able to apply statistical techniques and interpret the results; (3) professional statisticians who help others apply statistical techniques to a particular problem; and (4) mathematical statisticians who develop new statistical techniques and discover new characteristics of old techniques.[6] If our hopes for this book are realized, by the end of this course you will be comfortable calling yourself a Category 2 statistician.

HOW TO USE THIS BOOK

At the beginning of this chapter, we mentioned the importance of your active participation in learning statistics. We have tried to organize this book in a way that will encourage active participation. Here are some of the features we hope you will find helpful.

Objectives

At the beginning of each chapter, there is a list of the skills and abilities that the chapter is designed to help you acquire. Before reading each chapter, read this list of objectives to find out what you are to learn to do. Then, thumb through the chapter and look at the headings. Next, study the chapter and work all the problems. The last thing you should do is go back to the objectives to see if you have accomplished them.

Clues to the Future

Many times we will present an idea that will be used extensively in later chapters. You will find each idea, or "Clue to the Future," separated in a box from the rest of the text. You have already met two of these "Clues" in this chapter. Pay particularly close attention to the concepts the clues describe. This will save you time and trouble later in the course.

Error Detection

We have also boxed in, at various points in the book, some ways to detect errors in your working of a problem. These "Error-Detection" tips will often be useful in providing you with a better understanding of the overall statistical concepts as well. Since some of these checks can be made early in a statistical problem, their use can prevent the frustrating experience of making an error early and carrying it all the way through to the final answer.

Problems and Answers

You have already been introduced to our method of presenting problems and answers. Some additional no-nos should be mentioned. Do not skip any assigned problems. Never start with the answer in front of you. Do not simply correct a wrong answer without discovering why you made the error. Sometimes you will find minor differences between your answer and ours (in the second

---

[6]See any edition of the journal, *Psychological Bulletin.*

decimal place or beyond). Most of these will probably be the result of rounding errors and do not deserve your worry and anxiety.

Many of the problems are conceptual questions that do not require you to do any arithmetic. Think these through and write your answer. The arithmetic associated with an experiment is of no value if you cannot tell what it means in English.

Transition Pages

At various points in the book, there will be major differences between the material you have been studying and that which you are about to begin. At each of these points, we have provided a "Transition Page," which points out those major differences. These pages should help you to shift gears in your thinking processes in preparation for the new kinds of problems.

Glossaries

We have compiled three glossaries that you can use to jog your memory, if necessary. Two of them are in the back of the book and one is printed on the inside cover of the book.

1. A glossary of words (p. 333). This is a glossary of statistical words and phrases used in the text. When these items are first introduced in the text, they will appear in boldface type.
2. A glossary of formulas (p. 338). The formulas for all the statistical techniques discussed in the text are printed here, in alphabetical order according to the name of the technique.
3. A glossary of symbols (inside cover of this book). Statistical symbols are defined here.

Some General Thoughts

Statistics is the sort of subject that requires you to use old knowledge in the learning of new topics. You must know the material in Chapter 3 before you can learn the material in Chapter 4. You must understand both Chapters 3 and 4 before you can learn what is presented in Chapter 5, and so on throughout the book.

Skipping anything would be a serious mistake that would haunt you throughout the course. Not quite so obvious, but very true, is that looking back at chapters you have already completed and looking ahead at chapters you have yet to do can be valuable aids to understanding the chapter you are presently working on.

We wish to emphasize that this book is a fairly complete introduction to elementary statistics. There is more to the study of statistics—lots, lots more; but there is a limit to what you can do in one term. Some of you, however, will learn the material in this book and want to know more. For those students (may they live long and joyful lives!), we have provided references in footnotes. With the help of your instructor, you can read and learn this intermediate-level material.

Finally, most students will find that this text becomes a helpful reference book. In future courses, many students find themselves pulling out their copy to look up some forgotten definition or vaguely recalled procedure. We hope you not only learn from this book but that you, too, join that group of students who count this book as part of their personal library.

We find the study of statistics very satisfying. We hope you do too.

Problems

4. Name the four kinds of scales identified by S. S. Stevens.
5. Give the properties of each of the four kinds of scales.
6. Identify which kind of scale is being used in each of the following cases.
   a. Geologists have a "hardness scale" for identifying different rocks, called the Moh scale. The hardest rock (diamond) has a value of 10 and will scratch all others. The second hardest will scratch all but the diamond, and so on, down to a rock such as talc, with a value of 1, which can be scratched by any other rock. (On this scale, a fingernail has a value of between 2 and 3.)
   b. Three different cubes were measured with a ruler and their volumes computed to be 40, 65, and 75 cubic inches, respectively.
   c. Three different highways were identified by their numbers: 40, 65, and 75.
   d. Republicans, Democrats, Independents, and Others were identified on the voters' list with the numbers 1, 2, 3, and 4.
   e. The pages of a book were numbered 1 through 150.
   f. The winner of the Miss America contest was Miss Rhode Island; runners-up were Miss Kentucky and Miss Kansas.
7. Identify the independent and dependent variable in the following statements.
   a. The ability to solve new problems depends on previous experience.
   b. The more you eat, the better you feel.
   c. The more you eat, the worse you feel.
   d. For some problems, the more anxious subjects solved the problems more quickly.
   e. These plants are growing poorly, since they do not get much sunlight.
   f. Experience is the best teacher.
   g. April showers bring May flowers.
8. For each of the four studies described, identify
   a. the independent variable,
   b. the dependent variable,
   c. a controlled extraneous variable,
   d. the population,
   e. a sample,
   f. a statistic that might be used,
   g. a parameter that might be estimated,
   h. a variable measured with a nominal scale, and
   i. a variable measured with a ratio scale.
      i. Twenty-five high-anxious and 25 low-anxious female college students read the same paragraph while having their eye movements photographed. Films revealed that high-anxious subjects reread words and lines more often than did low-anxious subjects.
      ii. A litter of ten rats was divided into two groups at birth. Group 1 rats were handled and petted daily. Group 2 rats were not handled at all. At the age of five months, all rats were given a task to learn. The mean number of trials to learn the task was lower for Group 1 rats.

iii. An industrial psychologist was hired by a manufacturer of lawn mowers to determine which of four colors would be best. For one of her tests she obtained a rating from 40 participants. Each rated one color. The rating required the participant to judge "the suitability of this color for a lawn-mower." Ratings were given on a scale of one to ten.

iv. A social psychologist wanted to know whether college students in the United States differed from their fathers in their political attitudes. He collected attitude-questionnaire data from ten college students at each of four colleges and answers to the same questionnaire from the fathers of the college students. He concluded that American college students and their fathers do differ in political attitudes.

2 REVIEW AND MORE INTRODUCTION

Objectives for Chapter 2: After studying the text and working the problems in this chapter, you should be able to:

1. work with decimals, fractions, negative numbers, exponents, absolute values, simple algebra, proportions, and percents,
2. round off numbers,
3. estimate answers, and
4. define a few statistical concepts and pronounce the symbols that stand for the concepts.

This chapter gives you a review of the basic mathematical skills you will need in order to work the problems in this course. The math is not complex. For this course, you will need no mathematical sophistication beyond simple algebra.

A basic statistics course is not as much of a mathematics course as it first appears to be. Although numbers are manipulated in solving problems, logical reasoning is more important. A course in statistics develops your ability to reason tightly and provides you with a way of thinking scientifically.

However, in order to be good at statistics, you need to be good at arithmetic. A wrong answer is wrong, whether the error is arithmetical or logical. Most people are better at arithmetic when they use an electronic calculator with a square root key and a memory. This will speed up your work and will result in fewer computational errors.

Numerical answers to all of the computational problems in this book and in the accompanying Student Guide can be found using a computer. Many can be found using a calculator hard-wired to compute such statistics as standard deviations and correlation coefficients. The question for you is, "At what stage of learning should I use these aids?"

Remember that the major goal of this course is for you to develop an understanding of the logic of statistical reasoning. To develop this understanding, we believe you have to work through each process step-by-step. Thus, we believe that you should not use aids that produce a final statistic for you until after you have gone through the step-by-step, part-by-part calculations several times. In the words of one of the reviewers of this book, "The world already has too many people using computer packages without an adequate 'personal knowledge' of the statistical operations performed (and not performed)." Become a person with "personal knowledge."

This chapter is divided into three parts.

- Part 1 is a self-administered, self-scored test of your skills in arithmetic and algebra. If you do well on this, you may skip Part 2. Note that, even if you skip Part 2, you should read and do the problems in Part 3.
- Part 2 is a review of arithmetic and algebra for those whose pretest indicates a need for review. There are problems and worked-out answers for this part.
- Part 3 is a description of statistical rules, symbols, and shortcuts, which everyone should read.

The pretest asks you to work problems that are similar to those you will have to work later in the course. Take the pretest and then check your answers against the answers in the back of the book. If you find that you made any mistakes, work through those sections of Part 2 that explain the problems you missed. If you made no mistakes, proceed to Part 3, page 30. The decision to work through all or part of Part 2 is up to you.

As we stated earlier, to be good at statistics, you need to be good at arithmetic. To be good at arithmetic, you need to know the rules and to be careful in your computations. Part 2 is designed to refresh your memory of the rules. It is up to you to be careful in your computations.

PART 1: PRETEST

Decimals

Add:
1. 3.12, 6.3, 2.004   *11.424*
2. 12, 8.625, 2.316, 4.2   *27.141*

Subtract:
3. 28.76 − 8.91   *19.85*
4. 3.2 − 1.135   *2.065*

Multiply:
5. 6.2 × 8.06
6. .35 × .162

Divide:
7. 64.1 ÷ 21.25
8. .065 ÷ .0038

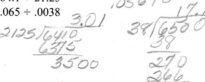

Fractions

Add:
9. $\dfrac{1}{8} + \dfrac{3}{4} + \dfrac{1}{2}$ *1 3/8*

10. $\dfrac{1}{3} + \dfrac{1}{2}$ *5/6*

Subtract:
11. $\dfrac{11}{16} - \dfrac{1}{2}$ *3/16*

12. $\dfrac{5}{9} - \dfrac{1}{2}$ *1/18*

Multiply:
13. $\dfrac{4}{5} \times \dfrac{3}{4}$ *3/5*

14. $\dfrac{2}{3} \times \dfrac{1}{4} \times \dfrac{3}{5}$ *1/10*

Divide:
15. $\dfrac{3}{8} \div \dfrac{1}{4}$ *1 1/2*

16. $\dfrac{7}{9} \div \dfrac{1}{4}$ *3 1/9*

Negative Numbers

Add:
17. −5, 16, −1, −4   *6*
18. −11, −2, −12, 3   *−22*

Subtract:
19. (−10) − (−3)   *−7*
20. (−8) − (−2)   *−6*

Multiply:
21. (−5) × (−5)   *25*
22. (−8) × (3)   *−24*

Divide:
23. (−10) ÷ (−3)   *3 1/3*
24. (−21) ÷ 4   *−5 1/4*

Percents and Proportions

25. 12 is what percent of 36?   *33 1/3%*
26. What percent of 84 is 27?   *32%*
27. What proportion of 112 is 21?   *.19*

28. A proportion .40 of the tagged birds was recovered. 150 were tagged in all. How many were recovered?   *60*

Absolute Value

29. |−5| *5*
30. |8 − 12| *4*

± Problems

31. $8 \pm 2 =$ *6 - 10*

32. $13 \pm 9 =$ *4 - 22*

Exponents

33. 4^2 *16*

34. 2.5^2 *6.25*

35. $.35^2$ *.1225*

.35
.35
175
105

Complex Problems (Round all answers to two decimal places)

36. $\dfrac{3+4+7+2+5}{5} =$ *4.20*

37. $\dfrac{(5-3)^2 + (3-3)^2 + (1-3)^2}{3-1} =$ *4.00*

38. $\dfrac{(8-6.5)^2 + (5-6.5)^2}{2-1} =$ *4.50*

39. $\left(\dfrac{8+4}{8+4-2}\right)\left(\dfrac{1}{8}+\dfrac{1}{4}\right) =$ *$\frac{12}{10} \times \frac{3}{8} = \frac{9}{20}$*   *=0.45*

40. $\left(\dfrac{3.6}{1.2}\right)^2 + \left(\dfrac{6.0}{2.4}\right)^2 =$ *$9 + 6.25 = 15.25$*

41. $\dfrac{(3)(4) + (6)(8) + (5)(6) + (2)(3)}{(4)(4-1)} =$ *$\frac{12+48+30+6}{12}$*   *$\frac{96}{12} = 8$*

42. $\dfrac{190 - (25^2/5)}{5-1} =$ *$\frac{65}{4} = 16.25$*

43. $\dfrac{12(50-20)^2}{(8)(10)(12)(11)} =$ *$\frac{900}{880} = 1.02$*

44. $\dfrac{6[(3+5)(4-1) + 2]}{5^2 - (6)(2)} =$ *$\frac{6 \times 26}{(25-12)} = \frac{6 \times 26}{13} = 12.00$*

Simple Algebra (Solve for x)

45. $\dfrac{x-3}{4} = 2.5$ *13*

46. $\dfrac{14-8.5}{x} = .5$ *11*

47. $\dfrac{20-6}{2} = 4x - 3$ *2.5*

48. $\dfrac{6-2}{3} = \dfrac{x+9}{5}$ *-2⅓*

PART 2: REVIEW OF FUNDAMENTALS

This section is designed to provide you with a quick review of the rules of arithmetic and simple algebra. We recommend that you work the problems as you come to them, keeping the answers covered while you work. We assume that you once knew all these rules and procedures but that you need to refresh your memory. Thus, we do not include much explanation. For a textbook that does include basic explanations, see Helen M. Walker, *Mathematics Essential for Elementary Statistics* (Holt, Rinehart and Winston, revised edition, 1951).

Definitions

1. *Sum.* The answer to an addition problem is called a sum. In Chapter 10, you will calculate a *sum of squares*, a quantity that is obtained by adding together some squared numbers.

2. *Difference.*   The answer to a subtraction problem is called a difference. Much of what you will learn in statistics deals with differences and the extent to which they are significant. In Chapter 8, you will encounter a statistic called the *standard error of a difference.* Obviously, this statistic involves subtraction.

3. *Product.*   The answer to a multiplication problem is called a product. Chapter 5 is about the *product-moment correlation coefficient,* which requires multiplication. Multiplication problems are indicated either by a × or by parentheses. Thus, 6 × 4 and (6)(4) call for the same operation.

4. *Quotient.*   The answer to a division problem is called a quotient. The IQ or *intelligence quotient* is based on the division of two numbers. The two ways to indicate a division problem are ÷ and −. Thus, $9 \div 4$ and $\frac{9}{4}$ (or 9/4) call for the same operation. It is a good idea to think of any common fraction as a division problem. The numerator is to be divided by the denominator.

Decimals

1. *Addition and Subtraction of Decimals.*   There is only one rule about the addition and subtraction of numbers that have decimals: *keep the decimal points in a vertical line.* The decimal point in the answer goes directly below those in the problem. This rule is illustrated in the five problems below.

Add: Subtract:

	.004	6.0		
1.26	1.310	18.0	14.032	16.00
10.00	4.039	.5	8.26	4.32
11.26	5.353	24.5	5.772	11.68

2. *Multiplication of Decimals.*   The basic rule for multiplying decimals is that the number of decimal places in the answer is found by adding up the number of decimal places in the two numbers that are being multiplied. To place the decimal point in the product, count from the right.

1.3	.21	1.47
× 4.2	× .4	× 3.12
26	.084	294
52		147
5.46		441
		4.5864

3. *Division of Decimals.*   Two methods have been used to teach division of decimals. The older method required the student to move the decimal in the divisor (the number you are dividing by) enough places to the right to make the divisor a whole number. The decimal in the dividend was then moved to the right the same number of places, and division was carried out in the usual way. The new decimal places were identified with carets, and the decimal place in the

quotient was just above the caret in the dividend. For example,

$$\overset{20.}{.016_\wedge \overline{\smash{)}.320_\wedge}}$$

decimal moved three places in
both the divisor and the dividend

$$\overset{38.46}{.39_\wedge \overline{\smash{)}15.00_\wedge}}$$

decimal moved two places in
both the divisor and the dividend

$$\overset{2.072}{6\overline{\smash{)}12.432}}$$

divisor is already a whole
number

$$\overset{.004}{9.1_\wedge \overline{\smash{)}.0_\wedge 369}}$$

decimal moved one place in
both the divisor and the dividend

The newer method of teaching the division of decimals is to multiply both the divisor and the dividend by the number that will make both of them whole numbers. (Actually, this is the way the caret method works also.) For example:

$$\frac{.32}{.016} \times \frac{1000}{1000} = \frac{320}{16} = 20.00;$$

$$\frac{.0369}{9.1} \times \frac{10,000}{10,000} = \frac{369}{91,000} = .004;$$

$$\frac{15}{.39} \times \frac{100}{100} = \frac{1500}{39} = 38.46;$$

$$\frac{12.432}{6} \times \frac{1000}{1000} = \frac{12,432}{6000} = 2.072.$$

Both of these methods work. Use the one you are more familiar with.

Problems

1. Define (a) sum, (b) quotient, (c) product, and (d) difference.
   Perform the operations indicated:
2. $.001 + 10 + 3.652 + 2.5 =$
3. $14.2 - 7.31 =$
4. $.04 \times 1.26 =$
5. $143.3 + 16.92 + 2.307 + 8.1 =$
6. $3.06 \div .04 =$
7. $\dfrac{24}{11.75} =$
8. $152.12 - 127.4 =$
9. $(.5)(.07) =$

Fractions

In general, there are two ways to deal with fractions.

1. Convert the fraction to a decimal and perform the operations on the decimals.

2. Work directly with the fractions, using a set of rules for each operation. The rule for addition and subtraction is: convert the fractions to ones with common denominators, add or subtract the numerators, and place the result over the common denominator. The rule for multiplication is: multiply the numerators together to get the numerator of the answer, and multiply the denominators together for the denominator of the answer. The rule for division is: invert the divisor and multiply the fractions.

For statistics problems, it is usually easier to convert the fractions to decimals and then work with the decimals. Therefore, this is the method that we will illustrate. However, if you are a whiz at working directly with fractions, by all means continue with your method. To convert a fraction to a decimal, divide the lower number into the upper one. Thus, $3/4 = .75$, and $13/17 = .765$.

1. *Addition and Subtraction of Fractions.*
   Addition:

   $$\frac{1}{2} + \frac{1}{4} = .50 + .25 = .75.$$

   $$\frac{13}{17} + \frac{21}{37} = .765 + .568 = 1.33.$$

   $$\frac{2}{3} + \frac{3}{4} = .667 + .75 = 1.42.$$

   Subtraction:

   $$\frac{1}{2} - \frac{1}{4} = .50 - .25 = .25.$$

   $$\frac{11}{12} - \frac{2}{3} = .917 - .667 = .25.$$

   $$\frac{41}{53} - \frac{17}{61} = .774 - .279 = .50.$$

2. *Multiplication of Fractions.*

   $$\frac{1}{2} \times \frac{3}{4} = (.5)(.75) = .38.$$

   $$\frac{10}{19} \times \frac{61}{90} = .526 \times .678 = .36.$$

   $$\frac{1}{11} \times \frac{2}{3} = (.09)(.67) = .06.$$

3. *Division of Fractions.*

   $$\frac{9}{21} \div \frac{13}{19} = .429 \div .684 = .63.$$

   $$14 \div \frac{1}{3} = 14 \div .33 = 42.$$

   $$\frac{7}{8} \div \frac{3}{4} = .875 \div .75 = 1.17.$$

Problems   Perform the operations indicated:

10. $\dfrac{9}{10} + \dfrac{1}{2} + \dfrac{2}{5} =$

11. $\dfrac{9}{20} \div \dfrac{19}{20} =$

12. $\left(\dfrac{1}{3}\right)\left(\dfrac{5}{6}\right) =$

13. $\dfrac{4}{5} - \dfrac{1}{6} =$

14. $\dfrac{1}{3} \div \dfrac{5}{6} =$

15. $\dfrac{3}{4} \times \dfrac{5}{6} =$

16. $18 \div \dfrac{1}{3} =$

Negative Numbers

1. *Addition of negative numbers.* (Any number without a sign is understood to be positive.)

a. To add a series of negative numbers, add the numbers in the usual way and attach a negative sign to the total.

$$
\begin{array}{r}
-3 \\
-8 \\
-12 \\
-5 \\
\hline
-28
\end{array}
\qquad (-1) + (-6) + (-3) = -10.
$$

b. To add two numbers, one positive and one negative, subtract the smaller number from the larger number and attach the sign of the larger number to the result.

$$
\begin{array}{r}
140 \\
-55 \\
\hline
85
\end{array}
\qquad
\begin{array}{r}
-14 \\
8 \\
\hline
-6
\end{array}
\qquad
\begin{aligned}
(28) + (-9) &= 19. \\[4pt]
74 + (-96) &= -22.
\end{aligned}
$$

c. To add a series of numbers, of which some are positive and some negative, add all the positive numbers together, all the negative numbers together (see 1a), and then combine the two sums (see 1b).

$$(-4) + (-6) + (12) + (-5) + (2) + (-9) = 14 + (-24) = -10.$$

$$(-7) + (10) + (4) + (-5) = 14 + (-12) = 2.$$

2. *Subtraction of negative numbers.* To subtract a negative number, change it to positive and add it. Thus

$$
\begin{array}{r}
(-14) \\
-(-2) \\
\hline
-12
\end{array}
\qquad
\begin{array}{r}
5 \\
-(-7) \\
\hline
12
\end{array}
\qquad
\begin{array}{r}
8 \\
-(-3) \\
\hline
11
\end{array}
\qquad
\begin{array}{r}
(-7) \\
-(-5) \\
\hline
-2
\end{array}
$$

3. *Multiplication of negative numbers.* When the two numbers to be multiplied are both negative, the product is positive.

$$(-3)(-3) = 9. \qquad (-6)(-8) = 48.$$

When one of the numbers is negative and the other is positive, the product is negative.

$$(-8)(3) = -24. \qquad 14 \times -2 = -28.$$

4. *Division of negative numbers.* The rule in division is the same as the rule in multiplication. If the two numbers are both negative, the quotient is positive.

$$(-10) \div (-2) = 5. \qquad (-4) \div (-20) = .20.$$

If one number is negative and the other positive, the quotient is negative.

$$(-10) \div 2 = -5. \qquad 6 \div (-18) = -.33.$$
$$14 \div (-7) = -2. \qquad (-12) \div 3 = -4.$$

Problems

17. Add the following numbers.
   a. $-3, 19, -14, 5, -11$
   b. $-8, -12, -3$
   c. $-8, 11$
   d. $3, -6, -2, 5, -7$

18. $(-8)(5) =$	19. $(-4)(-6) =$	20. $(4)(12) =$
21. $(11)(-3) =$	22. $(-18) - (-9) =$	23. $14 \div (-6) =$
24. $12 - (-3) =$	25. $(-6) - (-7) =$	26. $(-9) \div (-3) =$
27. $(-10) \div 5 =$	28. $4 \div (-12) =$	29. $(-7) - 5 =$

Proportions and Percents

A **proportion** is a part of a whole and can be expressed as a fraction or as a decimal. Usually, proportions are expressed as decimals. If eight students in a class of 44 received A's, we may express 8 as a proportion of the whole (44). Thus, we have 8/44, or .18. The proportion of the class who received A's is .18.

To convert a proportion to a percent (per one hundred), multiply by 100. Thus: $.18 \times 100 = 18$; 18 percent of the students received A's. As you can see, proportions and percents are two ways to express the same idea.

If you know a proportion (or percent) and the size of the original whole, you can find the number that the proportion represents. If .28 of the students were absent due to illness and there are 50 students in all, then .28 of the 50 were absent. $(.28)(50) = 14$ students who were absent. Here are some more examples.

a. 26 out of 31 completed the course. What proportion completed the course?

$$\frac{26}{31} = .83.$$

b. What percent completed the course?

$$.83 \times 100 = 83 \text{ percent.}$$

c. What percent of 19 is 5?

$$\frac{5}{19} = .26. \qquad .26 \times 100 = 26 \text{ percent.}$$

d. If 90 percent of the population agreed and the population consisted of 210 members, how many members agreed?

$$.90 \times 210 = 189 \text{ members.}$$

Absolute Value

The **absolute value** of a number ignores the sign of the number. Thus, the absolute value of -6 is 6. This is expressed with symbols as $|-6| = 6$. It is expressed verbally as "the absolute value of negative six is six." In a similar way, the absolute value of $4 - 7$ is 3. That is, $|4 - 7| = |-3| = 3$.

± Problems

A $\pm$ sign ("plus or minus" sign) means to both add *and* subtract. A $\pm$ problem *always* has two answers.

$$10 \pm 4 = 6, 14.$$

$$8 \pm (3)(2) = 8 \pm 6 = 2, 14.$$

$$\pm(4)(3) + 21 = \pm 12 + 21 = 9, 33.$$

$$\pm 4 - 6 = -10, -2.$$

$$\pm 7 - 2 = -9, 5.$$

Exponents

In the expression 5^2, 2 is the exponent. The 2 means that 5 is to be multiplied by itself. Thus, $5^2 = 5 \times 5 = 25$.

In elementary statistics, the only exponent used is 2, but it will be used frequently. When a number has an exponent of 2, the number is said to be squared. The expression 4^2 (pronounced "four squared") means 4×4, and the product is 16.

$$8^2 = 8 \times 8 = 64. \qquad \left(\frac{3}{4}\right)^2 = (.75)^2 = .5625.$$

$$1.2^2 = (1.2)(1.2) = 1.44. \qquad 12^2 = 12 \times 12 = 144.$$

Problems

30. Six is what proportion of 13?
31. Six is what percent of 13?
32. What proportion of 25 is 18?
33. What percent of 115 is 85?
34. $|-31| =$
35. $|21 - 25| =$
36. $12 \pm (2)(5) =$

37. $\pm(5)(6) + 10 =$
38. $\pm(2)(2) - 6 =$
39. $(2.5)^2 =$
40. $9^2 =$
41. $26^2 =$
42. $\left(\dfrac{1}{4}\right)^2 =$

Complex Expressions

Two rules will suffice for the kinds of complex expressions encountered in statistics.

1. Perform the operations within the parentheses first. If there are brackets in the expression, perform the operations within the parentheses and then the operations within the brackets.
2. Perform the operations in the numerator separately from those in the denominator, and, finally, carry out the division.

$$\frac{(8-6)^2 + (7-6)^2 + (3-6)^2}{3-1} = \frac{2^2 + 1^2 + (-3)^2}{2} = \frac{4+1+9}{2} = \frac{14}{2} = 7.00.$$

$$\left(\frac{10+12}{4+3-2}\right)\left(\frac{1}{4}+\frac{1}{3}\right) = \left(\frac{22}{5}\right)(.25+.333) = (4.40)(.583) = 2.57.$$

$$\left(\frac{8.2}{4.1}\right)^2 + \left(\frac{4.2}{1.2}\right)^2 = (2)^2 + (3.5)^2 = 4 + 12.25 = 16.25.$$

$$\frac{6[(13-10)^2 - 5]}{6(6-1)} = \frac{6(3^2-5)}{6(5)} = \frac{6(9-5)}{30} = \frac{6(4)}{30} = \frac{24}{30} = .80.$$

$$\frac{18 - \dfrac{6^2}{4}}{4-1} = \frac{18 - \dfrac{36}{4}}{3} = \frac{18-9}{3} = \frac{9}{3} = 3.00.$$

Simple Algebra

To solve a simple algebra problem, isolate the unknown (x) on one side of the equal sign and combine the numbers on the other side. To do this, remember that you can multiply or divide both sides of the equation by the same number without affecting the value of the unknown. For example,

a. $\dfrac{x}{6} = 9,$

$(6)\left(\dfrac{x}{6}\right) = (6)(9),$

$x = 54.$

b. $11x = 30,$

$\dfrac{11x}{11} = \dfrac{30}{11},$

$x = 2.73.$

c. $\dfrac{3}{x} = 14,$

$x\left(\dfrac{3}{x}\right) = (14)(x),$

$3 = 14x,$

$\dfrac{3}{14} = \dfrac{14x}{14},$

$.21 = x.$

In a similar way, the same number can be *added to* or *subtracted from* both sides of the equation without affecting the value of the unknown. For example:

a. $x - 5 = 12,$

$x - 5 + 5 = 12 + 5,$

$x = 17.$

b. $x + 7 = 9,$

$x + 7 - 7 = 9 - 7,$

$x = 2.$

We will combine some of these steps in the problems we will work for you. Be sure you see what operation is being performed on both sides in each step.

a. $\dfrac{x - 2.5}{1.3} = 1.96,$

$x - 2.5 = (1.3)(1.96),$

$x = 2.548 + 2.5,$

$x = 5.048.$

b. $\dfrac{21.6 - 15}{x} = .04,$

$6.6 = .04x,$

$\dfrac{6.6}{.04} = x,$

$165 = x.$

c. $4x - 9 = 5^2,$

$4x = 25 + 9,$

$x = \dfrac{34}{4},$

$x = 8.50.$

d. $\dfrac{14 - x}{6} = 1.9,$

$14 - x = (1.9)(6),$

$14 = 11.4 + x,$

$2.6 = x.$

Problems   Reduce these complex expressions to a single number rounded to two decimal places.

43. $\dfrac{(4-2)^2 + (0-2)^2}{6} =$

44. $\dfrac{(12-8)^2 + (8-8)^2 + (5-8)^2 + (7-8)^2}{4-1} =$

45. $\left(\dfrac{5+6}{3+2-2}\right)\left(\dfrac{1}{3}+\dfrac{1}{2}\right) =$

46. $\left(\dfrac{13+18}{6+8-2}\right)\left(\dfrac{1}{6}+\dfrac{1}{8}\right) =$

47. $\dfrac{8[(6-2)^2 - 5]}{(3)(2)(4)} =$

48. $\dfrac{[(8-2)(5-1)]^2}{5(10-7)} =$

49. $\dfrac{6}{1/2} + \dfrac{8}{1/3} =$

50. $\left(\dfrac{9}{2/3}\right)^2 + \left(\dfrac{8}{3/4}\right)^2 =$

51. $\dfrac{10 - (6^2/9)}{8} =$

52. $\dfrac{104 - (12^2/6)}{5} =$

Find the value of x in the problems below.

53. $\dfrac{x-4}{2} = 2.58.$

54. $\dfrac{x-21}{6.1} = 1.04.$

55. $x = \dfrac{14-11}{2.5}.$

56. $x = \dfrac{36-41}{8.2}.$

PART 3: RULES, SYMBOLS, AND SHORTCUTS

Rounding Numbers

There are two parts to the rule for rounding a number. If the first digit of the part that is to be dropped is less than 5, simply drop it. If the first digit of the part to be dropped is 5 or greater, increase the number to the left of it by one. These rules are built into most electronic calculators. Here are some illustrations of these rules.

rounding to the nearest whole number	6.2 = 6.	4.5 = 5.
	12.7 = 13.	163.5 = 164.
	6.49 = 6.	9.5 = 10.

rounding to the nearest hundredth	13.614 = 13.61.	12.065 = 12.07.
	.049 = .05.	4.005 = 4.01.
	1.097 = 1.10.	.675 = .68.
	3.6248 = 3.62.	1.995 = 2.00.

A reasonable question is "How many decimal places should an answer in statistics have?" A good rule of thumb in statistics is to carry all operations to three decimal places and then, for the final answer, round back to two decimal places.

Sometimes this rule of thumb can get you into trouble, though. For example, if halfway through some work you had a division problem of .0016 ÷ .0074, and you dutifully rounded those four decimals to three (.002 ÷ .007), you would get an answer of .2857, which becomes .29. However, division without rounding gives you an answer of .2162 or .22. The difference between .22 and .29 may be quite substantial. We will often give you cues if more than two decimal places are necessary but you will always need to be alert to the problems of rounding.

Square Roots

Statistics problems often require that a square root be found. Three possible solutions to this problem are

1. a calculator with a square root key,
2. a table of square roots, and
3. the paper-and-pencil method.

Our recommendation is that you use a calculator with a square root key. If this is not possible, find a book with a table of square roots. The third solution, the paper-and-pencil method, is so tedious and error-prone that we do not recommend it.

Problems

57. Round the following numbers to the nearest whole number.

a. 13.9	b. 126.4	c. 9.0
d. 8.5	e. 127.5	f. 12.62
g. 125.5	h. 12.5	i. 9.48

58. Round the following numbers to the nearest hundredth.

a. 6.333	b. 12.967	c. .050
d. .965	e. 2.605	f. .3445
g. .005	h. .015	i. .9949

59. Find the square root of the following numbers to two decimal places.
   a. 625   b. 231   c. 26
   d. 6.25   e. 4.91   f. 16,641
   g. 181,476   h. 4,920   i. 18.5
   j. 26.4

Estimating Answers

It is a very good idea to just look at a problem and make an estimate of the answer before you do any calculating. This is referred to as "eyeballing the data" and Edward Minium (1978) has captured its importance with Minium's First Law of Statistics: "The eyeball is the statistician's most powerful instrument."

Estimating answers should keep you from making gross errors, such as misplacing a decimal point. For example, 31.5/5 can be estimated as a little more than 6. If you make this estimate before you divide, you are likely to recognize that an answer of 63 or .63 is incorrect.

The estimated answer to the problem $(21)(108)$ is 2000, since $(20)(100) = 2000$. The problem $(.47)(.20)$ suggests an estimated answer of .10, since $(1/2)(.20) = .10$. With .10 in mind, you are not likely to write .94 for the answer, which is .094. Estimating answers is also important if you are finding a square root. You can estimate that $\sqrt{95}$ is about 10, since $\sqrt{100} = 10$; $\sqrt{1.034}$ is about 1.

To calculate a mean, eyeball the numbers and estimate the mean. If you estimate a mean of 30 for a group of numbers that are primarily in the 20s, 30s, and 40s, a calculated mean of 60 should arouse your suspicion that you have made an error.

Statistical Symbols

Although as far as we know, there has never been a clinical case of neoiconophobia,[1] some students show a mild form of this behavior. Symbols like σ, μ, and Σ may cause a grimace, a frown, or a droopy eyelid. In more severe cases, the behavior involves avoiding a statistics course entirely. We're sure that you don't have such a severe case, since you have read this far. Even so, if you are a typical beginning student in statistics, symbols like σ, μ, Σ, and $\bar{X}$ are not very meaningful to you, and they may even elicit feelings of uneasiness. We also know from our teaching experience that, by the end of the course, you will know what these symbols mean and be able to approach them with an unruffled psyche—and perhaps even approach them joyously. This section should help you over that initial, mild neoiconophobia, if you suffer from it at all.

Definitions and pronunciations of the symbols used in the next two chapters follow. Additional symbols will be defined as they occur. Study this list until you know it.

X X stands for a score or measurement. The scores 2, 3, 3, 5, and 6 may each be represented by X.

---

[1] An extreme and unreasonable fear of new symbols.

$\sum$ Pronounced "sum," $\sum$ is a capital Greek letter sigma. It stands for the instruction "add." Thus, $\sum X$ means to add the X's. For the numbers in the first example, $\sum X = 19$.

N N is an abbreviation for "number of scores." For our previous example, $N = 5$ because there are 5 scores.

$\bar{X}$ Pronounced "ex bar" or "mean," $\bar{X}$ is the mean of a *sample*. $\bar{X} = \sum X / N$. In our example, $\sum X = 19$ and $N = 5$, so $\bar{X} = \sum X / N = 19/5 = 3.80$. You probably recognize the mean as the arithmetic average.

f f is an abbreviation for frequency—the number of times a score occurs. In the example, scores 2, 5, and 6 have a frequency of 1; the score 3 has a frequency of 2, and the score 4, zero. Knowing the frequency of each score, you can express N. Thus, $N = \sum f$. For our example, $N = \sum f = 5$.

$>$ Pronounced "greater than." Thus $4 > 3$; $21 > 8$.

$<$ Pronounced "less than." Thus $18 < 21$; $2 < 4$.

μ Pronounced "mew," μ is the Greek letter mu; it is the symbol for the mean of a *population* of scores.

σ Pronounced "sig-mah" or "standard deviation." This is a lowercase sigma and is the symbol for the standard deviation of a *population*. You will learn to calculate σ in Chapter 4.

s This lowercase s is the symbol for the standard deviation of a *sample*, which is used as an estimate of σ. You will learn to calculate s in Chapter 4.

σ^2 Pronounced "sigma squared" or "variance." This is a symbol for the variance of a *population*.

s^2 This lowercase s is the square of the standard deviation of a *sample*. s^2 is the sample variance and is used as an estimate of σ^2.

S This capital S is the standard deviation of a sample. S is used to describe a sample. It is not used as an estimate of σ.

Pay careful attention to symbols. They serve as shorthand notations for the ideas and concepts you are learning. So, each time a new symbol is introduced, concentrate on it—learn it—memorize its definition and pronunciation. The more meaning a symbol has for you, the better you understand the concepts it represents and, of course, the easier the course will be.

Sometimes we will need to distinguish between two different σ's or two X's. We will use subscripts, and the results will look like σ_1 and σ_2, or X_1 and X_2. Later, we will use subscripts other than numbers to identify a symbol. You will see σ_X and $\sigma_{\bar{X}}$. The point to learn here is that subscripts are for identification purposes only; they never indicate multiplication. $\sigma_{\bar{X}}$ does not mean $(\sigma)(\bar{X})$.

Two additional comments—to encourage and to caution you. We encourage you to do more in this course than just read the text, work the problems, and pass the tests, however exciting that may be. We encourage you to occasionally get beyond this elementary text and read journal articles or short portions of other statistics textbooks. We will indicate our recommendations with footnotes at appropriate places. The word of caution that goes with this encouragement is that reading statistics texts is like reading a Russian novel—the same characters have different names in different places. For example, the mean of a sample in some texts is symbolized M rather than $\bar{X}$, and, in some texts, S.D., $\hat{\sigma}$, and $\bar{\sigma}$ are used as symbols for the standard deviation. If you expect such differences, it will be less difficult for you to make the necessary translations.

3 CENTRAL VALUES AND THE ORGANIZATION OF DATA

Objectives for Chapter 3: After studying the text and working the problems in this chapter, you should be able to:

1. arrange data into simple and grouped frequency distributions,
2. describe the characteristics and uses of frequency polygons, histograms, bar graphs, and line graphs, and be able to interpret them,
3. recognize some distributions by their shapes and name them,
4. find the mean, median, and mode of raw scores of a simple frequency distribution, and of a grouped frequency distribution,
5. determine whether a central value measure is a statistic or a parameter and interpret it,
6. determine which central value measure is most appropriate for a set of data, and
7. find the mean of a set of means.

Now that the preliminaries are out of the way, you are ready to start on the basics of descriptive statistics. The starting point is a group of scores or measures, all obtained by administering the same test or procedure to a group of subjects. In the case of an experiment, the scores are measurements of the dependent variable. Often, this results in an unorganized mass of numbers such as those in Table 3.1, which lists 100 college students' scores on a self-esteem inventory.

Table 3.1

Scores of 100 College Students on a Self-Esteem Inventory

40	40	39	36	41	23	25	34	25	29
23	61	42	55	29	30	39	29	34	30
40	32	38	42	55	25	36	29	36	53
51	41	38	25	23	34	44	44	44	30
55	30	34	46	44	39	34	42	39	49
47	55	58	44	41	36	36	47	39	32
34	39	41	53	42	27	47	42	46	27
51	39	51	39	41	47	39	34	40	49
32	39	41	42	65	53	46	34	51	51
34	41	38	30	41	32	40	38	27	29

In this chapter you will learn to (1) organize data like those in Table 3.1 in meaningful ways, (2) present the data graphically, and (3) calculate and interpret central values and determine whether they are statistics or parameters.

ARRANGING SCORES IN DESCENDING ORDER

Look again at Table 3.1. If you knew that a score of 40 were considered average, could you tell just by looking that this group is about average? Probably not. Often, in research, so many measurements are made on so many subjects that just looking at all those numbers is a mind-boggling experience. Although you can do many computations using unorganized data, it is often very helpful to organize the numbers in some way. Meaningful organization will permit you to get some general impressions about characteristics of the scores by simply "eyeballing" the data (looking at it carefully).

One way of making some order out of the chaos in Table 3.1 is to rearrange the numbers into a list, from highest to lowest. Table 3.2 presents this rearrangement of the self-esteem scores. (It is usual in statistical tables to put the high numbers at the top and the low numbers at the bottom.) Compare the unorganized data of Table 3.1 with the rearranged data of Table 3.2. The ordering from high to low permits you to quickly gain some insights that would have been difficult to glean from the unorganized data. For example, by simply looking at the center of the table, you get an idea of what the central value is. The highest and lowest scores are readily apparent, and you get the impression that there are large differences in the self-esteem levels of these students. You can easily see that some scores occurred often (such as 44) and that some were not made by anyone (for example, 60, 50, 33). This information is gleaned by quickly eyeballing the rearranged data.

Table 3.2   **Scores on a Self-Esteem Inventory, Arranged in Descending Order**

65	51	47	42	41	39	38	34	30	27
61	51	46	42	41	39	38	34	30	27
58	51	46	42	41	39	36	34	30	27
55	51	46	42	40	39	36	34	30	25
55	51	44	42	40	39	36	34	30	25
55	49	44	41	40	39	36	34	29	25
55	49	44	41	40	39	36	32	29	25
53	47	44	41	40	39	34	32	29	23
53	47	44	41	39	38	34	32	29	23
53	47	42	41	39	38	34	32	29	23

SIMPLE FREQUENCY DISTRIBUTIONS

A much more useful method of organizing data is to construct a **simple frequency distribution.** Table 3.3 is a simple frequency distribution for the self-esteem data in Table 3.1.

The X column shows the scores and the f column shows how frequently each score appeared. The tally marks are used only to construct a first draft of a frequency distribution. (Later we will describe a final form for formal presentation.)

A simple frequency distribution is a very common way to present the results of a study or experiment. We will tell you some of the reasons why it is so useful; you will learn others later in the chapter. At just a glance, you can find the highest and lowest scores and it is also easier to see runs of scores with zero frequencies.

Table 3.3   **Simple Frequency Distribution of Self-Esteem Scores (A Rough Draft)**

X	Tally Marks	f	X	Tally Marks	f
65	/	1	43		0
64		0	42	⁄⁄⁄⁄ /	6
63		0	41	⁄⁄⁄⁄ ///	8
62		0	40	⁄⁄⁄⁄	5
61	/	1	39	⁄⁄⁄⁄ ⁄⁄⁄⁄	10
60		0	38	////	4
59		0	37		0
58	/	1	36	⁄⁄⁄⁄	5
57		0	35		0
56		0	34	⁄⁄⁄⁄ ////	9
55	////	4	33		0
54		0	32	////	4
53	///	3	31		0
52		0	30	⁄⁄⁄⁄	5
51	⁄⁄⁄⁄	5	29	⁄⁄⁄⁄	5
50		0	28		0
49	//	2	27	///	3
48		0	26		0
47	////	4	25	////	4
46	///	3	24		0
45		0	23	///	3
44	⁄⁄⁄⁄	5			

In addition, the general shape or form of the distribution can be ascertained (although you may find that a little practice is necessary for this). In the case of Table 3.3, the distribution is fat in the middle and skinny on both ends since the most frequently occurring scores are grouped near the middle.

You will have a lot of opportunities to construct simple frequency distributions so here is a set of steps to follow:

1. Find the highest and lowest scores. In Table 3.1, the highest score is 65, and the lowest score is 23.
2. In column form, write down in descending order all possible scores between the highest score (65) and the lowest score (23). Head this column with the letter X.
3. Start with the number in the upper-left-hand corner of the unorganized scores (a score of 40 in Table 3.1), draw a line under it, and place a tally mark beside 40 in your frequency distribution.
4. Continue this process through all the scores.
5. Count the number of tallies by each score and place that number beside the tallies in a column headed f. Add up the numbers in the f column to be sure they equal N. You have now constructed a simple frequency distribution.

A formal presentation of a simple frequency distribution does not include either the tally marks or the scores for which the frequency is zero. A final form would be used for presentation to other people such as colleagues, professors, supervisors, or editors. Table 3.4 shows the data of Table 3.1, properly arranged for formal presentation.

Table 3.4   **Simple Frequency Distribution of Self-Esteem Scores, for Formal Presentation**

Self-Esteem Scores (X)	f
65	1
61	1
58	1
55	4
53	3
51	5
49	2
47	4
46	3
44	5
42	6
41	8
40	5
39	10
38	4
36	5
34	9
32	4
30	5
29	5
27	3
25	4
23	3

Problems

1. The following scores are the points earned by each of 50 participants in a dart-throwing contest. Organize them into a simple frequency distribution, using the rough-draft format.

19	27	12	16	21
31	7	11	20	21
22	34	24	5	18
29	19	25	10	14
17	23	16	18	26
8	20	15	12	29
17	11	4	32	16
19	23	12	24	10
13	25	19	20	28
7	15	12	21	18

2. The simple frequency distribution you will construct for this problem is a little different. The "scores" are simply names, making this a frequency distribution of nominal data.

   A political science student covered a voting precinct and noted the yard signs for the five candidates. Her plan was to find the relationship between yard signs and actual voting behavior in the city-wide election. The five candidates were Granger (G), Bassett (B), Clark (C), Allen (A), and Demosthenes (D). Below are the data she collected.

A	G	B	B	B	A	A	C	G	A	D	G	G	A	A	D	C	B	B
G	A	A	D	A	C	B	G	G	C	B	A	A	B	A	C	A	G	G
D	C	A	A	G	A	G	B	C	B	A	D	G	A	C	A	B	G	

GROUPED FREQUENCY DISTRIBUTIONS

Most researchers would condense Table 3.4 even more. The result would be a **grouped frequency distribution**, and Table 3.5 is a rough draft example.[1] Grouping data is often necessary when you want to construct a graph or to present your results in the form of a table. In the days before computers and cheap calculators, grouping saved time by simplifying computations.

In a grouped frequency distribution, X values are grouped into equal-sized ranges called **class intervals**. In Table 3.5, the entire range of scores, from 65 to 23, has been reduced to 15 class intervals. Each interval covers three scores, and the size of the interval (number of scores covered) is indicated by i. For Table 3.5, $i = 3$. The midpoint of each interval represents all scores in that interval. For example, there were nine students who had scores of 33, 34, or 35. The midpoint of the class interval 33–35 is 34. All nine are represented by 34.

---

[1]As is the case with a simple frequency distribution, a grouped frequency distribution that is to be presented formally does not include tally marks.

Table
3.5   **A Grouped Frequency Distribution of Self-Esteem Scores ($i = 3$), Rough-Draft Version**

Class Interval	Midpoint, X	Tally Marks	f
63–65	64	/	1
60–62	61	/	1
57–59	58	/	1
54–56	55	////	4
51–53	52	++++ ///	8
48–50	49	//	2
45–47	46	++++ //	7
42–44	43	++++ ++++ /	11
39–41	40	++++ ++++ ++++ ++++ ///	23
36–38	37	++++ ////	9
33–35	34	++++ ////	9
30–32	31	++++ ////	9
27–29	28	++++ ///	8
24–26	25	////	4
21–23	22	///	3

Class intervals have upper and lower limits, much like simple scores obtained by measuring a quantitative variable. A class interval of 33–35 has a lower limit of 32.5 and an upper limit of 35.5. Similarly, a class interval of 40–49 has a lower limit of 39.5 and an upper limit of 49.5. You will use this later when you calculate the median.

The only difference between grouped frequency distributions and simple frequency distributions is class intervals, which is our next topic.

Establishing Class Intervals

There are three conventions that are usually followed in establishing class intervals. We call them conventions because they are customs rather than hard-and-fast rules. There are two justifications for these conventions. First, they allow you to get maximum information from your data with minimum effort. Second, they provide some standardization of procedures, which aids in communication among scientists. These conventions are

1. *Data should be grouped into not fewer than 10 and not more than 20 class intervals.*

A primary purpose of grouping data is to provide a clearer picture of trends in the data. For example, Table 3.5 shows that there are many frequencies near the center of the distribution with fewer and fewer as the upper and lower ends of the distribution are approached. If the data are grouped into fewer than 10 intervals, such trends are not as apparent. In Table 3.6, the same scores are grouped into only five class intervals. The concentration of frequencies in the center of the distribution is not nearly so apparent.

On the other hand, the use of more than 20 class intervals puts you back into the simple frequency distribution situation—it is difficult to get an overall picture of the form of the distribution.

2. *Choose a convenient size for the class interval (i).*

Three and five are often chosen for i because if i is odd, the midpoint of the interval will be a whole number. This midpoint will be an important part of

Table
3.6

A Grouped Frequency Distribution of Self-Esteem Scores ($i = 10$) with Too Few Intervals

Class Interval	Midpoint X	f
60–69	64.5	2
50–59	54.5	13
40–49	44.5	33
30–39	34.5	32
20–29	24.5	20

any graph of the data and whole numbers are cleaner than decimal numbers in a graph. If an $i = 5$ will not cover all the scores within the 10–20 class intervals rate, data groupers usually jump to 10 or some multiple of 10. Sometimes, however, an interval size of 25 works; this is more commonly used than 30. You will occasionally find that an $i = 2$ is necessary in order to stay within 10–20 class intervals.

3. *Begin each class interval with a multiple of i.*

For example, if the lowest score is 44 and $i = 5$, the first class interval should be 40–44 because 40 is a multiple of 5. This convention is violated more often than the other two. A violation that seems to be justified occurs when $i = 5$. When the interval size is 5, it may be more convenient to begin the interval such that multiples of 5 will fall at the midpoint, since multiples of 5 make graphs easier to read. For example, an interval 23–27 has 25 as its midpoint, while an interval 25–29 has 27 as its midpoint.

In addition to these three conventions, remember that the highest scores go at the top and the lowest scores at the bottom.

Converting Unorganized Data into a Grouped Frequency Distribution

Now that you know the conventions for establishing class intervals, we will go through the steps for converting a mass of data like that in Table 3.1 into a grouped frequency distribution like Table 3.5:

1. Find the highest and lowest scores. In Table 3.1, the highest score is 65, and the lowest score is 23.
2. Find the range the scores cover by subtracting the lower limit of the lowest score from the upper limit of the highest score. The use of lower and upper limits is necessary in order to include the full range of scores.[2]
3. Determine i by a trial-and-error procedure. Remember that there are to be 10 to 20 class intervals and that the interval size should be convenient (3, 5, 10, or a multiple of 10). Dividing the range by a potential i value tells the number of class intervals that will result. For example, dividing the range of 43 by 5 provides a quotient of 8.60. Thus, $i = 5$ produces 8.6 or 9 class intervals. That does not satisfy the rule calling for at least 10 intervals, but it is close and might be acceptable. In most such cases, however, it is better to use a smaller i and get a larger number of intervals. Dividing the range by 3 ($43 \div 3$) gives

[2]For a more complete explanation, with a picture, see page 44.

you 14.33 or 15 class intervals. It sometimes happens that this process results in an extra class interval. This occurs when the lowest score is such that extra scores must be added to the bottom of the distribution to start the interval with a multiple of i. For the data in Table 3.1, the most appropriate interval size is 3, resulting in 15 class intervals.

4. Begin the bottom interval with the lowest score, if it is a multiple of i. If the lowest score is not a multiple of i, begin the interval with the next lower number that is a multiple of i. In the data of Table 3.1, the lowest score, 23, is not a multiple of i. Begin the interval with 21, since it is a multiple of 3. The lowest class interval, then, is 21–23. From there on, it's easy. Simply begin the next interval with the next number and end it such that it includes three score values (24–26). Look at the class intervals in Table 3.5. Notice that each interval begins with a number evenly divisible by 3.

5. The rest of the process is the same as for a simple frequency distribution. For each score in the unorganized data, put a tally mark beside its class interval and underline the score. Count the tally marks and put the number into the frequency column. Add the frequency column to be sure that $\sum f = N$.

Problems

3. A psychology professor who was trying to decide how much statistics to present in her Introduction to Psychology class developed a 65-item test of statistical knowledge that covered concepts such as the median, grouped frequency distributions, and the standard deviation. She gave the test to one class of 50 students, and on the basis of the results she planned a course syllabus for that class and the other four sections being taught that year. Arrange the data into an appropriate rough-draft frequency distribution.

20	56	48	13	30	39	25	41	52	44
27	36	54	46	59	42	17	63	50	24
31	19	38	10	43	31	34	32	15	47
40	36	5	31	53	24	31	41	49	21
26	35	28	37	25	33	27	38	34	22

4. A psychology instructor read 60 related statements to a General Psychology class. He then asked the students to indicate which of the next 20 statements they had heard among the first 60. Due to the relationships among the concepts in the sentences, many seemed familiar but, in fact, none of the 20 had been read before. The following scores represent the number (out of 20) that each student had "heard earlier." Arrange them into an appropriate rough-draft frequency distribution.

14	11	10	8	12	13	11	10	16	11
11	9	9	7	14	12	9	10	11	6
13	8	11	11	9	8	13	16	10	11
9	9	8	12	11	10	9	7	10	

5. As his project for a sociology class, a student decided to survey 99 randomly selected homes to determine the average number of pairs of women's shoes in

the homes of Midwesternville. Arrange the data into an appropriate rough-draft frequency distribution.

```
10   7  14  15  22  18   5  12   8  17
23   5   3   7   6  14  11  13  20  19
 5  16   7   9  29   4   5   9   6   8
11  14   6  20  11   4   8  13  17   7
 5   6   9   5   8  25  19   6   2   4
 8   5   7  10   6  11  21   4   7   3
 9  32  10  16  12   6  33  18  15  19
17   6  10  12   7   5  20   9  13   4
13  10   3  11  28   7   9  18   5  12
 5  17   8  23   4  16   8  15  19
```

GRAPHIC PRESENTATION OF DATA

In order to communicate your research findings to colleagues better (and to understand them better yourself), you will often find it useful to present the results in the form of a graph. It has been said, with considerable truth, that one picture is worth a thousand words; and a graph is a type of picture. Almost any data can be presented graphically. The major purpose of a graph is to get a clear overall picture of the data.

Graphs are composed of a horizontal axis (variously called the baseline, X axis, or **abscissa**) and a vertical axis called the Y axis or **ordinate.** We will take what seems to be the simplest course and use the terms X and Y.

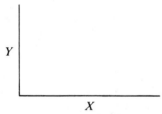

We will describe two kinds of graphs. The first kind is used to present frequency distributions like those you have been constructing. **Frequency polygons, histograms,** and **bar graphs** are examples of the first kind of graph. The second kind we will describe is the **line graph,** which is used to present the relationship between two different variables.

Frequency Distributions

The kind of variable you have measured determines whether you use a frequency polygon, a histogram, or a bar graph to present data. A frequency polygon or histogram is used for quantitative data, and the bar graph is used for qualitative data. It is not wrong to use a bar graph for quantitative data, but most researchers follow the rule just given. Qualitative data, however, *should not* be presented with a frequency polygon or a histogram. The self-esteem scores

(Table 3.1) are an example of quantitative data; the data on yard signs in Problem 2 are an example of qualitative data.

The frequency polygon.   Figure 3.1 shows a frequency polygon based on the grouped frequency distribution in Table 3.5. We will use it to demonstrate the characteristics of all frequency polygons. On the X axis we placed the midpoints of the class intervals, and named the variable self-esteem scores. Notice that the midpoints are spaced at equal intervals, with the smallest midpoint at the left and the largest midpoint at the right. The Y axis is labeled "Frequencies" and is also marked off into equal intervals.

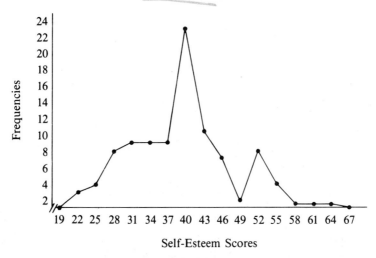

Figure   Frequency polygon of self-esteem scores of 100 college students.
3.1

Graphs are designed to "look right." They look right if *the height of the figure (Y axis) is 60 percent to 75 percent of its length (X axis)*. Since the midpoints must be plotted along the X axis, you must divide the Y axis into units that will satisfy this rule. Usually, this requires a little juggling on your part. Darrell Huff (1954) offers an excellent demonstration of the misleading effects that occur when this convention is violated.

The intersection of the X and Y axes is usually the zero point for both variables. For the Y axis in Figure 3.1, this is indeed the case. The distance on the Y axis is the same from zero to two as from two to four, and so on. On the X axis, however, that is not the case. Here, the scale jumps from zero to 19 and then is divided into equal units of three. It is conventional to indicate a break in the measuring scale by breaking the axis with slash marks between zero and the lowest score used, as we did in Figure 3.1. It is also conventional to close a polygon at both ends by connecting the curve to the X axis.

Each point of the frequency polygon represents two numbers: the class midpoint directly below it on the X axis and the frequency of that class directly across from it on the Y axis. By looking at the points in Figure 3.1, you can readily see that three people are represented by the midpoint 22; nine people by each of the midpoints 31, 34, and 37; 23 people by the midpoint 40, and so on.

The major purpose of the frequency polygon is to gain an overall view of the distribution of scores. Figure 3.1 makes it clear, for example, that the frequencies are greater for the lower scores than for the higher ones. It also illustrates rather dramatically that the greatest number of students scored in the center of the distribution.

The histogram.   Figure 3.2 is a histogram constructed from the same data that were used for the frequency polygon of Figure 3.1. Researchers may choose either of these methods for a given distribution of quantitative data, but the frequency polygon is usually preferred for several reasons: it is easier to construct, gives a generally clearer picture of trends in the data, and can be used to compare different distributions on the same graph. However, frequencies are easier to read from a histogram.

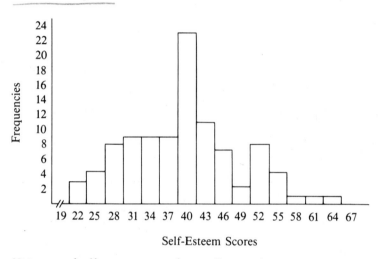

Self-Esteem Scores

Figure 3.2   Histogram of self-esteem scores of 100 college students.

Actually, the two figures are very similar. They differ only in that the histogram is made by raising bars from the X axis to the appropriate frequencies instead of plotting points above the midpoints. The width of a bar is from the lower to the upper limit of its class interval. Notice that there is no space between the bars.

The bar graph.   The third type of graph that presents frequency distributions is the bar graph. A bar graph presents frequencies of the categories of a qualitative variable. An example of a qualitative variable is laundry detergent; there are many different brands (types of the variable), but the brands don't tell you the order they go in.[3]

Figure 3.3 is an example of a bar graph. Notice that each bar is separated by a small space. This bar graph was constructed by a grocery store manager

---

[3]With quantitative variables, the measurements of the variable impose an order on themselves. Arithmetic achievement scores of 43 and 51 tell you the order they belong in. "Tide" and "Cheer" do not signify any order.

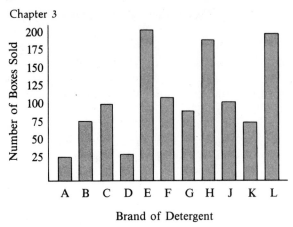

Brand of Detergent

**Figure
3.3**   Bar graph of number of boxes of laundry detergent sold in a grocery store over a one-week period.

who had a practical problem to solve. One side of an aisle in his store was stocked with laundry detergent, and he had no more space for this kind of product. How much of the available space should he allot for each brand? For one week, he kept a record of the number of boxes of each brand sold. From this frequency distribution of scores on a qualitative variable, he constructed the bar graph in Figure 3.3. (He, of course, used the names of the brands. We wouldn't dare!)

Brands E, H, and L are obviously the big sellers and should get the greatest amount of space. Brands A and D need very little space. The other brands fall between these. The grocer, of course, would probably consider the relative profits from the sale of the different brands in order to determine just how much space to allot to each. Our purpose here is only to illustrate the use of the bar graph to present qualitative data.

The Line Graph

Perhaps the most frequently used graph in scientific books and journal articles is the line graph. A line graph is used to present the relationship between two variables. A point on a line graph represents the two scores made by one person on each of the two variables. Often, the mean of a group is used rather than one person, but the idea is the same: a group with a mean score of X on one variable had a mean score of Y on the other variable. The point on the graph represents the means of that group on both variables.

Figure 3.4 is an example of a line graph of the relationship between subjects' scores on an anxiety test and their scores on a difficult problem-solving task. Many studies have discovered this general relationship. Notice that performance on the task is better and better for subjects with higher and higher anxiety scores up to the middle range of anxiety. However, as anxiety scores continue to increase, performance scores decrease. Chapter 5, "Correlation and Regression," will make extensive use of a version of this type of line graph.

A variation of the line graph places performance scores on the Y axis and some condition of training on the X axis. Examples of such training conditions are: number of trials, hours of food deprivation, year in school, and amount of

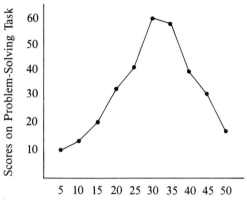

**Figure
3.4**

Hypothetical line graph of the relationship between anxiety scores and scores on a complex problem-solving task.

reinforcement. The "score" on the training condition is assigned by the experimenter. Figure 3.5 is a generalized learning curve with a performance measure (scores) on the Y axis and number of reinforced trials on the X axis. Early in training (after only one or two trials), performance is poor. As trials continue, performance improves rapidly at first and then more and more slowly. Finally, at the extreme right-hand portion of the graph, performance has leveled off; continued trials do not produce further changes in the scores.

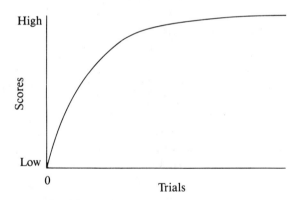

**Figure
3.5**

Generalized learning curve.

A line graph, then, presents a picture of the relationship between two variables. By looking at the line, you can tell what changes take place in the Y variable as the value of the X variable changes.

DESCRIBING DISTRIBUTIONS

There are three ways to describe the form or shape of a distribution: verbally, pictorially, and mathematically. In this section we will use the first two of these ways. We will not cover mathematical methods that describe the form of a distribution except for one method which appears in Chapter 12, the chapter on Chi Square.

Bell-Shaped Distributions

Look back at Table 3.5, graphed as Figure 3.1. Notice that the largest frequencies are found in the middle of the distribution. The same thing is true in Problem 1 of this chapter. These distributions are referred to as bell-shaped.

There is a special case of a bell-shaped distribution that you will soon come to know very well. It is called the **Normal Distribution** or **Normal Curve**. Figure 3.6 is an example of this distribution.

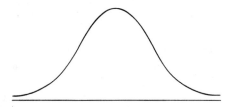

Figure 3.6   A normal distribution.

Skewed Distributions

In some distributions the scores with the greatest frequency are not found in the middle but near one end. Such distributions are said to be **skewed**.

The word *skew* is similar to the word *skewer*, the name of the cooking implement used in making shish kebab. A skewer is long and pointed and is thicker at one end than the other (not symmetrical). Although skewed distributions do not function like skewers (you would have a terrible time poking one through a chunk of lamb), the name does help you remember that a skewed distribution has a thin point on one side.

Figures 3.7 and 3.8 are illustrations of skewed distributions. Figure 3.7 is *positively skewed*; the thin point is toward the high scores, and the most frequent

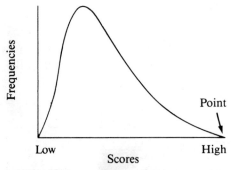

Figure 3.7   Example of a positively skewed distribution. Scores with the largest frequencies are concentrated among the low scores.

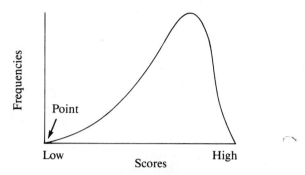

Figure
3.8

Example of a negatively skewed distribution. Scores with the largest frequencies are concentrated among the high scores.

scores are the low ones. Figure 3.8 is *negatively skewed*; the thin point or skinny end is toward the low scores, and the most frequent scores are the high ones.

Other Shapes

Curves of frequency distributions can, of course, take many different shapes. Some shapes are common enough to have names, and examples of these are presented in Figure 3.9.

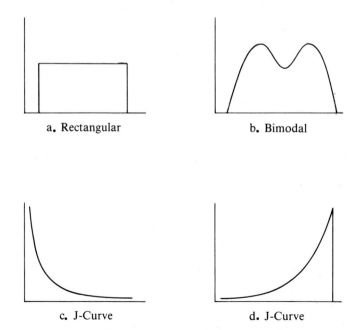

Figure
3.9

Examples of rectangular, bimodal, and J-curves.

Rectangular distribution. Figure 3.9a, the **Rectangular Distribution**, occurs when the frequency of each value on the X axis is the same. You will see this distribution again in Chapter 6.

Bimodal distribution. To understand the name of Figure 3.9b, you need to know that the mode of a distribution is the score that occurs most frequently. If two scores share this distinction, and there are scores with smaller frequencies between them, the distribution is said to be **bimodal**. Figure 3.9b is a picture of such a distribution.

J-curves. Finally, the c and d portions of Figure 3.9 are called **J-curves**. By looking at Figure 3.9c and using your imagination, the name will make sense to you. Both of these J-curves are examples of severely skewed distributions.

Problems

6. Answer the following questions for Figure 3.1.
   a. What is the meaning of the number 55 on the X axis?
   b. What is the meaning of the number 8 on the Y axis?
   c. How many students scored in the class interval 48–50?
   d. What information can you obtain from the point directly above the number 25 on the X axis?
7. Decide whether the following distributions should be graphed as frequency polygons or as bar graphs. Graph both distributions.

a. Class Interval	f
50–54	1
45–49	1
40–44	2
35–39	4
30–34	5
25–29	7
20–24	10
15–19	12
10–14	6
5–9	2

b. Class Interval	f
54–56	3
51–53	7
48–50	15
45–47	14
42–44	11
39–41	8
36–38	7
33–35	4
30–32	5
27–29	2
24–26	0
21–23	0
18–20	1

8. Determine the direction of the skew for the two curves in Problem 7 by examining the curves or the frequency distributions (or both).
9. Look at the simple frequency distribution that you constructed for Problem 2. Which kind of graph should be used to display these data? Graph the distribution.
10. Without looking at Figures 3.6 or 3.9 sketch the form of the Normal Distribution, a Rectangular Distribution, a Bimodal Distribution and two J-curves.
11. Is the point of a positively skewed distribution directed to the right or left?
12. Distinguish between a line graph and a frequency polygon.
13. For each distribution listed, tell whether it would be positively skewed, negatively skewed, or approximately symmetrical (bell-shaped or rectangular).

a. age of all people alive today
b. age in months of all first graders
c. number of children in families
d. wages in a large manufacturing plant
e. age at death of everyone who died in one year
f. shoe size

MEASURES OF CENTRAL VALUE

So far in this chapter you have learned some ways of condensing, summarizing, and clarifying data through the use of frequency distributions and graphs. These are valuable aids in understanding the meaning of your data and in making its meaning clear to others. In this section you will learn other (and even more valuable) ways of summarizing and giving meaning to data—the calculation of three measures of central value. Measures of **central value** (often called measures of central tendency) give you one score or measure that represents, or is typical of, an entire group of scores.

Recall from Chapter 1 that statistics are the numbers you get from samples and parameters are the numbers you get from populations. In the case of all three measures of central value, the formula for calculating the statistic is the same as that for calculating the parameter.[4] The interpretations, however, are different.

We will use the mean, which you are already somewhat familiar with, as an example. A sample mean, $\bar{X}$ (pronounced "ex-bar"), is only one of many possible sample means from a population. Suppose you have a sample from a population and you calculate the sample mean. Since other samples from that same population would probably produce somewhat different $\bar{X}$'s, there is some uncertainty that goes with the one $\bar{X}$ you have. However, if you measured the entire population, your mean, μ (pronounced "mew"), would be the only one and would carry no uncertainty with it. Clearly, the parameter μ is more desirable than the statistic $\bar{X}$, but, unfortunately, it is often impossible to measure the entire population. Most of the time you use $\bar{X}$ as your best estimate of μ.

The difference in interpretation, then, is that a statistic carries some uncertainty with it, a parameter does not.

The Mean

Let's begin with a very simple example. Suppose a college freshman arrives at school in the fall with a promise of a monthly allowance for spending money. Sure enough, on the first of each month the check comes in the mail. However, by Thanksgiving our student has discovered a monthly problem: there is too much month left at the end of the money.

---

[4]This is not true for the standard deviation (Chapter 4).

In pondering his problem it occurs to him that lots of money is escaping from his pocket at the Student Center. So, for a two-week period, he keeps a careful accounting of every cent spent at the Center on soft drinks, snacks, video games, coffee, and so forth. He then computes the mean amount spent per day. His data are presented in Table 3.7. You probably already know how to compute the mean of this set of scores. To find the mean, add the scores and divide that sum by the number of scores.

In terms of a formula:

$$\bar{X} = \frac{\sum X}{N}.$$

For the data in Table 3.7:

$$\bar{X} = \frac{\sum X}{N} = \frac{\$24.98}{14} = \$1.78.$$

Table 3.7

Amount of Money Spent Per Day at the Student Center Over a Two-Week Period

Day	Amount Spent
1	$2.56
2	.47
3	1.25
4	.00
5	3.25
6	1.15
7	.00
8	.00
9	6.78
10	2.12
11	.00
12	.00
13	3.78
14	3.62
	$\sum = \$24.98$

These are data for a two-week period but our freshman is interested in his expenditures for at least one month and more likely, for many months. Thus, this is a sample mean and the $\bar{X}$ symbol is appropriate. This $\bar{X}$ is an estimate of the mean amount per day that our friend spends in the Student Center.

Now we come to an important part of any statistical analysis, which is to answer the question, "So what?" Calculating numbers or drawing graphs is a part of almost every statistical problem, but unless you can tell the story of what the numbers or pictures mean, you won't find statistics worthwhile. We will deal with interpretation now, and again later.

The first interpretation you can make from Table 3.7 is to estimate monthly Student Center expenses. This is easy to do. Thirty days times $1.78 = $53.40.

Now, let's suppose that our student decides that this $53.40 is an important part of the "out of money before the end of the month" phenomenon. It strikes us that our student has three options to choose from. The first is to obtain more money. A second is to cut down on money spent at the Student Center. The third is to justify leaving things as they are. For this third option, our student might perform an economic analysis to determine what he gets in return for his $50 plus a month. His list might be pretty impressive: some excellent social times, several dates, occasional information about classes and courses and professors, a borrowed book that was just super, hundreds of calories, and more.

The point of all this is that in order to attack this money problem you need to know how to calculate a mean. However, an answer of $1.78 doesn't have much meaning by itself. Interpretation and comparisons are called for.

Characteristics of the mean.   There are two characteristics of the mean that we need to introduce you to now. Both characteristics will come up again later.

First, if the mean of a distribution is subtracted from each score in that distribution and the differences added algebraically, the sum will be zero. That is, $\sum (X - \bar{X}) = 0$. Each difference score is called a deviation and these will be explained more fully on p. 68 in Chapter 4.

Second, the mean is defined as the point about which the sum of the squared deviations is minimized. If the mean is subtracted from each score and each deviation is squared and all squared deviations are added together, the resulting sum will be smaller than if any number other than the mean had been used. That is, $\sum (X - \bar{X})^2$ is a minimum. This "least squares" characteristic of the mean will be referred to again.

The Median

The **median** is the point that divides a distribution of scores exactly in half. Half of the scores in the distribution are above the median and half are below it.

To find the median of the Student Center expense data, arrange the daily expenditures from highest to lowest as we have done in Table 3.8. Since there are 14 scores, the halfway point, or median, will have seven scores above it and seven scores below it. The seventh score from the bottom is $1.15. The seventh

Table 3.8	Data of Table 3.7 Arranged in Descending Order

X	
$6.78	
3.78	
3.62	
3.25	7 scores
2.56	
2.12	
1.25	
	Median = $1.20
1.15	
.47	
0	
0	7 scores
0	
0	
0	

score from the top is $1.25. The median, then, is halfway between these two, or $1.20. Remember, the median is a *point* in the distribution, and it may or may not be an actual score.

What interpretation can we make of a median of $1.20? The simplest would be that half the days our student spends less than $1.20 in the Student Center, and the other half he spends more. We will have more to say about the interpretation of the median later in the chapter.

What if there had been an odd number of days in our student's sample? Suppose he had chosen to sample half a month or 15 days. Then the median would be the 8th score. This would leave seven scores above and seven below. For example, had an additional day been included, during which $3.12 was spent, the median would be $1.25.

The Mode

The third central value statistic is the mode. As mentioned earlier, the mode is the most frequently occurring score—the score with the greatest frequency.

For the Student Center expense data, the mode is $0.00. This can be seen most easily in Table 3.8. The zero amount occurred 5 times and all other amounts occurred only once.

FINDING CENTRAL VALUES OF SIMPLE FREQUENCY DISTRIBUTIONS

Mean

Table 3.9 is an expanded version of Table 3.4, the distribution of self-esteem scores. We will use Table 3.9 to illustrate the steps for calculating the mean from a simple frequency distribution.

First, multiply each score in the X column by its corresponding f value, so that all the people making a particular score are included. Sum the fX values and divide by N. (N is the sum of the f values.) This will give you the mean of a simple frequency distribution. In terms of a formula,

$$\mu \text{ or } \bar{X} = \frac{\sum fX}{N}.$$

For the data in Table 3.9,

$$\mu \text{ or } \bar{X} = \frac{3943}{100} = 39.43.$$

To answer the question of whether this 39.43 is an $\bar{X}$ or a μ, we would need more information. If the 100 scores were a population, we would have a

Table
3.9

Simple Frequency Distribution of Self-Esteem Scores. Calculating the Mean

Self-Esteem Scores (X)	f	fX
65	1	65
61	1	61
58	1	58
55	4	220
53	3	159
51	5	225
49	2	98
47	4	188
46	3	138
44	5	220
42	6	252
41	8	328
40	5	200
39	10	390
38	4	152
36	5	180
34	9	306
32	4	128
30	5	150
29	5	145
27	3	81
25	4	100
23	3	69
Σ	100	3943

$$\mu \text{ or } \bar{X} = \frac{\Sigma fX}{N} = \frac{3943}{100} = 39.43$$

μ, but if they represent a sample of some larger population, $\bar{X}$ would be the appropriate symbol.

Median

Since there are 100 scores, the median will be a point that has 50 scores above it and 50 below it. If you begin adding frequencies in Table 3.9 from the bottom $(3 + 4 + 3 + 5 + \cdots)$ you will find a total of 42 when you include a score of 38. To include 39 would make your total 52, more than you need. So the median is somewhere among those 10 scores of 39.

Remember from Chapter 1 that any number actually stands for a range of numbers that has a lower and upper limit. This number, 39, has a lower limit of 38.5 and an upper limit of 39.5. To find the exact median somewhere within the range of 38.5–39.5, you use a procedure called **interpolation**. We will give you the procedure and the reasoning that goes with it at the same time. Study it until you understand it. It will come up again.

There are 42 scores below 39. You will need eight more $(50 - 42 = 8)$ scores to reach the median. Since there are ten scores of 39, you need 8/10 of them to reach the median. Assume that those ten scores of 39 are distributed evenly throughout the interval of 38.5 and 39.5 and that, therefore, the median is 8/10 of the way through the interval. Adding .8 to the lower limit of the interval, 38.5, gives you 39.3, which is the median for these scores.

When the number of scores is odd, the same reasoning applies. However, the halfway point will be a number with .5 at the end of it. For example, had the number of self-esteem scores been 99, the median would be the point with 49.5 scores above it. In such a case you might find that the median was 8.5/10 of the way through an interval. You will have a chance to practice on this in Problem 16.

Error Detection

Calculating the median by starting from the top of the distribution will produce the same answer as calculating it by starting from the bottom.

Mode

It is easy to find the mode from a simple frequency distribution. In Table 3.9, more people had a score of 39 than any other score, so 39 is the mode. You will note, however, that it was close. Ten people scored 39, but nine scored 34 and eight scored 41. Just a little random variation could have caused significant changes in the mode. This instability is one reason the mode is not a popular central value statistic.

Problems

14. Find the median for the following sets of scores.
    a. 2, 5, 15, 3, 9
    b. 9, 13, 16, 20, 12, 11
    c. 8, 11, 11, 8, 11, 8
15. Which of the following distributions would be described as bimodal?
    a. 10, 12, 9, 11, 14, 9, 16, 9, 13, 20
    b. 21, 17, 6, 19, 23, 19, 12, 19, 16, 7
    c. 14, 18, 16, 28, 14, 14, 17, 18, 18, 6
16. For Problem 4, find the mean, median, and mode.
17. For Problem 2
    a. Determine which of the measures of central value is appropriate and find it.
    b. Is this central value measure a statistic or a parameter?
    c. Write a sentence of interpretation.
18. Examine Problems 3 and 5 to determine whether the numbers are samples or populations.
19. Find the median of each distribution below.

a. X	f	b. 9, 4, 3, 6, 5, 3, 7, 5, 2
12	4	3, 6, 5, 7, 4, 2, 5, 6, 4
11	3	
10	5	
9	4	
8	2	
7	1	

FINDING CENTRAL VALUES OF GROUPED FREQUENCY DISTRIBUTIONS

As we mentioned earlier, the principal reason for constructing a grouped frequency distribution is to draw a graph or to present the data as a table. Sometimes, however, you need to find central values from such presentations. This involves only a step or two more than finding such values from a simple frequency distribution.

Mean

The procedure for finding the mean of a grouped frequency distribution is similar to that for determining the mean of a simple frequency distribution. In the grouped distribution, however, the midpoint of each interval represents all the scores in the interval. Look at Table 3.10, which is an expansion of Table 3.5. Assume that the scores in the interval are evenly distributed throughout the interval. Thus, X is the mean for all scores within the interval. Multiply each X by its f value in order to include all frequencies in that interval. Place the product in the fX column. Summing the fX column provides $\sum fX$, which, when divided by N, yields the mean.

In terms of a formula,

$$\mu \text{ or } \bar{X} = \frac{\sum fX}{N}.$$

Table 3.10   **A Grouped Frequency Distribution of Self-Esteem Scores, Expanded to Include Columns for Calculating the Mean**

Class Interval	Midpoint X	f	fX
63–65	64	1	64
60–62	61	1	61
57–59	58	1	58
54–56	55	4	220
51–53	52	8	416
48–50	49	2	98
45–47	46	7	322
42–44	43	11	473
39–41	40	23	920
36–38	37	9	333
33–35	34	9	306
30–32	31	9	279
27–29	28	8	224
24–26	25	4	100
21–23	22	3	66
		$\sum$ 100	3940

$$\mu \text{ or } \bar{X} = \frac{\sum fX}{N} = \frac{3940}{100} = 39.40.$$

For Table 3.10,

$$\mu \text{ or } \bar{X} = \frac{3940}{100} = 39.40.$$

Note that grouping introduces minor inaccuracies into your results. The mean self-esteem score from the raw data or from the simple frequency distribution is 39.43. For this grouping the mean is 39.40. (The mean for Table 3.6, which has coarser grouping, is off even more. That mean is 39.00.)

Median

Finding the median of a grouped distribution usually requires interpolation within the interval containing the median. That is the case for Table 3.10. As before, we are looking for the point that has 50 frequencies above it and 50 frequencies below it. Adding frequencies from the bottom of the distribution, you find that there are 42 who scored below the interval 39–41. You need 8 more frequencies $(50 - 42 = 8)$ to find the median. Since 23 people scored in the interval 39–41, you need 8 of these 23 frequencies or 8/23. Again, *assume that the 23 people in the interval are evenly distributed through the interval.* Thus, you need the same proportion of score points in the interval as you have frequencies—that is, 8/23 or .35 of the 3 score points in the interval. Since .35 × 3 = 1.05, you must go 1.05 score points into the interval to reach the median. Since the lower limit of the interval is 38.5, add 1.05 to find the median, which is 39.55. Figure 3.10, which will require careful study, illustrates this procedure.

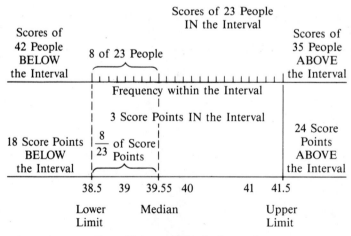

Figure 3.10   Finding the median within a class interval.

In summary, the steps for finding the median in a grouped frequency distribution are as follows.

1. Divide N by 2.
2. Starting at the bottom of the distribution, add the frequencies until you find the interval containing the median.
3. Subtract from $N/2$ the total frequencies of all intervals below the interval containing the median.
4. Divide the difference found in step 3 by the number of frequencies in the interval containing the median.
5. Multiply the proportion found in step 4 by i.
6. Add the product found in step 5 to the lower limit of the interval containing the median. That sum is the median.

Mode

The mode is the midpoint of the interval which has the highest frequency. In Table 3.10, the interval 39–41 has the highest frequency—23. The midpoint of that interval, 40, is the mode.

Error Detection

Eyeballing data is a valuable means of avoiding gross errors. Begin a problem by glancing at it and making a quick estimate of the answers. If the answers you then calculate differ from what you expect on the basis of eyeballing, wisdom dictates that you try to reconcile the difference. You have either overlooked something when eyeballing or made a mistake in your computations.

THE MEAN, MEDIAN, AND MODE COMPARED

A common question is "Which measure of central value should I use?" The general answer is "Given a choice, use the mean." Sometimes, however, the data give you no choice. Here are three considerations that limit your choice.

Scale of Measurement

A mean is appropriate for ratio or interval scale data, but not for ordinal or nominal distributions. A median is appropriate for ratio, interval, and ordinal data, but not for nominal. The mode is appropriate for any of the four scales of measurement.

You have already thought through part of this in working Problem 17. In it you found that the yard sign names (very literally, a nominal variable) could be characterized with a mode, but that it would be impossible to try to add up the names and divide by N or to find the median of the names.

For an ordinal scale like class standing in college, either a median or a mode would make sense. The median would probably be part of the way through the classification, sophomore, and the mode would be freshman.

Skewed Distributions

Even if you have interval or ratio data, there are two situations in which the mean is inappropriate because it gives an erroneous impression of the distribution. The first situation is the case of a severely skewed distribution. The following story demonstrates why the mean is inappropriate for severely skewed distributions. The story also presents an example of a population of data.

The developer of Swampy Acres Retirement Homesites is attempting, with a computer-selected mailing list, to sell the lots in his southern paradise to northern buyers. The marks express concern that flooding might occur. The developer reassures them by explaining that the average elevation of his lots is 78.5 feet and that the water has never exceeded 25 feet in that area. On the average, he has told the truth; but this average truth is misleading. Look at the actual lay of the land in Figure 3.11 and examine the frequency distribution in Table 3.11, which summarizes the picture.

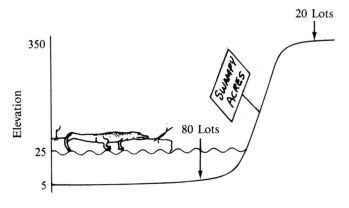

Figure 3.11   Elevation of Swampy Acres.

Table 3.11   **Frequency Distribution of Lot Elevations at Swampy Acres**

Elevation in feet (X)	Number of Lots (f)	fX
348–352	20	7000
13–17	30	450
8–12	30	300
3–7	20	100
	100	7850

$$\mu = \frac{\sum fX}{N} = \frac{7850}{100} = 78.50 \text{ feet.}$$

The mean elevation, as the developer said, is 78.5 feet; however, only 20 lots, all on a cliff, are out of the flood zone. The other 80 lots are, on the average, under water. The mean, in this case, is misleading. In this instance, the central value that describes the typical case is the median because it is unaffected by the size of the few extreme lots on the cliff. The median elevation is 12.5 feet, well below the high-water mark. Darrell Huff's delightful and informative book *How to Lie with Statistics* (1954) gives a number of such examples. We heartily recommend this book to you. It provides many cautions concerning misinformation conveyed through the use of the inappropriate statistic. A more recent and equally delightful book is *Flaws and Fallacies in Statistical Thinking* by Stephen Campbell (1974).

Open-Ended Categories

There is another instance that requires a median, even though you have a symmetrical distribution. This is when the class interval with the largest (or smallest) scores is not limited. In such a case, you do not have a midpoint and, therefore, *cannot* compute a mean. For example, age data are sometimes reported with the highest category as "75 and over." The mean cannot be computed. Thus, when one or both of the extreme class intervals is not limited, the median is the appropriate measure of central value.

To reiterate: given a choice, use the mean.

DETERMINING SKEWNESS FROM THE MEAN AND MEDIAN

Examining the relationship of the mean to the median is a way of determining the direction of skew in a distribution without having to draw a graph. When the mean is smaller than the median, there is some amount of negative skew. When the mean is larger than the median, there is positive skew. The reason for this is that the mean is affected by the size of the numbers and is pulled in the direction of the extreme scores. The median is not influenced by the size of the scores. The relationship between the mean and the median is illustrated by Figure 3.12. The size of the difference between the mean and the median gives you an indication of how much the distribution is skewed. There is a mathematical way of measuring the *degree* of skewness that is more precise than eyeballing, but it is beyond the scope of this book.

THE MEAN OF A SET OF MEANS ($\bar{\bar{X}}$)

Occasions arise in which means are available from several samples taken from the same population. If these means are combined, the mean of the set of means will give you the best estimate of the population parameter, μ. If every sample has the same N, you can compute the average mean simply by adding the means and dividing by the number of means. If, however, the means to be averaged have varying N's, it is essential that you take into account the various sample sizes by multiplying each mean by its own N before summing.

Here is a story that illustrates the right way and the wrong way to calculate the mean of a set of means. After his sophomore year one of your authors found

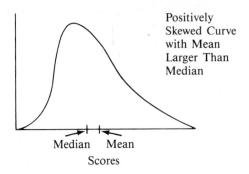

Positively
Skewed Curve
with Mean
Larger Than
Median

Median   Mean

Scores

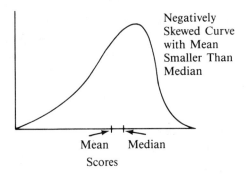

Negatively
Skewed Curve
with Mean
Smaller Than
Median

Mean   Median

Scores

**Figure
3.12**   Illustration of the effect of skewness on the relative position of the mean and median.

that if a person had a cumulative grade point average of 3.25 in the middle of his junior year, he could begin a program to "graduate with honors." Calculating a cumulative GPA seemed easy enough to do: four semesters had produced GPA's of 3.41, 3.63, 3.37, and 2.16. Given another GPA of 3.80, the sum of the five semesters would be 16.37 and dividing by five gave an average of 3.27, well above the required 3.25.

Graduating with honors seemed like a neat ending for college, so our hero embarked on a goal-oriented semester: a B in German and an A in everything else, a GPA of 3.80. And, at the end of the semester the goal had been accomplished. Unfortunately, "graduating with honors" was not to be.

There was a flaw in the method of calculating the cumulative GPA. The method assumed that all of the semesters were equal in weight, that they had been all based on the same number of semester hours. The calculations based on this assumption are shown on the left side of Table 3.12.

However, all five semesters were not the same. For example, the semester with the GPA of 2.16 was based on 19 semester hours, rather than the usual 16 or so.[5] Thus, this semester must be weighted more heavily than semesters in which only 16 hours were taken.

The right side of Table 3.12 shows a correct way to calculate a cumulative GPA. Each semester's GPA is multiplied by its number of semester hours. These

[5]The semester was more educational than a GPA of 2.16 would indicate. A great deal of American literature was consumed that spring, but unfortunately, I was not registered for any courses in American Literature.

products are summed and that total is divided by the sum of the semester hours. As you can see from the figures on the right, the actual cumulative GPA was 3.24, not high enough to qualify for the honors program.

Table 3.12 **Two Methods of Calculating a Mean of a Set of Means from Five Semesters' GPA's. The Method on the Left is Correct Only If All Semesters are Equal in Weight**

Semester GPA's	Semester GPA's	Hours Credit	GPA's × Hours
3.41	3.41	17	58
3.63	3.63	16	58
3.37	3.37	19	64
2.16	2.16	19	41
3.80	3.80	16	61
$\sum = 16.37$		$\sum = 87$	$\sum = 282$

$$\bar{\bar{X}} = \frac{16.37}{5} = 3.27 \qquad\qquad \bar{\bar{X}} = \frac{282}{87} = 3.24$$

More generally, to find the mean of a set of means, multiply each separate mean by its N, add these products together, and divide by the sum of the N's. Thus, means of 2.0, 3.0, and 4.0 based on N's of 6, 3, and 2 produce an overall mean (mean of a set of means) of 2.64 ($29 \div 11 = 2.64$).

Clue to the Future

The distributions that you have been working with in this chapter are **empirical distributions** based on scores actually gathered in experiments. This chapter and the next two are about these empirical frequency distributions. Starting with Chapter 6, and throughout the rest of the book, you will also make use of **theoretical distributions**—distributions based on mathematical formulas and logic rather than on actual observations.

Problems

20. Find the mean, median, and mode of the grouped frequency distribution you constructed for Problem 3.

21. Find the mean, median, and mode of the grouped frequency distribution you constructed for Problem 5.

22. Suppose you overheard the following statement: "My grade-point average for three semesters is 3.0. I made a 3.5 average for the six credits I took in summer school, a 3.0 for 12 credits in the fall, and a 2.5 for 16 credits last spring. That averages out to 3.0." What do you think of such reasoning? If you are dissatisfied with it, explain how you would reason differently.

23. For the following situations, tell which central value is appropriate and why.
   a. As part of a study on prestige, an investigator sat on a corner in a high-income residential area and classified passing autombiles according to producers: General Motors, Ford, Chrysler, American Motors, and foreign.
   b. In a study of helping behavior, an investigator pretended to have locked himself out of his car. Passersby who stopped were classified on a scale of

1 to 5 as (1) very helpful, (2) helpful, (3) slightly helpful, (4) neutral, and (5) discourteous.

c. In a study of per capita income in a city, the following income categories were established: $0–$5000, $5001–$10,000, $10,001–$15,000, $15,001–$20,000, $20,001–$25,000, $25,001–$30,000, $30,001–$35,000, $35,001–$40,000, $40,001–$45,000, and $45,001–up.

d. In a study of per capita income in a city, the following income categories were established: $0–$5000, $5001–$10,000, $10,001–$15,000, $15,001–$20,000, $20,001–$25,000, $25,001–$30,000, $30,001–$35,000, $35,001–$40,000, $40,001–$45,000, and $45,001–$50,000.

e. First admissions to a state mental hospital for the previous five years were classified as schizophrenic, depressed, paranoid, senile, and other.

f. A teacher gave her class an arithmetic test, and most of the children scored in the range 70–79. A few were above this, and a few were below.

g. The frequency distribution of pairs of shoes in Problem 5.

24. A senior psychology major performed the same experiment on three groups, obtaining means of 74, 69, and 75. The groups consisted of 12, 31, and 17 subjects respectively. What is the overall mean for all subjects?

25. In Problem 8, you eyeballed two distributions to determine the direction of the skew. Verify that judgment now with a comparison of the means and medians of the distributions.

26. A three-year veteran of the local baseball team was figuring out his lifetime batting average. During the first year he played half the season and batted .350 (28 hits in 80 at-bats). The second year he had about twice the at-bats and his average was .325. The third year, although he played even more regularly, he was in a slump and batted only .275. By adding the three season averages and dividing by three, he found his lifetime batting average to be .317. Is this correct? Justify your answer.

4 VARIABILITY

Objectives for Chapter 4: After studying the text and working the problems in this chapter, you should be able to:

1. explain the concept of variability,
2. find the range of distribution,
3. distinguish among the standard deviation of a population, the standard deviation of a sample used to describe the sample, and the standard deviation of a sample used to estimate the population standard deviation,
4. calculate a standard deviation from grouped or ungrouped data,
5 calculate the variance of a distribution,
6. convert raw scores to z scores, and
7. use z scores to compare a score in one distribution with a score in a second distribution.

You learned in Chapter 3 how to compute some central value statistics that give you *one number* to represent all the numbers in a set of scores. A central value statistic is a typical value for the entire distribution. However, it is also important to know about the **variability** of the scores—that is, how much the scores differ from the typical value. Are the scores clustered closely together, or are they spread apart?

The importance of knowing about variability is illustrated by the story of two brothers who went water skiing on Christmas day (far north of Miami). On the *average*, each skier finished his turn 5 feet from the shoreline (where one may step off the ski into only 1 foot of very cold water). This bland average of the two actual stopping places does not convey the excitement of the day. In fact, the first brother, determined to avoid the cold water, overshot the beach so much that he finally stopped rolling, scraped and bruised, up on the rocky shore 35 feet from the water. The second brother, determined not to share that fate, stopped too soon. Although he swam the 45 feet to shore very quickly, his lips were very blue. No, the average stopping place of 5 feet from shore doesn't convey the excitement of the day, but knowledge of the variability around that average does.

To take another example of the importance of variability, pretend for a moment that you are an elementary school teacher who has just been given a description of the children who will be your charges for the coming school year. You are delighted to discover that the mean IQ of this group is 115—well above the IQ of 100 that is the average for the general population. You decide that you can use more complex teaching materials and begin to plan some rather sophisticated projects for this year. But wait just a bit, you don't have all the information you need. The fact that the *mean* is 115 doesn't mean that all the children in the room are bright. You need to know the variability of the group. Consider just the two possibilities illustrated by the graphed frequency distributions in Figure 4.1. In one group, the lowest IQ is 110 and the highest is 120. In the other group, however, the lowest IQ is 75 and the highest is 155. Obviously, although the mean is the same for both groups, what will work with one group will not work

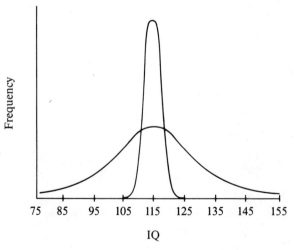

Figure 4.1   Frequency distributions of the IQs of two groups with the same mean but different variability.

with the other. One group is made up of above-average youngsters of about equal ability. The other group ranges from borderline retardates to budding geniuses.

What you need then, is a measure that will give you information about the variability of distributions. This is a chapter that is primarily about statistics and parameters that measure the variability of a distribution. The **range** is the first one we will describe. The second, the **standard deviation**, is the most important. Most of this chapter will be about the standard deviation. A third way to measure variability is with the **variance**. Finally, at the end of this chapter, **z scores** will be described. The standard deviation is necessary in order to calculate z scores, which are used to compare the relative standing of scores in two different distributions. A z score is not a measure of variability.

THE RANGE

The **range** is the upper limit of the highest score minus the lower limit of the lowest score; that is,

$$\text{range} = X_H - X_L$$

where:

X_H = the upper limit of the highest score in the distribution
and X_L = the lower limit of the lowest score in the distribution.

Thus, the range from 5 to 10 is 6. You can check this for yourself by counting the spaces in the following illustration. In terms of our formula, $10.5 - 4.5 = 6$.

Applying this same logic to a distribution with a high score of 3.7 and a low score of 2.0, the range is 1.8 $(3.75 - 1.95)$. Similarly, the range is .26 when the high score is 0.70 and the low score is .45.

As a matter of fact, you have already been exposed to the range as a measure of variability. You made use of it in Chapter 3 when you learned to set up frequency distributions. You will recall that this same procedure was used to find out how many class intervals to establish.

Although it is very easy to compute, the range has a flaw. It depends on only the two *extreme* scores. Because of this, it is easy to picture two very *different* distributions with the *same* mean and *same* range. Figure 4.2 illustrates such a situation.

What is needed is a measure of variability that will differentiate the two distributions in Figure 4.2. The range does not. Fortunately, the standard deviation, which depends on every score, clearly differentiates the two distributions in Figure 4.2.

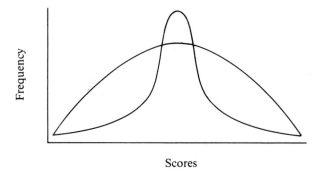

Frequency

Scores

**Figure
4.2**   Two frequency distributions with the same mean and range but with different variability.

THE STANDARD DEVIATION

A most useful and informative measure of the variability is the **standard
deviation**. Deviation is part of its name because its computation takes into account
the distance that *each* score lies away from, or deviates from, the mean. The
standard deviation is used in measuring distances along the baseline of a theoretical distribution called the normal curve—something you will learn about in
Chapter 6 and then use for the rest of your statistical life. The standard deviation
will be used throughout this book. It is difficult to overemphasize the importance
of the concept of variability and its yardstick, the standard deviation.

There are three different situations in which variability is measured with a
standard deviation. We will use three different symbols for these three situations.
As you will see, one of these situations calls for a modified formula for the
standard deviation. The three symbols and their respective situations are

σ This is the lowercase Greek letter sigma. It is the symbol for the standard
deviation of a population. σ is a parameter and is used to *describe* variability
when all the data in the population are at hand.

s s is an *estimate* of σ, in the same way that $\bar{X}$ is an estimate of μ. As you
know, parameters are rarely known. The best we can do is draw a sample
from the population and use statistics such as $\bar{X}$ and s as estimates of
parameters such as μ and σ. s, then, is an estimate of σ, and is the variability
statistic you will be using most often in this book.

S There are some occasions when you want the standard deviation of a sample
but have no interest in trying to estimate σ. In such instances, S is the
statistic to use. The computation of S is the same as the computation of σ
except that the formula for S contains $\bar{X}$ rather than μ. You will encounter
S in this book only in this chapter and in places where correlation is stressed
(principally in Chapter 5).

Distinguishing among these three standard deviations and clearly understanding them is sometimes a problem for beginning students. Our advice is to
memorize a definition of each symbol; then be alert to the situations in the text

where a standard deviation is used. With each situation you can add more meaning to your understanding. We will first discuss the calculation of σ and S (which are computed in the same way) and then deal with s.

THE STANDARD DEVIATION AS A DESCRIPTIVE INDEX OF VARIABILITY

Both σ and S are used to *describe* the variability of some data on hand. σ is a parameter based on a population; S is a statistic based on a sample. The two are calculated with the same formula. We will show you two ways to arrange the arithmetic of this formula—the raw-score method and the deviation-score method. The quick and accurate raw-score method is the one used by experienced researchers and statisticians and it is the method you will use after you complete this chapter. The deviation-score method, however, gives you a much better appreciation of what a standard deviation is measuring, and this leads to better understanding. Algebraically, the two methods are identical.

So, we will begin your introduction to the standard deviation with the deviation-score method. But before you can use this method, you need to be introduced to a common statistic, the deviation score which will also be used in situations other than the calculation of standard deviations.

Deviation Scores

A **deviation score** is a raw score minus the mean, either $X - \bar{X}$ or $X - \mu$. It is simply the difference between a score in the distribution and the mean of that distribution. Deviation scores are encountered so often that they have a special symbol, the lowercase x. Note that a capital X is used for a score and a lowercase x is used for a deviation score. Not only must you know this; you must also be sure that you (and your instructor) can tell the difference between your written versions of X and x.

Since $x = X - \bar{X}$, raw scores that are larger than the mean will have positive deviation scores, raw scores that are smaller than the mean will have negative deviation scores, and raw scores that are equal to the mean will have a deviation score equal to zero.

Table 4.1

The Computation of Deviation Scores from Raw Scores

Name	Score	$X - \mu$	x
John	14	14–8	6
Shelly	10	10–8	2
Joshua	8	8–8	0
Melissa	5	5–8	−3
Jeri	3	3–8	−5
	$\sum X = 40.$		$\sum x = 0.$

$$\mu = \frac{\sum X}{N} = \frac{40}{5} = 8.$$

Consider a small set of numbers to see how these deviation scores work. Table 4.1 gives you a brief demonstration of how to compute deviation scores for a small population of data.

In Table 4.1, we first computed the mean by adding the scores and dividing by N. We then subtracted the mean from each score to obtain the deviation scores, which appear in the right-hand column.

What a deviation score tells you is the number of points that a particular score deviates from, or differs from, the mean. In Table 4.1 the x value for John tells you that he scored six points above the mean, Joshua scored at the mean, and Jeri was five points below the mean.

Error Detection

Notice that the sum of the deviation scores is zero. This will always be the case, so summing the deviation scores serves as a useful check on computations. If the sum is not zero, you've made an error.

Problems

For each of the four distributions below, find the range and the deviation scores. Check to see that $\sum x = 0$.

1. 15, 13, 12, 10, 8, 7, 5
2. 17, 5, 1, 1
3. 3.4, 3.1, 2.7, 2.7, 2.6
4. .45, .30, .30
5. Give the symbol and purpose of the three standard deviations.

The Deviation Method of Computing σ and S

The deviation-score formula for computation of the standard deviation as a descriptive index is

$$\sigma = \sqrt{\frac{\sum(X - \mu)^2}{N}} = \sqrt{\frac{\sum x^2}{N}} \quad \text{or} \quad S = \sqrt{\frac{\sum(X - \bar{X})^2}{N}} = \sqrt{\frac{\sum x^2}{N}},$$

where: σ = the standard deviation of a population,
S = the standard deviation of a sample,
$\sum x^2$ = the sum of the squared deviations,
and N = the number of deviations (which is the same as the number of scores).

As you can see from the formulas, the standard deviation is a sort of average of the deviations from the mean;[1] thus, the standard deviation can be viewed as the average amount that scores deviate from the mean. It is sometimes called a root-mean-square because it is the square root of the mean of the squared deviation scores.

---

[1]The mean, $\sum X/N$, is an average of the scores. $\sum x^2/N$ is an average of the squared deviations.

Table 4.2 shows the results when a word processing problem was given to 10 secretaries who work for a law firm. The scores represent the time in minutes necessary to complete the problem and print the results. We will use these scores to illustrate the computation of σ. If the secretaries had been a sample and your only interest were in that sample, the standard deviation computed would be identified by S instead of σ. The computations, however, would be the same.

Table
4.2

The Deviation-Score Method of Computing σ for the Time Scores of a Population of 10 Secretaries

deviation
from the Mean

Ind'l Scores

X	x	x^2
16	5	25
14	3	9
13	2	4
12	1	1
11	0	0
11	0	0
9	−2	4
9	−2	4
8	−3	9
7	−4	16
Σ 110	0	72

$$\mu = \frac{\Sigma X}{N} = \frac{110}{10} = 11.$$

$$\sigma = \sqrt{\frac{\Sigma (X - \mu)^2}{N}} = \sqrt{\frac{\Sigma x^2}{N}} = \sqrt{\frac{72}{10}} = \sqrt{7.20} = 2.68.$$

To compute σ for the data in Table 4.2, first compute the mean by summing the scores and dividing by N. Then subtract the mean from each of the scores to obtain x for each score. Add these deviation scores to be certain that their sum is zero. Next, square each x and sum the x^2 values to obtain Σx^2. Finally, divide Σx^2 by N and extract the square root. $\sigma = 2.68$ minutes.

Now, what does this mean? Knowing that $\sigma = 2.68$ minutes, what can we say? In subsequent chapters you will learn new concepts that will lead to a more thorough understanding of the standard deviation, but at this point, we have a simple definitional interpretation for you.

The figure 2.68 is a measure of the variability of the time taken to work the word processing problem by the 10 secretaries. If σ had been 0, we would know that each person completed the task in exactly the same amount of time. (There would have been no variability.) The closer σ is to 0, the more confidence we would have in predicting that any one secretary's time was the mean for the group. Conversely, the further σ is from 0, the less confidence we have that the mean will be descriptive for a particular secretary.

Now, look back at Table 4.2 and the formula for σ. Notice what is happening. The mean is being subtracted from each score. This difference, whether positive or negative, is squared and these squared differences are added together. This sum is divided by N and the square root is found. Every score in the distribution contributes to the final answer.

Notice the contribution made by a score far from the mean: it is large. This makes sense because the standard deviation is a measure of variability and if there are scores far from the mean, they cause the standard deviation to be larger. Take a moment to think through the contribution to the standard deviation made by a score near the mean.[2]

Error Detection

All standard deviations are positive numbers. If you find yourself trying to take the square root of a negative number, you've made an error.

Problems

Compute σ or S for each distribution, using the deviation-score method.

6. 7   6   5   2
7. 14   11   10   8   8
8. 107   106   105   102
9. Compare the standard deviation of Problem 6 with that of Problem 8. What conclusion can you draw about the effect of the size of the numbers on the standard deviation?
10. Does the size of the numbers in a distribution have any effect on the mean?
11. The temperatures in this problem are averages for the months of March, June, September, and December. Calculate the mean and standard deviation for each city. Summarize your results as a sentence.

San Francisco	54°	59°	62°	52°
Albuquerque	46°	75°	70°	36°

12. No computation is needed for this one. Eyeball the following pairs of distributions to determine which of the two has the greater variability if they are not equal.
    a. 1, 2, 4, 1, 3, and 9, 7, 3, 1, 0
    b. 9, 10, 12, 11, and 4, 5, 7, 6
    c. 1, 3, 9, 6, 7, and 14, 15, 14, 13, 14
    d. 114, 113, 114, 112, 113, and 14, 13, 14, 12, 13
    e. 8, 4, 6, 3, 5, and 4, 5, 7, 6, 15

The Raw-Score Method of Computing σ and S

The deviation-score method of computing the standard deviation requires that you "cook" the raw scores by converting them to deviation scores. This

---

[2]Playing with a formula is an effective way to become comfortable with it and to understand it better. Make up a small set of numbers and calculate a standard deviation. See what happens when you change one of the numbers, add a number, or leave out a number.

"cooking" usually requires that each deviation score be rounded off, which introduces some error into the final answer. With the raw-score method, no such "cooking" is required. The raw-score method, then, is always exact and the deviation-score method is not.

Although the raw-score formula may appear to be somewhat more forbidding at first glance, it is actually easier to use than the deviation-score formula. With the raw-score formula, you don't have to go through the intermediate step of computing deviation scores. The numbers you will be working with will be larger, but your calculator probably won't mind (unless you exceed its capacity).

We will illustrate the calculation of σ and S by the raw-score method using the data of Table 4.3. These data are the same as those of Table 4.2—the word processing time scores of the 10 secretaries. Notice that the answers for σ in Table 4.3 and Table 4.2 are the same. In this case, the deviation scores did not have to be rounded, so no rounding errors were introduced.

Table 4.3   **The Raw-Score Method of Computing σ for Time Scores of a Population of 10 Secretaries**

X	X^2
16	256
14	196
13	169
12	144
11	121
11	121
9	81
9	81
8	64
7	49

$\sum = 110$ 1282

$\sum X = 110$ $\sum X^2 = 1282$ $(\sum X)^2 = (110)^2 = 12,100$

$$\sigma = \sqrt{\frac{1282 - \dfrac{(110)^2}{10}}{10}} = \sqrt{\frac{1282 - 1210}{10}} = \sqrt{\frac{72}{10}} = \sqrt{7.2} = 2.68$$

The formula for the raw-score method is

$$\sigma \text{ or } S = \sqrt{\frac{\sum X^2 - \dfrac{(\sum X)^2}{N}}{N}},$$

where: $\sum X^2 = $ the sum of the squared scores,
        $(\sum X)^2 = $ the square of the sum of the raw scores,
and $N = $ the number of scores.

For this formula, follow the steps in Table 4.3. Square the sum of the X column and divide by N. Subtract this quotient from the sum of the X^2 column.

Divide this difference by N and extract the square root. The result is σ or S.[3]

Be sure you understand the difference between the terms $\sum X^2$ and $(\sum X)^2$. The term $\sum X^2$ is obtained by squaring each score to obtain X^2 and then summing the X^2 column to obtain $\sum X^2$. The term $(\sum X)^2$ is obtained by summing the scores and then squaring $\sum X$. Playing with some numbers here should make this distinction clearer.

Error Detection

The range is usually two to five times larger than the standard deviation when $N = 100$ or less. The range (which can be calculated quickly) serves as a useful check on any large errors you may have made in calculating the standard deviation.

Problems

13. Look at the two sample distributions below. Without any calculation (just eyeball the data), decide which one has the larger standard deviation. Guess the size of S for each distribution. (You may wish to calculate the range before you choose a number.) Compute S for each distribution. Compare your computation with your guess.
   a. 6 5 4 3 2
   b. 6 6 6 2 2

14. For each of the distributions in Problem 13, divide the range by the standard deviation. Is the result between 2 and 5?

15. By now you can look at the two distributions below and see that (a) is more variable than (b). The difference in the two distributions is in the lowest scores (2 and 6). Calculate σ for each distribution, using the raw-score formula. Notice that the difference in σ is due to how extreme the lowest score is.
   a. 9 8 8 7 2
   b. 9 8 8 7 6

---

[3]Some textbooks and statisticians prefer the algebraically equivalent formula

$$\sigma \text{ or } S = \sqrt{\frac{N \sum X^2 - (\sum X)^2}{N^2}}$$

because it introduces less rounding error. (In this formula you divide only once, but in the formula in the text there are two divisions to do.) We are using the formula in the text because the same form, or parts of it, will be used in other procedures.

Yet another formula is often used in the field of testing. This formula, which requires you to begin by calculating the mean, is

$$\sigma = \sqrt{\frac{\sum X^2}{N} - \mu^2} \quad \text{and} \quad S = \sqrt{\frac{\sum X^2}{N} - \bar{X}^2}.$$

All three of these arrangements of the arithmetic are algebraically equivalent.

Clue to the Future

The term $\sum x^2$, or $\sum X^2 - \dfrac{(\sum X)^2}{N}$ is referred to as the **sum of squares**. It will be used in Chapters 10 and 11. When the sum of squares is divided by $N - 1$, producing $\dfrac{\sum x^2}{N-1}$ or

$$\dfrac{\sum X^2 - \dfrac{(\sum X)^2}{N}}{N - 1}$$ you have the **variance** or **mean square**. These terms are used later in this chapter and again in Chapters 10 and 11.

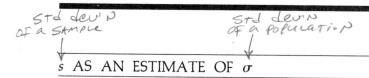

STD dev'N OF a SAMPLE

STD dev'N OF a POPULATION

s AS AN ESTIMATE OF σ

We want to emphasize again that s is the principal statistic you will learn in this chapter. It will be used again and again throughout the rest of this text.

As we explained in Chapters 1 and 3, the purpose of a sample is usually to find out something about a population. That is, a statistic based on a sample is used to estimate the parameter of the population. The most obvious pitfall in this reasoning is that two samples from the same population are often slightly different, yielding two different statistics. Which one of the statistics is closest in value to the parameter? There is no way of knowing other than by measuring the entire population, which is impossible. The best you can do is to calculate the statistic in such a way that, *on the average*, its value is equal to the parameter. In the language of the mathematical statistician, you want a statistic that is the "best estimate" of the corresponding population parameter.

It turns out that if you have sample data and you want to calculate the best estimate of σ, you should use the statistic $\sqrt{\dfrac{\sum (X - \bar{X})^2}{N - 1}}$. This statistic is symbolized with a lowercase s.[4] Note that the difference between s and σ is that s has $N - 1$ in the denominator where σ has N.

This issue of dividing by N or by $N - 1$ can leave students shrugging their shoulders and muttering, "O.K., I'll memorize it and do it however you want." We'd like, however, to explain why you use N or $N - 1$. You will recall from Chapter 3 that for any set of scores the numerical value of $\sum (X - \text{mean})^2$ is minimized. Thus, if you have a *population* of scores, the mean is μ and $\sum (X - \mu)^2$ is minimized (compared to using any value other than μ). Similarly, if you have a *sample* of scores, the mean is $\bar{X}$ and $\sum (X - \bar{X})^2$ is minimized. Finally, if you have a sample of scores but you also know μ and you calculated $\sum (X - \mu)^2$, the value will be greater than $\sum (X - \bar{X})^2$, assuming that $\bar{X}$ and μ are not exactly the same.

[4]Here is a technical point that you should know if you plan to learn more about statistics: Even with $N - 1$ in the denominator, s is not an unbiased estimate of σ. Since the bias is not very serious, s is usually used as the best estimate of σ. However, s^2 (see the next section) *is* an unbiased estimate of σ^2.

Now, faced with the problem of having a sample and wanting to estimate population variability, you would like to calculate

$$\sqrt{\frac{\sum (X - \mu)^2}{N}},$$

but of course you don't have μ. If you substitute $\bar{X}$ for μ then the numerator will be too small unless some correction is made. The correction, though, is simple: divide the numerator by a smaller amount, $N - 1$ instead of N. Note also that as N gets larger, the subtraction of 1 from N has less and less effect on the size of the estimate of variability. This makes sense because the larger the sample size is, the closer $\bar{X}$ will come to μ.

There is one new task that comes with the introduction of s. It is the decision of whether to calculate σ, S, or s on a given set of data. Your decision will be based on the purpose of the standard deviation. If the purpose is to describe, use σ or S depending on whether or not you have population data. If the purpose is to estimate a population σ from sample data, use s.

Calculating s

As in the previous section on calculating σ, a raw score formula is recommended for s.

$$s = \sqrt{\frac{\sum X^2 - \frac{(\sum X)^2}{N}}{N - 1}}.$$

This is practically the same as the formula for σ, so you will have no trouble calculating s (assuming you mastered the calculation of σ).[5]

This raw score formula will be the one you will probably use for your own data. You may be confronted, though, with someone else's frequency distribution for which you want to calculate a standard deviation. For a simple frequency distribution or a grouped frequency distribution, the formula is

$$s = \sqrt{\frac{\sum fX^2 - \frac{(\sum fX)^2}{N}}{N - 1}},$$

where all terms are defined as before.

Here is a problem that illustrates the calculation of s both for raw scores and a frequency distribution. Consider puberty. As you know, females reach puberty earlier than males (about 2 years earlier on the average). Is there any difference between the sexes in the *variability* of reaching this developmental milestone? Comparing standard deviations will give you an answer.

---

[5]Calculators that have a standard deviation function differ. Some use N in the denominator and some use $N - 1$. You will have to check yours to see how it is programmed.

We have only a sample of ages for both sexes and since the interest is in all females and males, s is the appropriate standard deviation. Table 4.4 shows the calculation of s for the females. A value of 2.19 years was obtained for the eight scores.

Table
4.4

Calculation of s for Ages at Which Females Reach Puberty (Raw Score Method)

Age (X)	X^2
17	289
15	225
13	169
12	144
12	144
11	121
11	121
11	121
$\sum X = 102$	$\sum X^2 = 1334$

$(\sum X)^2 = 10{,}404$

$$s = \sqrt{\frac{\sum X^2 - \frac{(\sum X)^2}{N}}{N-1}} = \sqrt{\frac{1334 - \frac{(102)^2}{8}}{7}} = \sqrt{\frac{1334 - 1300.50}{7}} = 2.19$$

We will illustrate the calculation of s for a simple frequency distribution using the data for males in Table 4.5. Note that the grouping causes two extra columns in the calculations. The standard deviation for males for these data is 1.44.

Calculation of s for Ages at Which Males Reach Puberty (Simple Frequency Distribution)

Age (X)	f	fX	fX^2
18	1	18	324
17	1	17	289
16	2	32	512
15	4	60	900
14	5	70	980
13	3	39	507
	$N = 16$	$\sum = 236$	$\sum = 3512$

$$s = \sqrt{\frac{\sum fX^2 - \frac{(\sum fX)^2}{N}}{N-1}} = \sqrt{\frac{3512 - \frac{(236)^2}{16}}{15}} = \sqrt{\frac{3512 - 3481}{15}} = 1.44$$

Thus, we may conclude that, based on these sample data, there is more variability among females in reaching puberty than there is among males.

Here is a final point on calculating s (or S or σ). For a grouped frequency distribution use the midpoints of the class intervals as the X values. This technique is the same as the one you used in Chapter 3 when you worked with grouped frequency distributions there. (Refer to Table 3.10 for a quick review.)

Problems

16. A researcher had a sample of scores from the freshman class on a test that measured attitudes toward authority. She wished to estimate the standard deviation of the entire freshman class, since she believed that the current group of students was more homogeneous than students in the past. Given the summary statistics below, calculate the standard deviation.

$$N = 21, \qquad \sum X = 304, \qquad \sum X^2 = 5064.$$

17. A statistics instructor wanted to know the standard deviation on the last test for a 12-member class. Use the data below and explain why you chose S or s.

$$
\begin{array}{cccccc}
12 & 15 & 9 & 8 & 14 & 10 \\
11 & 12 & 9 & 17 & 5 & 12
\end{array}
$$

18. A high-school English teacher measured the attitudes of 11th-grade students toward poetry. After a nine-week unit on poetry, she measured the students' attitudes again. She was disappointed to find that the mean change was 0. Below are some representative scores. (High scores indicate favorable attitudes.) Calculate s for both before and after, and write a conclusion based on the standard deviations.

$$
\begin{array}{lcccccccc}
\text{Before} & 7 & 5 & 3 & 5 & 5 & 4 & 5 & 6 \\
\text{After} & 9 & 8 & 2 & 1 & 8 & 9 & 1 & 2
\end{array}
$$

THE VARIANCE

In a "Clue to the Future," you were alerted to the fact that a characteristic of a distribution called the *variance* emerged in the process of computing the standard deviation. The variance is symbolized as the square of the standard deviation—that is, σ^2 (population variance) and s^2 (estimate of the population variance). In all formulas for the standard deviation, the final step is to take the square root of a number. That number is the variance. For example,

$$s = \sqrt{\frac{\sum (X - \bar{X})^2}{N - 1}} \qquad \text{and} \qquad s^2 = \frac{\sum (X - \bar{X})^2}{N - 1}$$

As is the case with the standard deviation, the difference between s^2 and σ^2 is that s^2 has $N - 1$ in the denominator while σ^2 has N. s^2 is the variance of a sample used to estimate the population variance, and σ^2 is the population variance.

The variance is used in many statistical methods. Fortunately, you already know how to compute it.[6] You will find it widely used in psychological testing

---

[6]Some calculators have a variance function, which typically uses N in the denominator. One calculator we know of uses $N - 1$ for the standard deviation and N for the variance. Given a square key and a square root key, this arrangement covers all possibilities.

and in the **analysis of variance**. The analysis of variance will be covered in Chapters 10 and 11 of this book.

z SCORES

You have used measures of central value and measures of variability to describe a *distribution* of scores. The next statistic, z, is used to describe *a single score.*

Suppose one of your friends tells you he made a 95 on a math exam. What does that tell you about his mathematics ability? Due to your previous experience with tests, 95 may seem like a pretty good score, but unless you make a couple of assumptions, a score of 95 is meaningless. To illustrate, suppose he tells you 200 points were possible. Suddenly, a score of 95 seems like something to hide. Next, he tells you the highest score on that difficult exam was 101. Now 95 looks pretty good. But what if you discover that the mean score was 98? A 95 takes a nosedive in respectability. As a final blow, you find out that 95 was the lowest score, that nobody was worse than your friend. What this example illustrates is that 95 acquires meaning only when it is compared with the rest of the test scores. In particular, a score gets its meaning from its relation to the mean and the variability of its fellow scores. A z score is a mathematical way to change a raw score so that it reflects its relationship to the mean and standard deviation of its fellow scores. The formula is

$$z = \frac{X - \bar{X}}{S}.$$

Any distribution of raw scores can be converted to a distribution of z scores; for each raw score, there is a z score. Raw scores above the mean will have positive z scores; those below the mean will have negative z scores. Since $X - \bar{X}$ is our old friend the deviation score (x), we can also write the formula as

$$z = \frac{x}{S}.$$

A z score is also called a **standard score** because it is a deviation score expressed in standard deviation units. It is the number of standard deviations a score is above or below the mean.

What a z score describes is the relation of an X to $\bar{X}$ with respect to general width (variability) of the distribution. For example, you can say that a distance of five units from X to $\bar{X}$ ($X - \bar{X} = 5$) is large or small only if you know how wide the entire distribution is. If the distribution is ten units wide, and $\bar{X} = 50$, then $X = 55$ is the largest score. If the distribution is 100 units wide, and $\bar{X} = 50$, then $X = 55$ is very similar to the mean, compared with all the other scores. This means that raw scores from different distributions cannot be compared unless the means and standard deviations of the distributions are equal. Figure 4.3 is a picture of the ideas in this paragraph.

When used as a descriptive measure of one distribution of scores, z scores range from -3 to $+3$ for $N = 100$ or less. Thus, a z score of -2.5 is one that is at or near the bottom of the distribution. However, for some distributions, the highest z score may be as low as 1, and the lowest z score may be -1. The point is that a z score tells you the relative position of a raw score in the distribution.

You will discover in later chapters that z scores are also used for inferential purposes. Much larger z scores may occur then.

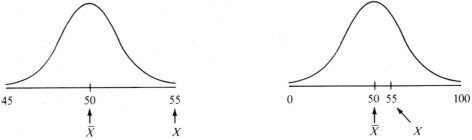

Figure 4.3

A comparison of $X - \bar{X} = 5$ in two distributions with different standard deviations.

Use of z Scores

z scores are used to compare two scores in the same distribution. They are also used to compare two scores from different distributions, even when the

Table 4.6

A Comparison of the Attitudes of 15 Fifth-Grade Students toward Arithmetic and toward Reading

Student	Attitude toward Arithmetic			Attitude toward Reading		
	Raw Score (X)	X^2	$z = \dfrac{x}{S}$	Raw Score (X)	X^2	$z = \dfrac{x}{S}$
1	45	2025	$-.02$	41	1681	$-.86$
2	57	3249	$.94$	57	3249	$.55$
3	32	1024	-1.05	46	2116	$-.42$
4	15	225	-2.41	63	3969	1.08
5	43	1849	$-.18$	18	324	-2.89
6	41	1681	$-.33$	43	1849	$-.68$
7	40	1600	$-.41$	48	2304	$-.24$
8	56	3136	$.86$	57	3249	$.55$
9	63	3969	1.42	60	3600	$.82$
10	62	3844	1.34	54	2916	$.29$
11	44	1936	$-.10$	53	2809	$.20$
12	44	1936	$-.10$	56	3136	$.47$
13	61	3721	1.26	61	3721	$.91$
14	38	1444	$-.57$	62	3844	1.00
15	37	1369	$-.65$	42	1764	$-.77$
Σ	678	33008	$.00$	761	40531	$.01$

$$\bar{X} = \frac{678}{15} = 45.20.$$

$$S = \sqrt{\frac{33008 - \dfrac{(678)^2}{15}}{15}} = 12.55.$$

$$\bar{X} = 50.73.$$

$$S = \sqrt{\frac{40531 - \dfrac{(761)^2}{15}}{15}} = 11.32.$$

distributions are measuring different things. (If this sounds like trying to compare apples and oranges, see Problem 23.) We will illustrate these uses with the data in Table 4.6.

Fifteen fifth-grade students were measured on their attitude toward arithmetic and their attitude toward reading. The highest possible score was 77 (extremely positive attitude), and the lowest possible score was 11 (extremely negative attitude). The means, standard deviations, and z scores are computed in Table 4.6.

From the two group means, you can see that these students had a more positive attitude toward reading than toward arithmetic. The standard deviations show that attitudes toward arithmetic were more variable than those toward reading.

The z scores, each calculated by subtracting the mean from the raw score and dividing by S, show each student's relative attitude. For example, Student 8 scored 56 on attitude toward arithmetic and slightly higher (57) on attitude toward reading. However, relative to the total group, this student's attitude toward arithmetic is more positive ($z = .86$) than his or her attitude toward reading ($z = .55$). The z scores can be compared, raw scores cannot (unless the $\bar{X}$'s are equal and the S's are equal).

Other comparisons are possible. Student 9, who had the most positive attitude toward arithmetic ($z = 1.42$), also had a positive attitude toward reading ($z = .82$). On the other hand, the one with the most positive attitude toward reading (Student 4, $z = 1.08$) had the most negative attitude toward arithmetic ($z = -2.41$).[7]

Besides comparing scores in two distributions, you can use z scores to compare two scores within a distribution. For example, Students 3 and 12 had raw scores of 46 and 56, respectively, on attitudes toward reading. z scores show Student 3's attitude to be .42 below the mean and Student 12's to be .47 above the mean.

Clue to the Future

z is an important statistic and will be used again in later chapters. Be sure you understand it and know how to compute it.

Problems

19. The following scores are the number of eggs found by each of five children on an Easter egg hunt. No attempt is being made to generalize from this sample to a population.

    1   3   4   7   10

    a. Convert each score to a z score.
    b. Determine the variance of the egg scores.

---

[7]Chapter 5 describes a technique (correlation) with which you can determine the *degree* to which positive attitudes toward reading go with negative attitudes toward arithmetic.

20. What conclusion can you come to about $\sum z$?

21. For any distribution, what is the z score of a raw score that is equal to the mean?

22. In a hypothetical experiment, a group of male rats was trained to run a maze. Two measures of performance were taken. The first measure was the number of seconds taken to run the maze on the last trial. The second measure was the number of entries into blind alleys during learning. Below are the data.

Rat	Running Time X	Errors X
1	20	22
2	17	23
3	16	25
4	15	20
5	15	14
6	13	19
7	12	16
8	12	17
9	10	19
10	8	10

Convert both distributions of raw scores to z scores.

a. Which rat was the best maze performer?

b. Was his running time or his error performance better?

c. Which rats were better at running time than at staying out of blind alleys?

23. Tobe grows apples and Zeke grows oranges. In the local orchards, the mean weight of apples is 5 oz. with $S = 1.0$ oz. For oranges the mean weight is 6 oz. with $S = 1.2$ oz. At harvest time, each enters his largest specimen in the County Fair. Tobe's apple was 9 oz. and Zeke's orange was 10 oz.

This particular year Tobe fell ill on the day of judgment so he sent his friend Hamlet to inquire who had won. Adopt the role of Judge and use z scores to determine the winner. Hamlet's query to you as judge must be, "Tobe, or not Tobe; that is the question."

24. Milquetoast's anthropology professor announced that he would drop the poorest exam grade for each student. Milquetoast scored 79 on the first anthropology exam. The mean was 67 and the standard deviation 4. On the second exam, he made 125. The class mean was 105, the standard deviation 15. On the third exam, the mean was 45, the standard deviation 3. Milquetoast made 51. On which test was Milquetoast's performance poorest?

TRANSITION PAGE

Thus far in our exposition of the wonders of statistical methods, we have presented data on only one variable at a time. On such data you now know how to perform basic kinds of manipulations. You can graph data and compute some essential statistics, such as means and standard deviations. In the next chapter and most of the remaining chapters in this book, these statistics will be vital tools.

As we explained in Chapter 1, most research involving statistical analysis attempts to answer questions about how variables are related or how groups differ on certain variables. In the next chapter, you will take up the first of these questions—the question of how variables are related. To do this, you will have to deal with two variables at the same time. You might want to know, for example, to what extent the following variables are related.

- Number of reinforced responses and number of trials to extinction
- Anxiety scores and introversion scores
- Mathematics achievement scores and language achievement scores
- IQ scores and height
- Level of income and probability of being psychotic

The next chapter, "Correlation and Regression," explains methods for determining the degree of such relationships (correlation) and methods of making predictions about measurements on one of the variables when you know the measurements on the other variable (regression).

5 CORRELATION AND REGRESSION

Objectives for Chapter 5. After studying the text and working the problems in this chapter, you should be able to:

1. explain the concept of correlation,
2. explain the difference between positive and negative correlation,
3. compute a Pearson product-moment correlation coefficient (r),
4. interpret the meaning of correlation coefficients,
5. identify situations in which the Pearson r will not accurately reflect the degree of relationship,
6. explain the meaning and use of the regression equation,
7. compute regression coefficients and fit a regression line to a set of data, and
8. make predictions about one variable from measurements of another variable.

Some of the earliest investigations making use of statistical analysis were conducted by Sir Francis Galton in England. Galton was concerned with the general question of whether people of the same family were more alike than people of different families. Specifically, he wanted to know whether genius, height, reaction time, and musical talent tended to run in families. A clue to the position Galton took on the matter was that he and his illustrious cousin, Charles Darwin, shared the same illustrious grandfather, Erasmus Darwin (although they had no grandmother in common).

Galton asked such questions as, "Are fathers and their adult sons more alike than men who are of different families?" To answer this question, he measured fathers and sons for weight, strength of grip, and height. He also measured many unrelated men. His subjects, people stopping at his booth at a fair, paid him three pence each to participate in his research. His subjects left with new self-knowledge; Galton left with a mass of data and jingling pockets. Modern psychologists often pay subjects for their participation. Where have we gone wrong?

In order to analyze this mass of data, Galton needed a method that would describe the degree to which, for example, heights of fathers and their sons were alike. The method he invented for this purpose is called correlation (co-relation). With it, Galton could also measure the degree to which the heights of unrelated men were alike. He could then compare these two results and answer his question.

Galton's student Karl Pearson, with Galton's aid, later developed a formula that yielded a statistic known as a correlation coefficient. Pearson's product-moment correlation coefficient, and other correlation coefficients based on Pearson's work, have been widely used in statistical studies in psychology, education, sociology, medicine, and many other areas.

It is the basic idea of correlation that we will explain next. Then you will learn to compute and interpret a Pearson product-moment correlation coefficient. Finally, you will learn a technique for predicting the value of one variable given a value of a second variable. This technique is called regression and it is closely related to correlation.

THE CONCEPT OF CORRELATION

In order to compute a correlation coefficient, you must have two variables, with values of one variable (X) paired in some *logical* way with values of the second variable (Y). Such an organization of data is referred to as a *bivariate* (two-variable) *distribution*. Paired scores for a bivariate distribution can occur in a number of different ways. The same group of people may take two tests, for example. Then each person's score on the first test is paired with his or her score on the second test. The result is a bivariate distribution like Table 5.1.

Family relationships may be organized as bivariate distributions. Height of fathers is one variable, X. Height of sons is another variable, Y. There is a logical pairing, which is to put a son's height with his father's height. Table 5.2 is an example of the pairing of these two variables into a bivariate distribution.

Table 5.1	A Bivariate Distribution of Scores on Two Tests Taken by the Same Group of People		
Person	*Test 1* *X*	*Test 2* *Y*	
Elgin	50	55	
Greg	40	41	
Debbie	30	33	
Ursula	20	19	

Notice in Table 5.2 that there is a tendency for tall fathers to have tall sons and for short fathers to have short sons. As variable X (father's height) increases, variable Y (son's height) also increases.

Table 5.2	Data on Two Variables That Can Be Correlated: The Heights of Fathers and Their Sons			
Father	*Height* *X*	*Son*	*Height* *Y*	
Mark Smith	5'9"	Mark, Jr.	5'10"	
Kenneth Johnson	6'3"	Kenneth, Jr.	6'2"	
Bruce Brown	5'7"	Bruce, Jr.	5'8"	
Paul Williams	5'10"	Paul, Jr.	5'10"	

Two variables may be positively correlated, be negatively correlated, or have no relationship to each other (zero correlation). We will discuss each of these possibilities in turn.

Positive Correlation

In the case of a *positive correlation* between two variables, high measurements on one variable tend to be associated with high measurements on the other variable and low measurements on one variable with low measurements on the other. In other words, the two variables vary together in the same direction. This is the case in Table 5.2. Tall fathers (as Galton found) tend to have sons who grow up to be tall men. Short fathers tend to have sons who grow up to be short men. If this relationship were so undeviating that every son grew to be exactly his father's height, the correlation would be perfect, and the correlation coefficient would be 1.00. A graph plotting the relationship would look like Figure 5.1. If such were the case (which, of course, is ridiculous; mothers and environments have their say, too), then it would be possible to predict perfectly the adult height of an unborn son simply by measuring his father.

Examine Figure 5.1 carefully. Each point represents a pair of scores—the height of a father and the height of his son. Such an array of points is called a **scatterplot** or scattergram. These scatterplots will be used throughout this chapter. Incidentally, it was when Galton cast his data in the form of a scatterplot that he conceived the idea of a correlationship between the variables.

The line that runs through the points in Figure 5.1 (and in Figures 5.2, 5.3, 5.4, and 5.5) is called a **regression line** or "line of best fit." When there is perfect

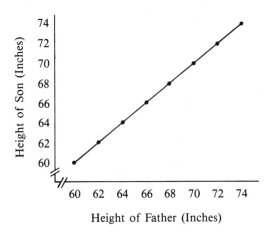

Figure 5.1   Hypothetical scatterplot if correlation were perfect positive.

correlation, $(r = 1.00)$ all points fall exactly on the line. When the points are scattered away from the line as in Figure 5.3, correlation is less than perfect and the correlation coefficient falls between .00 and 1.00. The term *regression* was chosen by Galton because it described a phenomenon that he discovered in his data on inheritance. He found that tall fathers had sons who were somewhat shorter than themselves and that short fathers had sons who were somewhat taller than themselves. From such data, he conceived his "law of universal regression," which states that there exists a tendency for each generation to regress, or move toward, the mean of the general population. It is from Galton's use of the term *regression* that we get the symbol *r* for correlation.

Today, the term regression also has a second meaning. It refers to a statistical method that is used to fit a straight line to bivariate data and to predict scores on one variable from scores on a second variable. The second definition is the topic covered in the last part of this chapter. Be certain that you understand these two different meanings of the term "regression." Now, let's get on with our explanation of positive correlation, using Galton's topic of heights.

In the past generation or so, with changes in nongenetic factors such as nutrition and medical care, sons tend to be somewhat taller than their fathers, be the father average, short, or tall (but not extremely tall). If it were the case that every son grew to be not the same height but 2 inches taller than his father (or 1 inch or 6 inches, or 5 inches shorter), the correlation would still be perfect, and the coefficient would be 1.00. Figure 5.2 demonstrates this point. It is not necessary that the numbers on the two variables be exactly the same in order to have perfect correlation. The only requirement is that the differences between pairs of scores all be the same. Another way of stating this requirement is to say that the relationship must be such that all points in a scatterplot will lie on the regression line. If this requirement is met, correlation will be perfect, and an exact prediction can be made.

Nature, of course, is not so accommodating as to permit such perfect prediction, at least at science's present state of knowledge. People cannot predict their sons' heights precisely. The points do not all fall on the regression line;

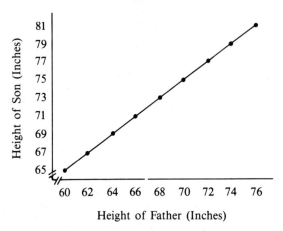

**Figure
5.2**
Hypothetical scatterplot of perfect positive correlation, where every son is 5 inches taller than his father.

some miss it badly. However, as Galton found so long ago, there is some positive relationship; the correlation coefficient is about .50. The points do tend to cluster around the regression line. In only rare cases does a very short man sire a very tall son or a very tall man sire a very short one.

Another example of a less-than-perfect correlation is presented in Figure 5.3. This is a scatterplot of some data collected to examine the relationship between reading achievement test scores (X) and mathematics achievement test scores (Y) for one class of ninth graders. We found the actual correlation between these two sets of scores to be far from perfect, but sufficient to indicate that there is a positive relationship. Thus, if you knew a student's score for reading achievement, you could predict his or her score for mathematics achievement. This prediction, although not perfectly accurate, would be far better than a random

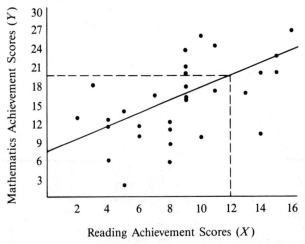

**Figure
5.3**
Scatterplot and regression line for reading and mathematics achievement scores for 30 students. Each point represents both the reading score and the mathematics score of one person ($r = .54$).

guess. One way of making this prediction is to draw a vertical line from the reading achievement score (X) to the regression line and then a horizontal line from the intersection of those lines to the Y axis. The score predicted for mathematics achievement can then be read at the point where the Y axis is intersected. For example, a mathematics achievement score of about 20 would be predicted for persons scoring 12 on reading achievement, as demonstrated by the dotted lines in Figure 5.3. Later in this chapter the method of accurately drawing the regression line on a scatterplot of data, and a more precise way of making predictions, will be discussed.

Negative Correlation

As was mentioned earlier, some sets of data may be negatively correlated. With *negative correlation*, high scores of one variable are associated with low scores of the other. The two variables thus tend to vary together but in opposite directions. Where correlation is negative, the regression line runs from the upper left of the graph to the lower right, as in Figure 5.4.

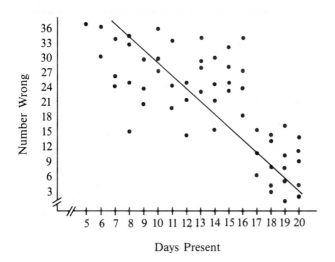

Figure 5.4   Hypothetical scatterplot of the negative correlation between days present in class and number of items answered incorrectly on an examination $(r = -.63)$.

The hypothetical scores in Figure 5.4 are the days present in class during a four-week period (X) and the number of items answered incorrectly on a test of material covered during that period (Y). The figure shows that there is a tendency for those more often present to miss fewer items: as attendance increases, errors decrease.

It might be noted that this negative correlation could be changed to positive simply by changing the type of score plotted on one of the variables. You could correlate days absent, instead of days present, with number wrong, and the correlation would be positive. The same is true if days present were correlated with number right instead of number wrong.

Perfect negative correlation exists, as does perfect positive correlation, when all points are on the regression line. The correlation coefficient in such a case is -1.00. For example, there is a perfect negative relationship between the amount of money in your checking account and the amount of money you have written checks for (if you ignore that nasty little service charge and assume you make no deposits). As the amount of money you write checks for increases, your balance decreases by exactly the same amount.

Other examples of negative correlation (less than perfect) are

1. temperature and inches of snow at the top of a mountain, measured at noon each day in May,
2. hours of sunshine and inches of rainfall per day at Miami, Fla.,
3. number of pounds lost and number of calories consumed per day by a person on a strict diet.

Negative correlation permits prediction in the same way that positive correlation does. With correlation, positive is not better than negative. In both cases, the size of the correlation coefficient indicates the strength of the relationship—the larger the absolute value of the number, the stronger the relationship. The algebraic sign ($+$ or $-$) indicates the direction of the relationship.

Zero Correlation

A *zero correlation* means that there is no relationship between the two variables. High and low scores on the two variables are not associated in any predictable manner.

Figure 5.5 shows a scatterplot in which the correlation is zero. Although a regression line can be drawn, it is not useful for making predictions. In the case of zero correlation, the best prediction from any X score is the mean of the Y scores. The regression line, then, runs parallel to the X axis at a height of $\bar{Y}$ on the Y axis.

Clue to the Future   Correlation will come up again in future chapters. If there is correlation between two sets of measurements, and you want to compare the *mean* of one set with the mean of the other set, you will have to take into account the existing correlation (Chapter 9).

Problems

1. What is meant by the statement "Variable X and variable Y are correlated?"
2. Describe the difference between positive and negative correlation.
3. Can the following variables be correlated, and, if so, would you expect the correlation to be positive or negative?
   a. heights and weights of a group of adults
   b. driving speed and gasoline mileage

c. height of oak trees and height of pine trees
d. average daily temperature and the cost of heating a home
e. IQ and reading comprehension
f. scores made by two sections of a math course on the same quiz

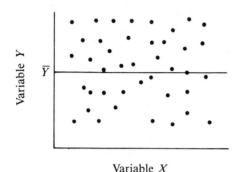

Variable X

Figure 5.5   Hypothetical scatterplot demonstrating zero correlation.

THE CORRELATION COEFFICIENT

A correlation coefficient gives you a quantitative way to express the *degree* of relationship that exists between two variables. As we mentioned before, computational procedures for the most widely used correlation coefficient were developed by Karl Pearson. The resulting coefficient is known as the Pearson product-moment correlation coefficient. In terms of a formula:

$$r = \frac{\sum (z_X z_Y)}{N},$$

where: r = Pearson product-moment correlation coefficient,
   z_X = a z score for variable X,
   z_Y = a corresponding z score for variable Y,
and N = number of pairs of X and Y values.

You can think through the formula above and discover what happens when high scores on one variable are paired with high scores on the other variable (positive correlation). The large positive z scores are multiplied together, and the large negative z scores are multiplied together. Both of these products are positive and when added together make a large positive numerator. The result is a large, positive value of r. Think through for yourself what happens in the formula when there is a negative correlation or a zero correlation.

This definitional formula with z scores is cumbersome if you want to calculate a value for r. Other formulas have been developed that require less computation (a boon for all). We will use these latter formulas to illustrate computational procedures.

Computational Formulas

We will illustrate two formulas for r, a blanched (partially cooked) formula and a raw-score formula. The blanched formula requires you to calculate means and standard deviations from the summary data before you calculate r. It is useful if, in addition to calculating r, you are going to do a regression analysis (draw a regression line and predict Y scores) because one formula for the regression line requires means and standard deviations. The blanched formula also gives you a better "feel" for the data (because you know means and standard deviations) and is more convenient if your calculator has a small capacity. The raw-score formula, however, does not require any intermediate steps between the summary data and r. If all you want is r, and you have a calculator with large capacity, the raw-score formula is the quicker method.

Blanched formula. This procedure requires you to "cook out" the means and standard deviations of both X and Y before computing r. Researchers often want to know them, so this formula is used by many.

$$r = \frac{\dfrac{\sum XY}{N} - (\bar{X})(\bar{Y})}{(S_X)(S_Y)},$$

where: X and Y are paired observations,

$\qquad XY$ = the product of each X value multiplied by its paired Y value,

$\qquad \bar{X}$ = the mean of variable X,

$\qquad \bar{Y}$ = the mean of variable Y,

$\qquad S_X$ = the standard deviation of variable X,

$\qquad S_Y$ = the standard deviation of variable Y,

and $\qquad N$ = the number of pairs of observations.

The only really new term in this formula is $\sum XY$. To obtain this value, it is necessary to multiply each of the X values by its paired Y value and then sum those products. Do *not* try to obtain $\sum XY$ by multiplying $(\sum X)(\sum Y)$. *It won't work.* The next example demonstrates this fact. Work through it.

X	Y	XY
2	5	10
3	4	12
$\sum = 5.$	$\sum = 9.$	$\sum = 22.$

$$\sum XY = 22.$$
$$(\sum X)(\sum Y) = (5)(9) = 45.$$

The term $\sum XY$ is called "the sum of the cross-products." Some form of the sum of the cross-products is a necessary part of all procedures for computing the Pearson r. One other caution about formulas for a correlation coefficient: remember that N is the number of *pairs* of observations.

Table 5.3 is a demonstration of the steps necessary to compute r by the blanched procedure. The data were collected from a group of 20 fifth-grade pupils. Each child took a reading achievement test and a mathematics achievement test. Reading achievement was arbitrarily designated as the X variable, and mathematics achievement was called Y. Scores on the two tests were then paired for each child.

Table 5.3   **Illustration of the Calculation of r for Reading Achievement and Mathematics Achievement by the Blanched Method**

Subject	Reading Ach. (X)	Math Ach. (Y)	X^2	Y^2	XY
1	66	36	4356	1296	2376
2	63	61	3969	3721	3843
3	62	65	3844	4225	4030
4	53	66	2809	4356	3498
5	53	63	2809	3969	3339
6	52	40	2704	1600	2080
7	50	42	2500	1764	2100
8	47	44	2209	1936	2068
9	46	38	2116	1444	1748
10	43	51	1849	2601	2193
11	41	38	1681	1444	1558
12	41	49	1681	2401	2009
13	41	42	1681	1764	1722
14	39	38	1521	1444	1482
15	38	31	1444	961	1178
16	38	38	1444	1444	1444
17	36	34	1296	1156	1224
18	34	49	1156	2401	1666
19	34	40	1156	1600	1360
20	24	46	576	2116	1104
	Σ 901	911	42801	43643	42022

$$\bar{X} = \frac{\Sigma X}{N} = \frac{901}{20} = 45.05. \qquad \bar{Y} = \frac{\Sigma Y}{N} = \frac{911}{20} = 45.55.$$

$$S_X = \sqrt{\frac{\Sigma X^2 - \frac{(\Sigma X)^2}{N}}{N}} = \sqrt{\frac{42801 - \frac{(901)^2}{20}}{20}} = 10.51.$$

$$S_Y = \sqrt{\frac{\Sigma Y^2 - \frac{(\Sigma Y)^2}{N}}{N}} = \sqrt{\frac{43643 - \frac{(911)^2}{20}}{20}} = 10.36.$$

why don't use $\bar{\varsigma}$ to calc. std. dev?

$$r = \frac{\frac{\Sigma XY}{N} - (\bar{X})(\bar{Y})}{S_X S_Y} = \frac{\frac{42022}{20} - (45.05)(45.55)}{(10.51)(10.36)} = \frac{2101.10 - 2052.03}{108.88}$$

$$= \frac{49.07}{108.88} = .451.$$

$$r = .45.$$

As usual, for the computation of means and standard deviations you must sum X, sum Y, square each X value and each Y value, and obtain ΣX^2 and ΣY^2. For values in the XY column, you must multiply each X value by its paired Y value. $\sum XY$ is obtained by summing the XY column. You then compute $\bar{X}$ and $\bar{Y}$. Next, you compute S_X and S_Y. Now you have all the information necessary to compute r. Work through each step in Table 5.3.

Raw-score formula.   With this formula, you start with the raw scores and obtain r without having to compute means and standard deviations.

A word of caution here: calculators vary in capacity. Some of them do not have the capacity to carry some of the terms called for by this formula. If raw scores are many or large, it is best to go back to the blanched formula.

CORRELATION
Coefficient $=$

$$r = \frac{N\sum XY - (\sum X)(\sum Y)}{\sqrt{[N\sum X^2 - (\sum X)^2][N\sum Y^2 - (\sum Y)^2]}}.$$

You have already learned what all the terms of this formula mean. Remember that N is the number of *pairs* of values.

All calculators with two or more memory storage registers—and some with one—permit the simultaneous accumulation of values for $\sum X$ and $\sum X^2$. Then $\sum Y$ and $\sum Y^2$ may be computed simultaneously. This leaves only $\sum XY$ to be computed.

Some calculators have a function for r built in; by entering X and Y values and pressing the r key, the coefficient is displayed. If you have such a calculator, we recommend that you use this labor-saving device after you have used the computation formulas a number of times. Working directly with terms like $\sum XY$ leads to an understanding of what goes into r.

Table 5.4 illustrates the use of the raw-score procedure for computation of r. The data are the same as those used in Table 5.3, in order to demonstrate that the value of r is the same for both methods. You will notice that, although there is no direct computation of means and standard deviations for the X and Y variables, the terms necessary for their computation are included in the raw-score formula for r.

Table 5.4   **Illustration of the Calculation of r by a Raw-Score Procedure Using the Data of Table 5.3**

$\sum X = 901.$ $\sum Y = 911.$ $\sum X^2 = 42801.$ $\sum Y^2 = 43643.$ $\sum XY = 42022.$

$$r = \frac{N\sum XY - (\sum X)(\sum Y)}{\sqrt{[N\sum X^2 - (\sum X)^2][N\sum Y^2 - (\sum Y)^2]}}$$

$$= \frac{(20)(42022) - (901)(911)}{\sqrt{[(20)(42801) - (901)^2][(20)(43643) - (911)^2]}}$$

$$= \frac{840440 - 820811}{\sqrt{(856020 - 811801)(872860 - 829921)}}$$

$$= \frac{19629}{\sqrt{(44219)(42939)}} = \frac{19629}{43574.30} = .451.$$

Our final step is interpretation. What is the story that goes with a correlation coefficient of .45 between reading and arithmetic achievement scores? The story is that there is a moderate correlation. Students who do well in one *tend* to do well in the other, although there are exceptions. Had the correlation been near zero, we could say that the two abilities are independent or unrelated. Had the coefficient been strong and negative, we could say, "Good in one, poor in the other."

You may have noticed that we are describing r as a sample statistic; there are no σ_X, σ_Y, μ_X, or μ_Y symbols in this chapter. Of course, if a population of data is available that can be correlated, the parameter may be computed. The procedure is the same as for sample data. Most recent statistical texts use ρ (the Greek letter rho) as the symbol for this parameter, although you must be cautious in your outside reading; ρ has also been used to symbolize Spearman's correlation coefficient, which you will learn about in Chapter 13.

Correlation coefficients should be based on an "adequate" number of pairs of observations. As a general rule of thumb, "adequate" means 30 or more. We will ask you to work some problems with fewer than 30 pairs, but this is so you can spend your time on interpretation and understanding rather than "number crunching."

Error Detection

The correlation coefficient ranges between -1.00 and 1.00. Values smaller than -1.00 or larger than 1.00 are not possible.

Now it is time for you to try your hand at computing r. We suggest that you compute it using both the raw-score formula and the blanched formula. You should get the same answer both ways. We will use small numbers so that you can work with the raw-score formula even if your calculator has limited capacity.

Problems

4. Draw a scatterplot and calculate r for these hypothetical data on fraternal twins. Scores represent the number of errors made in recalling details of a short story they read.

Twin Pair	Twin A X	Twin B Y
1	9	7
2	8	8
3	7	6
4	6	4
5	5	3
6	4	5
7	3	5
8	2	2

5. This problem is based on actual data. By now you know how to obtain the summary values for $\sum X$, $\sum X^2$, $\sum Y$, $\sum Y^2$, and $\sum XY$, so we are providing them for you. The raw data are scores on two different tests that measure self-esteem. Subjects were seventh-grade students. Since the two tests were designed to measure the same trait, they should be positively correlated. Summary values are: $N = 38$, $\sum X = 1755$, $\sum Y = 1140$, $\sum X^2 = 87373$, $\sum Y^2 = 37592$, $\sum XY = 55300$. Compute r.

6. The X variable is the population of each of the states in the U.S. in millions. The Y variable is expenditure per pupil in public schools for each of the states. Calculate r, using either method.
$N = 50$, $\sum X = 202$, $\sum Y = 41048$, $\sum X^2 = 1740$, $\sum Y^2 = 35451830$, $\sum XY = 175711$.

THE MEANING OF r

The interpretation of r can be a tricky business. You must exercise caution to avoid pitfalls in interpretation. Keep firmly in mind that r is one of those summary index numbers, like the mean and standard deviation. All such numbers are used to describe a set of data.

A correlation coefficient is a measure of the relationship between two variables. It describes the tendency of two variables to vary together (covary); that is, it describes the tendency of high or low values of one variable to be regularly associated with either high or low values of the other variable. The absolute size of the coefficient (from 0 to 1.00) indicates the strength of that tendency to covary.

Look at Figure 5.6, which shows scatterplots of correlational relationships of $r = .20$, $.40$, $.60$, and $.80$. Notice that as the size of the correlation coefficient gets larger, the points cluster more and more closely to the regression line; that is, the envelope containing the points becomes thinner and thinner. This means that a stronger and stronger tendency to covary exists as r becomes larger and larger. It also means that predictions made about values of the Y variable from values of the X variable will be more accurate when r is larger.

The algebraic sign tells the direction of the covariation. When the sign is positive, high values of X are associated with high values of Y, and low values of X are associated with low values of Y. When the sign is negative, high values of X are associated with low values of Y, and low values of X are associated with high values of Y. (For pictures of these relationships, see Figures 5.3 and 5.4.) Knowledge of the size and direction of r, then, permits some prediction of the value of one variable if the value of the other variable is known. For example, if $r = .98$, then, for an individual with a low X score, it would be reasonable to predict a low Y score.

Correlation vs. Causation

A high correlation coefficient does not tell you that one of the variables is *causing* the variation in the other. Quite possibly some third variable is responsible

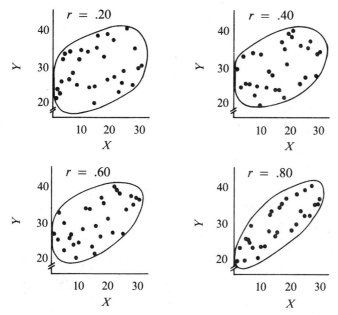

**Figure
5.6** Scatterplots of data in which r = .20, .40, .60, .80.

for the variation in both. For example, in most parts of the United States, there is a positive correlation between the number of birds in deciduous trees and the number of leaves on those trees; however, a change in one of these variables does not *cause* a change in the other variable. It is obviously foolish to think that falling leaves frighten away the birds or that the birds knock off leaves as they fly away. Variation in the number of birds does not cause variation in the number of leaves, nor does variation in leaves cause variation in birds. A third variable, seasonal changes, causes variation in both. Thus, r can be interpreted only in terms of relationship and prediction, not in terms of cause and effect.

Jumping to a cause-and-effect conclusion is easy to do; it usually seems so reasonable. And, of course, a cause-and-effect relationship may in fact exist. However, a correlation coefficient *alone* cannot establish a causal relationship. As another example, the early statements about cigarette smoking causing cancer were based on correlational data; persons with cancer were often heavy smokers. However, as careful thinkers (and the cigarette companies) pointed out, both cancer and smoking might have been caused by a third variable; stress and anxiety were often suggested as possible causes. What was required to establish the cause-and-effect relationship was experimental data, not correlational data. It was the experimental data, then, that established the cause-and-effect relationship between smoking and cancer. (Chapter 8, "Differences between Means," includes a discussion of experimental data.)

Coefficient of Determination

This entire chapter is about ways to express the relationship between two variables. Of course, you have already had a lot of experience with expressing relationships. For example, a statement that specifies corresponding values on the two variables is a common way to express a relationship. A statement like "If you correctly answer 90 percent of the questions, you'll get an 'A'" is an example. A second example is a prediction about one variable when you know a value on the second variable. "The predicted height of the son of a 5' 9" father is 5' 10"" is an example. (You will learn to make such predictions in the section of this chapter on regression.) In the present section you will learn a more complicated method of expressing the relationship between two variables. It is widely used because it gives, with one number, an overall index that specifies the proportion of variance the two variables have in common. The name of this index is the **coefficient of determination**. It is easy to calculate; just square r.

In the field of testing, experts usually say that the variance two tests have in common can be attributed to the same cause. Thus, a correlation of .70 between arithmetic and reading tests means that 49 percent of the variance of each test was caused by the same factors.[1] The primary factor was probably differences in the general intellectual ability of the persons taking the test. Factors such as the health of the people the day they took the tests, their ability to follow directions and to take tests, the pleasantness of the teacher the day the tests were given, and the distraction in the room that day may also enter into the common variance. These factors have their effect on both tests taken by the student. On the other hand, 51 percent of the variance was not held in common; 51 percent was independent and was not predictable from the other test. Fifty-one percent of the variance was specific to the test. This independent variance presumably was due to skills specific to arithmetic and reading.

Note what happens to a fairly strong correlation of .70 when it is interpreted in terms of variance. To the novice, such a correlation may seem to mean that the two tests are measuring pretty much the same things, but in fact they hold only about one-half the variance in common. To predict 70 percent of the variance requires a correlation of .84, a rather high correlation in much research.

Here is another example to help you conceptualize the notion of variance held in common by two variables. First-semester college grade-point averages usually correlate with academic-aptitude test scores, with an r equal to about .50. Squaring .50 gives a coefficient of determination of .25. If you now wished to predict grade-point average (G.P.A.) from an academic-aptitude test, you should recognize that you would be able to account for only 25 percent of the G.P.A. variance. The other 75 percent is independent of academic-aptitude scores and cannot be accounted for by the test. Figure 5.7 presents a picture of this concept. The shaded area represents the common variance—variance held in common by both variables. This common variability in the test scores and in G.P.A. is related to the same factors. As an example, intelligence and previous learning will influence both variables and thus enter into the common variance. The rest of the variance (75 percent of both variables) is independent. For example, part of the variance of G.P.A. may be due to such factors as illness, falling in love (or is that just another illness?), developing a taste for beer, or financial problems. Academic-aptitude tests are never likely to predict such factors.

[1]If you have some skill at algebra and some curiosity about this, you can find proofs in Minium (1978, p. 209), McNemar (1969, p. 142), or Edwards (1976, p. 45).

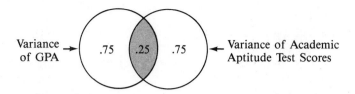

Figure
5.7

Illustration of the common variance in academic-aptitude test scores and first-semester freshman grade-point average.

The coefficient of determination is also useful in comparing correlation coefficients. When one compares an r of .80 with an r of .40, the tendency is to think of the .80 as being twice as high as .40; but that is not the case. Correlation coefficients are compared in terms of the amount of common variance, as shown:

$$.80^2 = .64,$$

$$.40^2 = .16,$$

$$.64 \div .16 = 4.$$

Thus, two variables that are correlated with $r = .80$ have four times as much common variance as two variables correlated with $r = .40$.

Problems

7. In Tables 5.3 and 5.4 we found a correlation of .45 between reading achievement-test scores and mathematics achievement-test scores. Compute the coefficient of determination and interpret its meaning.

8. You found a correlation of .57 between two measures of self-esteem. What is the coefficient of determination, and what does it mean?

9. What percent of variance in common do two variables have if their correlation is (a) .40? (b) .10?

10. Imagine that you were told that there is a correlation of 1.00 between anxiety and annual income. Could you conclude that anxiety leads people to make money? Why or why not?

11. Interpret the following statements:
    a. $r = .72$ was found between scores on a test of intolerance of ambiguity and scores on a test of authoritarianism.
    b. The correlation between vocational-interest scores at age 20 and at age 40 for the same subjects was found to be .70.
    c. The correlation between intelligence-test scores of identical twins is in the high .90s.
    d. The correlation between IQ and family size is about −.30.
    e. $r = .22$ between height and IQ for 20-year-old men.
    f. $r = −.83$ between income level and probability of psychosis.

Practical Significance of *r*

In this section, we will discuss practical (as opposed to statistical) significance.[2] That is, how high must a correlation coefficient be before it is of use? How low must it be before we conclude that it is useless? Remember that correlation is useful if it improves prediction over guessing. In this sense, any reliable correlation other than zero, whether positive or negative, is of some value because it will reduce to some extent the incorrect predictions that might otherwise be made. Recognize, however, that very low correlations allow little improvement over guessing in prediction. Such poor prediction usually is not worth the costs involved in practical situations.

For example, suppose that a personnel manager wants to be able to predict the job success of applicants for a particular job. Administering a mechanical-aptitude test to his current employees, he finds that the scores correlate .25 with actual job success. Squaring .25 gives him a coefficient of determination of .0625. This means that scores on the aptitude test and job success have only about 6 percent of their variance in common, so the aptitude test is not a great deal better than flipping a coin. Buying, administering, and scoring tests require time and money. The value to be gained by using the test is probably less than the costs involved.

There is no general sense in which we can answer the question of how high a correlation must be in order to be of value. The total situation must always be considered. How important are the predictions? How expensive in time and money will it be to include this correlation in making predictions? The answers to such questions may depend on the nature of the research. Generally, researchers are satisfied with lower correlations in theoretical work but require higher ones in practical situations. In developing theories, the theorist wants to account for as much variance as possible. In practical situations, this usually becomes too time consuming and expensive.

STRONG RELATIONSHIPS BUT LOW CORRELATIONS

One of the neat things about understanding something is that you come to know what's going on beneath the surface. Knowing the inner workings, you can judge whether the surface appearance is to be trusted or not. You are about to learn about two "inner workings" of correlation. These will help you evaluate the meaning of low correlations. Low correlations do not always mean that there is no relationship between two variables.[3]

Nonlinearity

For *r* to be a meaningful statistic, the best-fitting line through the scatterplot of points must be a *straight line.* If a curved regression line fits the data better

---

[2]Chapter 9 has a section on the statistical significance of *r*. Intervening chapters will provide you with the background needed to understand the concept of statistical significance.

[3]Correlations that do not reflect the true degree of relationship are said to be *spuriously* low or high.

than a straight line, r will be low, and will not reflect the true relationship between the two variables.

Figure 5.8 is an example of a situation in which r is inappropriate because the best-fitting line is curved. The X variable is chronological age. The Y variable is physical endurance for hard labor. Endurance increases with increasing age until the mid-20s; then it levels off until middle age, when it begins gradually to decrease with advancing age.

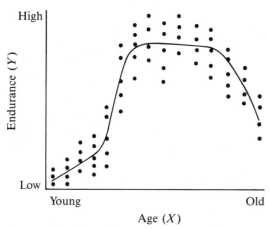

**Figure
5.8**   An example of curved regression.

There is obviously a strong relationship between age and endurance, but r for these two variables would be low, not reflecting the strong relationship. The product-moment correlation coefficient is just not appropriate as a measure of curved relationships. Special nonlinear correlation techniques for such relationships do exist and are described in texts such as McNemar (1969) and Guilford and Fruchter (1978).

Truncated Range

Besides nonlinearity, there is another situation that can give you a spuriously low Pearson coefficient, even though there is a strong relationship between the two variables.

This second situation we have in mind is called a *truncated range*. Suppose, for example, that you wanted to investigate the relationship between IQ scores and self-esteem scores and that you planned to use Introductory Psychology students as participants in your study. Since college students are in the higher ranges of IQ scores, your study will omit the lower range of scores in the population. The same is probably true of the self-esteem scores. Now look at Figure 5.9 for a picture of what a scatterplot of scores might look like for the population. Inset on this scatterplot is a pair of axes for the Introductory Psychology students in your study. The scatterplot for the population indicates a moderate relationship, but the scatterplot within the inset axes leads to the conclusion that the correlation between the two variables is near zero. If you did

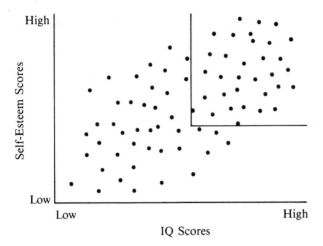

Figure 5.9   Illustration of the effect of a truncated range on r.

not recognize that your study had used a truncated range of scores, you would not discover the relationship that exists between IQ and self-esteem.

OTHER KINDS OF CORRELATION COEFFICIENTS

Correlation is a widely used statistical technique. You have learned about one of the several kinds of linear correlation—the Pearson product-moment coefficient. Here is a list of some other kinds of data for which a descriptive index number can be calculated. These numbers are also correlation coefficients, but they are not Pearson product-moment coefficients.

1. Correlations may be computed on data for which one or both of the variables are dichotomous (having only two possible values). An example is the correlation of the dichotomous variable sex and the quantitative variable grade-point average.
2. Several variables can be combined, and the resulting combination can be correlated with one variable. With this technique, called **multiple correlation**, a more precise prediction can be made. Performance in school or on the job usually can be predicted better by using several measures of a person rather than just one.
3. A technique called **partial correlation** allows you to separate or partial out the effects of one variable from the correlation of two other variables. For example, if we want to know the true correlation between achievement-test scores in two school subjects, it will probably be necessary to partial out the effects of intelligence since IQ and achievement are correlated.
4. Chapter 13 will teach you how to compute a statistic called Spearman's r_s. r_s is used when the data are ranks rather than raw scores.
5. If the relationship between two variables is curved rather than linear, the correlation ratio, *eta* (η) gives the degree of association.

These and other correlational techniques are covered in intermediate-level textbooks. For example, McNemar (1969) has over 100 pages on correlation and Downie and Heath (1983) have complete chapters on some of these and other correlational methods.

Problems

12. The correlation between number of older brothers and sisters and degree of acceptance of personal responsibility for one's own successes and failures is −.37.
    a. How would you interpret this correlation?
    b. What can you say about the cause of this correlation?
    c. What practical significance might it have?

13. A correlation of .97 has been found between anxiety and neuroticism.
    a. Interpret the meaning of this r.
    b. Are you justified in assuming that being neurotic is responsible in large part for manifestations of anxiety?

14. Plot the following set of data, and compute r if appropriate.

X	1	8	10	3	5	2	12	12	4	9	3	5
Y	2	11	4	3	7	2	3	1	5	8	5	8

X	7	11	2	8	1	9	11	4	8	7	5	4
Y	12	1	3	10	1	5	3	4	9	10	10	7

CORRELATION AND REGRESSION

Throughout this chapter we have emphasized the importance of the correlation coefficient in making predictions. We have used the term *regression line* many times. You have now arrived at the part of the chapter where you will learn how to make predictions through the use of a linear regression technique. We hope you will now see how correlation (the relationship between variables) is related to regression (making predictions).

We said earlier in this chapter that the correlation between scores on typical college-entrance examinations and first-semester-freshman grade-point averages is about .50. Knowing this fact permits you to predict that those who score high on college-entrance exams will have a somewhat higher probability of success as freshmen than those who score low. The size of the correlation coefficient gives you information about how much faith to put in these predictions. The higher the value of r, the closer the points come to the regression line, and the more faith you may put in your predictions.[4] All this is rather general though. Usually, what you want to do is to predict a *specific* value of Y for a given value of X. To do this, you must have the regression line.

---

[4]For an entire chapter on prediction, using regression, see Ferguson (1981, Chapter 9).

THE REGRESSION EQUATION

If you happen to be the lazy type who doesn't mind sloppy predictions, you can simply make a scatterplot of the data and fit the regression line by the "eyeball" method. This involves looking at the scatterplot of data and attempting to draw a line through the center of the plot of points. Predictions are then made by the method demonstrated in Figure 5.3. If you're getting the idea that we don't really approve of drawing lines by the eyeball method, you're right. The **regression equation** is much more accurate, and we highly recommend it.

The method for calculating the regression equation is called the *least-squares solution*. Recall from Chapters 3 and 4 that the sum of the squared deviations from the mean will be smaller than the sum of the squared deviations from any other value. Think of the regression line as a moving mean sliding through the scatterplot of points. The regression line minimizes the squared deviations from itself.

The regression equation is

$$Y' = a + bX,$$

where: Y' = the Y value predicted from a particular X value (Y' is pronounced "y prime"),

a = the point at which the regression line intersects the Y axis,

b = the slope of the regression line—that is, the amount Y is increasing for each increase of one unit in X,

and X = the X value for which you wish to predict a Y value.

The symbols X and Y can be assigned arbitrarily in correlation, but, in a regression equation, Y is assigned to the variable you wish to predict. To make predictions of Y using the regression equation, you need to calculate the values of the constants a and b, which are called **regression coefficients**.

If you have already computed r and the standard deviations for both X and Y,[5] b can be obtained very simply by the formula

$$b = r\frac{S_Y}{S_X},$$

where: r = correlation coefficient for X and Y,

S_Y = the standard deviation of the Y variable,

and S_X = the standard deviation of the X variable.

Notice that for positive correlation b will be a positive number. For negative correlation b will be negative.

You can compute a from the formula

$$a = \bar{Y} - b\bar{X},$$

[5]If you have not computed r, S_X, and S_Y, b can be obtained by the formula

$$b = \frac{N\sum XY - (\sum X)(\sum Y)}{N\sum X^2 - (\sum X)^2}.$$

where: $\bar{Y}$ = the mean of the Y scores,

         b = the regression coefficient computed previously,

and $\bar{X}$ = the mean of the X scores.

Figure 5.10 illustrates the meaning of the regression coefficients. The regression line in Figure 5.10 is the heavy line that goes up at an angle. The point marked a in Figure 5.10 is the point at which the regression line crosses the Y axis. It is called the "Y intercept." For these data, $a = 1.00$. The symbol b stands for the slope of the regression line. Regression slopes for positive correlations range from 0 to $+\infty$; for negative correlations, they range from 0 to $-\infty$. (A line horizontal to the X axis has a slope of 0; a vertical line has a slope that approaches $+\infty$.) The slope of a line is measured by the vertical distance the line rises divided by the horizontal distance the line covers. If you measure the line DE (vertical rise of the line FD) in Figure 5.10, you will find it is one-half the length of FE (the horizontal distance of line FD). Thus, the slope of the regression line (DE/FE) is .50 ($b = .50$). Put another way, the value of Y increases one-half point for every one-point increase in X. Expressed in a third way, the regression line is "going out" horizontally twice as fast as it is "going up" vertically.

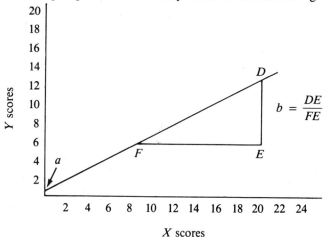

Figure 5.10   An illustration of the regression coefficients.

Two words of caution are in order here. The appearance of the slope of the regression line will depend on the units chosen for X and Y. Look at Figure 5.11. The two lines do not look the same even though the value of b (1.00) is the same for both. The difference in appearance is due to the difference in units plotted on the Y axis. In both figures the regression line is rising one unit of Y for each unit of X. The line rises less steeply in the right figure because the units on the Y axis are half as large as those of the left figure. In a similar way, a break in either axis will change the appearance of the line.

Drawing a Regression Line

In Table 5.3, we presented data concerning the relationship between mathematics achievement and reading achievement for a group of fifth graders. For

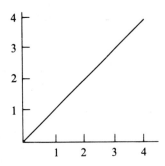
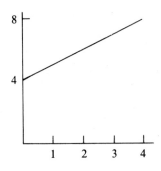

Figure 5.11   Two regression lines with the same slope ($b = 1.00$) but with different appearances. The difference is due to the units chosen for the Y axis.

those data, $r = .45$, $S_X = 10.51$, $S_Y = 10.36$, $\bar{X} = 45.05$, and $\bar{Y} = 45.55$. Using the formula

$$b = r\frac{S_Y}{S_X},$$

$$b = (.45)\frac{10.36}{10.51} = (.45)(.99) = .44.$$

And using the formula

$$a = \bar{Y} - b\bar{X},$$

$$a = 45.55 - (.44)(45.05) = 45.55 - 19.82 = 25.73.$$

We now have our two constants: $b = .44$ and $a = 25.73$. The b coefficient tells us that, when X (reading achievement scores) increases by one point, Y (mathematics achievement scores) increases by .44 points, or a little less than one-half point. The a coefficient tells us that the regression line will intersect the Y axis at a value of 25.73.

The formula for the regression line relating mathematics achievement and reading achievement is

$$Y' = 25.73 + .44X.$$

Now, all we need in order to draw the regression line are two points that fall on the line. Any two will do; here are the easiest. When $X = 0$, $Y' = a = 25.73$. This is the point at which the regression line intersects the Y axis. Thus, one point is $X = 0$, $Y = 25.73$. The means of both variables, $\bar{X}$ and $\bar{Y}$, are always on the line and they serve as a second point. Thus, a second point is $X = 45.05$, $Y = 45.55$. Both points are marked on Figure 5.12 with circles. To draw the

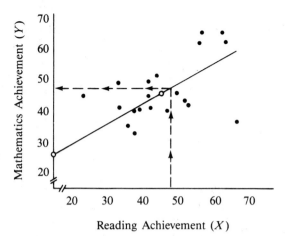

Figure
5.12
Scatterplot of the data from Table 5.3, with the regression line.

regression line, simply connect the two points with a straight line and extend the line on through the scatterplot.

Predicting a Y Score

If you are given a particular X score, what Y score should be predicted? If you have a regression line drawn accurately on a scatterplot, you may use it for gross predictions. Start with the given X score and draw a vertical line up to the regression line. Then draw a horizontal line to the vertical axis. The Y score there is the predicted Y for the X score you began with. This is demonstrated by the dotted lines in Figure 5.12 for a reading achievement score of 48. The projection to the Y axis shows a predicted mathematics achievement score of approximately 47.

A more accurate way to predict a Y score is to use the regression equation. If you wished to predict a mathematics achievement score from a reading achievement score of 48, the regression equation would be

$$Y' = a + bX = 25.73 + (.44)(48)$$

$$= 25.73 + 21.12$$

$$= 46.85.$$

Thus, given an X score of 48, the predicted Y score is 46.85 or 47.

Unless you wish to draw the regression line, it is usually more convenient to combine the formulas for a and b into one prediction equation. A little algebraic manipulation produces the following formula:

$$Y' = r\frac{S_Y}{S_X}(X - \bar{X}) + \bar{Y}.$$

This formula permits you to use the specific value of X for which you need a predicted Y without going through the intermediate step of computing a.

To demonstrate how this formula works, we'll again use the data from Table 5.3 to predict a mathematics achievement score from a reading achievement score of 48:

$$Y' = (.45)(.99)(2.95) + 45.55$$

$$= 1.31 + 45.55$$

$$= 46.86.$$

Thus, for students who have reading achievement scores of 48, we would predict math achievement scores of 47.

Now, you know how to make predictions. However, predictions are cheap; anyone can make them. Respect accrues only when the predictions come true. So far we have not discussed the accuracy of predictions made from a regression equation. Unfortunately, our discussion will be superficial, but we will point you in the direction of more complete information.

Both the size of the correlation, r, and the standard deviation of the variable you are predicting affect the accuracy of your prediction. The larger r is, the more faith you can put in your prediction; likewise, the smaller the standard deviation of Y, the greater the accuracy of your prediction. If $r = 1.00$, your predictions will be exact and each will come true. If $r = .00$, all your predictions will be $\bar{Y}$. Some of the observed Y's will be close but many will deviate widely from $\bar{Y}$.

To be able to *measure* the accuracy of a prediction made from a regression analysis, you need to know about the *standard error of estimate.* This statistic is discussed in most intermediate-level statistics textbooks and in books on psychological testing. After you have covered the material in Chapters 6 to 8 of this book, you will have the background necessary to understand discussions about the standard error of estimate.

Problems

15. In Problem 4, the twins problem, you computed r, using eight pairs of small numbers.
    a. Compute the regression coefficients a and b.
    b. Plot the points.
    c. Draw the regression line for those eight pairs of numbers. Use an X value of 9 to draw the line.

16. In Problem 5 you also computed an r.
    a. Compute a and b.
    b. What Y score would you predict for a child scoring 42 on X?

17. For the following data, compute the regression coefficients and plot the regression line.

X	1	3	3	4	5	5	5	6	7	7	9	10	10
Y	15	13	14	12	10	12	13	8	9	6	7	4	2

18. The correlation between Stanford-Binet IQ scores and Wechsler Adult Intelligence Scale (WAIS) IQs has been found to be .80. Both tests have a mean of

100. The standard deviation of the Stanford-Binet is 16. For the WAIS, $S = 15$. What WAIS IQ would you predict for a person scoring 65 on the Stanford-Binet? (An IQ score of 70 has been used by some schools as a cutoff point between regular classes and special education classes.)

19. The Trial 1 and Trial 20 scores below are the number of seconds of time on target in a 60-second tracking task (a pursuit rotor).

Person	Trial 1	Trial 20
1	3	21
2	6	25
3	2	20
4	6	33
5	5	23
6	8	34

a. Compute the correlation coefficient.
b. Explain what conclusions can be drawn about how people change when they are learning this tracking task.
c. Construct a scatterplot using Trial 1 as the X variable and carefully draw the regression line.
d. What score on Trial 20 would be predicted for a person whose Trial 1 score was 9?

*T*RANSITION PAGE

You are now through with the part of the book devoted to descriptive statistics. You should be able to describe a set of data with a graph and a few descriptive numbers, such as a mean, a standard deviation, and (if appropriate) a correlation coefficient. The remainder of the book is devoted to inferential statistics.

Inferential statistics are concerned with decision making. Usually, the decision is whether the difference between two samples is probably due to chance or probably due to some other factor. Inferential statistics help you make a decision by giving you the probability that the difference is due to chance. If the probability is very high, a decision that the difference is due to chance is supported. If the probability is very low, a decision that the difference is due to some other factor is supported. As you will see rather quickly, most of the descriptive statistics that you've learned will be used in these decision-making processes.

6 THEORETICAL DISTRIBUTIONS INCLUDING THE NORMAL DISTRIBUTION

Objectives for Chapter 6: After studying the text and working the problems in this chapter, you should be able to:

1. distinguish between a theoretical and an empirical distribution,
2. predict the probability of certain events from your knowledge of the theoretical distribution of those events,
3. list the characteristics of the normal distribution,
4. find the proportion of a normal distribution that lies between two scores,
5. find the scores between which a certain proportion of a normal distribution falls, and
6. find the number of scores associated with a particular proportion of a normal distribution.

In this chapter (and the following two chapters), your most important task is to understand the concepts. There are calculations for you to do, but the main purpose of the calculations is to help you understand some ideas that are at the heart of inferential statistics. Once you grasp these ideas and their logical progression, the rest of this book (as well as more advanced books) should be much easier for you to understand. You may, in fact, find these ideas stimulating and exciting. Let's begin now with the most basic of these ideas—the concept of theoretical distributions.

In Chapter 3, you learned how to arrange scores into frequency distributions. The scores you worked with there were either scores obtained from actual research or scores that we made up and pretended were obtained from actual research. Distributions of such observed scores are called **empirical distributions**.

This chapter is about theoretical distributions. Like the empirical distributions in Chapter 3, a theoretical distribution is a presentation of all the scores, arranged from highest to lowest. However, **theoretical distributions** are based on mathematical formulas and logic rather than on empirical observations. In this chapter, we will describe theoretical distributions that will be valuable to you as an applied statistician because they correspond closely to empirical distributions. When there is such a correspondence, you can accurately estimate the probability of future empirical observations from a theoretical distribution.[1]

In this chapter, you are going to learn about three theoretical distributions. The first two will be used to establish some points that are true for all theoretical distributions and to introduce the topic of probability. Then you will study the normal distribution in some detail.

This is the first chapter on the topic of inferential statistics. As you may recall, inferential statistics are used to decide whether some observed difference or observed relationship should or should not be attributed to chance. The probability that the event was due to chance is found by using a theoretical distribution. In this chapter and in Chapters 7 and 8, you will learn how to use the normal distribution to determine these probabilities. Understanding these probabilities is fundamental to inferential statistics and thus to the rest of this book.

The concept of probability is one you are somewhat familiar with. The probability of the occurrence of any event ranges from .00 (there is no possibility that the event will occur) to 1.00 (the event is certain to happen). Theoretical distributions are used to find the probability of an event or a group of events.

A RECTANGULAR DISTRIBUTION

We'll start with an example of a theoretical distribution that you are already familiar with to show you the relationship between theoretical distributions and probabilities. Figure 6.1 is a histogram that shows the distribution of types of cards in an ordinary deck of playing cards. There are 13 kinds of cards, and the

---

[1]Mathematical statisticians use the term probability density functions when they refer to these theoretical distributions.

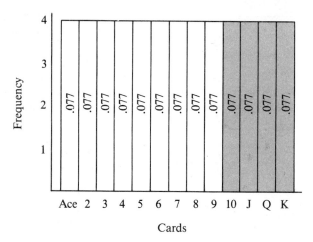

Figure 6.1   Theoretical distribution of 52 draws from a deck of playing cards.

frequency of each card is four. This theoretical curve is rectangular in shape. (The line that encloses a frequency polygon is called a curve, even if it is straight.) The number in the area above each card is the probability of obtaining that card in a chance draw from the deck. That probability (.077) was obtained by dividing the number of cards that represent the event (4) by the total number of cards (52).

Probabilities are often stated as "chances in a hundred." The expression $p = .077$ means that there are 7.7 chances in 100 of the event in question occurring. Thus, from Figure 6.1 you can tell at a glance that there are 7.7 chances in 100 of drawing an ace from a deck of cards. This knowledge might be helpful in a poker game.

With this theoretical distribution, you can determine other probabilities. Suppose you wanted to know your chances of drawing a face card or a 10. These are the shaded events in Figure 6.1. Simply add the probabilities associated with a 10, jack, queen, and king. Thus, $.077 + .077 + .077 + .077 = .308$. This knowledge might be helpful in a game of blackjack, in which a face card or a 10 is an important event.

One property of the distribution in Figure 6.1 that is *true for all theoretical distributions is that the total area under the curve is 1.00.* In Figure 6.1, there are 13 kinds of events, each with a probability of .077. Thus, $(13)(.077) = 1.00$. With this arrangement, any statement about area is also a statement about probability. Of the total area under the curve, the proportion that signifies "ace" is .077, and that is also the probability of drawing an ace from the deck.

Clue to the Future   The probability of an event or a group of events corresponds to the *area* of the theoretical distribution associated with the event or group of events. This idea will be used throughout the rest of this book.

Problems
1. What is the probability of drawing a card that falls between 3 and jack, excluding both?
2. If you drew a card at random, recorded the result, and replaced it, how many 7's would you expect in 52 draws?
3. What is the probability of drawing a card larger than a jack *or* smaller than a 3?
4. If you made 78 draws from a deck, replacing each card, how many 5's and 6's would you expect?

A BINOMIAL DISTRIBUTION

The binomial (two names) is another example of a theoretical distribution. Suppose you took three new quarters and tossed them in the air. What is the probability that all three would come up "heads?" As you may already know, the answer is found by multiplying together the probabilities of each of the independent events. For each coin, the probability of a head is $\frac{1}{2}$, so the probability that all three will be heads is $(\frac{1}{2})(\frac{1}{2})(\frac{1}{2}) = \frac{1}{8} = .1250$.

Here are two other questions about tossing those three coins. What is the probability of two heads? What is the probability of one head or zero heads? Since you could answer these questions easily if you had a theoretical distribution of the probabilities, we will construct one for you. We will start by listing in Table 6.1 the eight possible outcomes of tossing the three quarters in the air. Each of these eight outcomes is equally likely so the probability for any one of them is $\frac{1}{8} = .1250$. There are three outcomes in which two heads appear so the probability of two heads is $.1250 + .1250 + .1250 = .3750$. This is the answer to the first question. Based on Table 6.1, we can construct Figure 6.2, which is the theoretical distribution of probabilities we need. You can use it to answer the following questions.[2]

Table 6.1

All Possible Outcomes When Three Coins Are Tossed

Outcomes	Number of Heads	Probability of Outcome
Heads, heads, heads	3	.1250
Heads, heads, tails	2	.1250
Heads, tails, heads	2	.1250
Tails, heads, heads	2	.1250
Heads, tails, tails	1	.1250
Tails, heads, tails	1	.1250
Tails, tails, heads	1	.1250
Tails, tails, tails	0	.1250

Problems
5. What is the probability of one head or two heads?
6. What is the probability of obtaining all heads or all tails?

[2]Loftus and Loftus (1982), Chapter 4; Howell (1982), Chapter 5; and Downie and Heath (1983), Chapter 9, all discuss binomial distributions.

7. If you threw the three coins in the air 16 times, how many times would you expect to find zero heads?

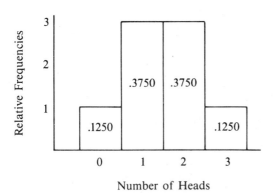

Number of Heads

Figure 6.2   A binomial distribution showing the number of heads when three coins are tossed.

COMPARISON OF THEORETICAL AND EMPIRICAL DISTRIBUTIONS

We have carefully called Figures 6.1 and 6.2 theoretical distributions. They may not, in fact, reflect exactly what would happen if you drew cards from an actual deck of playing cards or tossed quarters in the air. Actual results could be influenced by lost or sticky cards, sleight of hand, unbalanced coins, or chance deviations. Now we'll turn to the empirical question of what a frequency distribution of actual draws from a deck of playing cards would look like. Figure 6.3 is a histogram based on 52 draws from a used deck shuffled once before each draw.

As you can see, Figure 6.3 is not exactly like Figure 6.1. In this case, the differences between the two distributions are due to chance or worn cards and not to lost cards or sleight of hand (at least not conscious sleight of hand). Of course, if we made 52 more draws from the deck and constructed a new histogram, the picture would probably be different from both Figures 6.3 and 6.1. However, if we continued, drawing 520 or 5200 or 52,000 times,[3] and only chance were at work, we would get a curve that was practically flat on the top. That is, the empirical curve would look like the theoretical curve.

The major point here is that a theoretical curve represents the "best estimate" of how the events would actually occur. As with all estimates, the theoretical curve is somewhat inaccurate; but in the world of real events it is better than any other estimate.

In summary, then, a theoretical distribution is one based on logic and mathematics rather than on observations. It shows you the probability of each event that is part of the distribution. When it is similar to an empirical distribution,

---

[3]Statisticians describe this extensive sampling as "the long run."

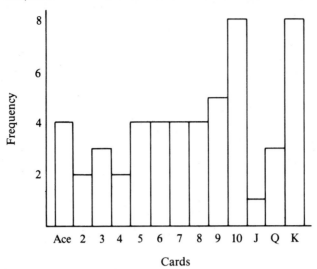

Figure 6.3   Empirical frequency distribution of 52 draws from a deck of playing cards.

the probability figures obtained from the theoretical distribution are accurate predictors of actual events.

There are a number of theoretical distributions that applied statisticians have found useful. In the rest of this chapter and the next two, you will work with a theoretical distribution called the normal distribution. In later chapters you will deal with others. At this point, we will simply name some of these other theoretical distributions: t distribution, F distribution, chi square distribution, and U distribution.

THE NORMAL DISTRIBUTION

One theoretical distribution that has proved to be extremely valuable is called the **normal distribution**. In this case, *normal* has few of its usual connotations. The name was established by early statisticians, who found that frequency distributions of data gathered from a wide variety of fields were similar. Thus, *normal* simply means that this distribution is frequently found.

The normal distribution is sometimes called the Gaussian distribution after Carl Friedrich Gauss (1777–1855), who developed the curve (about 1800) as a way to represent the random error in astronomy observations (Stewart, 1977). Because this curve was such an accurate picture of the effects of random variation, early writers referred to the curve as the *law* of error.[4]

Description of the Normal Distribution

Figure 6.4 is a normal distribution. It is a bell-shaped, symmetrical distribution, a theoretical distribution based on a mathematical formula rather than on

---

[4]This word, *error*, meaning random variations, remains in statistical terminology. In the next chapter you will learn about a statistic called the *standard error*.

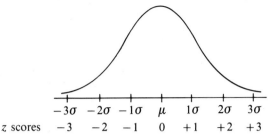

z scores   −3   −2   −1   0   +1   +2   +3

Figure 6.4   The normal distribution.

any empirical observations. (Even so, if you peek ahead to Figures 6.7, 6.8, and 6.9, you will see that empirical curves often look similar to this theoretical distribution.) When the theoretical curve is drawn, the Y axis is usually omitted. On the X axis, z scores are used as the unit of measurement for the standardized normal curve, where

$$z = \frac{X - \mu}{\sigma}$$

We will discuss the standardized curve first and later add the raw scores to the X axis.

There are a number of other things to note about the normal distribution. The mean, the median, and the mode are the same score—the score on the X axis at which the curve is at its peak. If a line were drawn from the peak to the mean score on the X axis, the area under the curve to the left of the line would be half the total area—50 percent—leaving half the area to the right of the line. The tails of the curve are **asymptotic** to the X axis; that is, they never actually cross the axis but continue in both directions indefinitely with the distance between the curve and the X axis becoming less and less. Although theoretically the curve never ends, it is convenient to think of (and to draw) the curve as extending from -3σ to $+3\sigma$. (The *table* for the normal curve, however, covers the area from -4σ to $+4\sigma$.)

Another point about the normal distribution is that the two inflection points in the curve are at exactly -1σ and $+1\sigma$. An inflection point is where a curve changes from bowed down to bowed up, or vice versa. (See the point above -1σ and $+1\sigma$ on Figure 6.4.)

Unfortunately, the antonym for *normal* is *abnormal*. The word *abnormal* has connotations that are not appropriate in statistics. *Curves that are not normal distributions are definitely not abnormal.* There is nothing abnormal about the distribution of playing cards in Figure 6.1, and it is not a normal curve. There is nothing abnormal about the distribution in Figure 6.5. It simply shows what numbers were picked when an instructor asked introductory psychology students to pick a number between one and ten. (This is an example of a bimodal distribution with modes at 3 and 7.)

Use of the Normal Distribution

The theoretical normal distribution is used to determine the probability of an event just as Figure 6.1 was. Figure 6.6 is a picture of the normal curve,

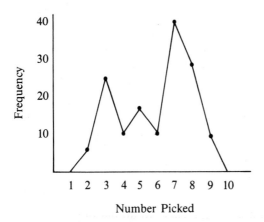

**Figure
6.5**   Frequency distribution of choices of numbers between 1 and 10.

showing the probabilities associated with certain areas. These probability figures were obtained from Table C in the Appendix. Look at Table C now. It is arranged so that you may begin with a z score (Column A) and find

1. the area between the mean and the z score (Column B)
2. the area beyond the z score (Column C)

In Column A find the z score of 1.00. The proportion of the curve between the mean and a z score of 1.00 is .3413. The proportion beyond the z score of 1.00 is .1587. Since the normal curve is symmetrical and since the area under the entire curve is 1.00, you will probably find it satisfying to add .3413 and .1587 together. Also, since the curve is symmetrical, these same proportions hold for $z = -1.00$. Thus, all the proportions in Figure 6.6 were derived by looking up the proportions associated with a z value of 1.00 in Table C. Don't just read this paragraph; do it. Understanding the normal curve *now* will pay you dividends throughout the book.

Notice that the proportions in Table C are carried to four decimal places and that we used all of them. This is customary practice in dealing with the normal curve.

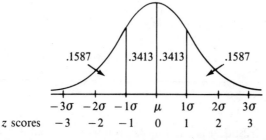

**Figure
6.6**   The normal distribution showing the probabilities of certain z scores.

Problem

8. What proportion of the normal distribution is found
    a. between the mean and $z = .21$
    b. between the mean and $z = -2.01$
    c. between the mean and $z = .55$

Many empirical distributions are approximately normally distributed. Figure 6.7 shows a set of 261 IQ scores, Figure 6.8 shows the diameter of 199 ponderosa pine trees, and Figure 6.9 shows the hourly wage rates of 185,822 union truck drivers in 1944. As you can see, these distributions from diverse fields are similar to Figure 6.4, the theoretical normal distribution. Please note that all of these empirical distributions are based on a "large" number of observations. Usually, 100 or more observations are required for the curve to fill out nicely.

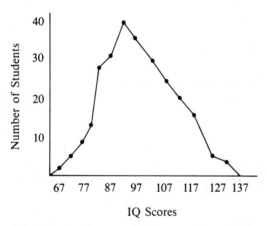

Figure 6.7 Frequency distributions of IQ scores of 261 fifth-grade students. Source: unpublished data of J. O. Johnston.

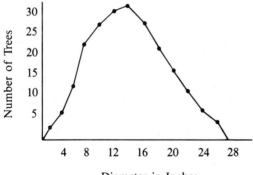

Figure 6.8 Frequency distribution of diameters of 100-year-old ponderosa pine trees on one acre. $N = 199$. Source: Forbs and Meyer, 1955, pp. 3–34.

So far in this section, we have made two points: first, that Table C can be used to determine areas (proportions) of a normal distribution, and, second, that

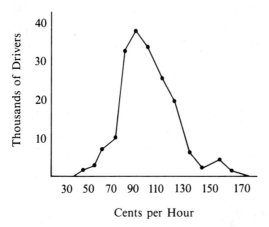

Figure 6.9   Frequency distribution of hourly wage rates of union truck drivers on July 1, 1944. $N = 185,822$. Source: *Monthly Labor Review*, December 1944.

many empirical distributions are approximately normally distributed. Our final point is that any normally distributed empirical distribution can be made to correspond to the standardized normal distribution (a theoretical distribution) by using z scores. Converting the raw scores of *any* empirical normal distribution to z scores will give the distribution a mean equal to zero and a standard deviation equal to 1.00, and that is exactly the scale used in the theoretical normal distribution. With this correspondence established, the theoretical normal distribution can be used to determine the probabilities of empirical events, whether they are IQ scores, tree diameters, or hourly wages.

In the following examples, the normal curve will be used with IQ scores. IQ scores are distributed normally (Figure 6.7 is evidence for that) with a mean of 100 and a standard deviation of 15.[5]

Finding What Proportion of a Population Has Scores of a Particular Size or Greater

Suppose you were interested in knowing the proportion of all people who have IQs of 120 or greater. The normal distribution shown in Figure 6.10 has IQs included on the X axis. You need to know the proportion under the curve to the right of IQ = 120 (the shaded area in Figure 6.10). For an IQ score of 120, the z score would be $(120 - 100)/15 = 20/15 = 1.33$. Table C shows that the proportion beyond $z = 1.33$ is .0918. Thus you would expect a proportion of .0918 or 9.18 percent of the population to have an IQ of 120 or greater. *Since the size of an area under the curve is also a probability statement about the events in that area, we can say that there are 9.18 chances in 100 that any randomly selected person will have an IQ of 120 or above.* Figure 6.11 shows the proportions just determined.

---

[5]This is true for the Wechsler intelligence scales (WAIS, WISC, and WPPSI). The Stanford-Binet has a mean of 100 also, but the standard deviation is 16.

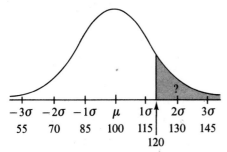

Figure 6.10   Theoretical distribution of IQ scores.

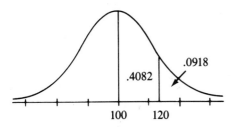

Figure 6.11   Proportion of the population with an IQ of 120 or greater.

Table C gives the proportions of the normal curve for positive z scores only. However, since the distribution is symmetrical, knowing that .0918 of the population has an IQ of 120 or greater tells you that .0918 has an IQ of 80 or less. An IQ of 80 has a z score of -1.33.

You can answer questions of "how many" as well as questions of proportions. Suppose there were 500 first graders entering school. How many would be expected to have IQs of 120 or greater? You just found that 9.18 percent of the population would have IQs of 120 or greater. If the population is 500, then calculating 9.18 percent of 500 would give you the number of children. Thus, $(.0918)(500) = 45.9$. So, 46 of the 500 first graders would be expected to have an IQ of 120 or greater.

There are 19 problems left in this chapter for you to do. You can do every one of them correctly by using just the z-score formula and Table C. However, if you will first sketch a normal curve for each problem, write in the givens and the unknown, and finally, as a last step, apply the z-score formula, you are more likely to get the correct answers and, as a bonus, you will be able to conceptualize the normal curve more quickly. Drawing pictures of the normal curve is very helpful.

Problems   **9.** Is the distribution in Figure 6.5 theoretical or empirical?

10. What proportion of the population has an IQ of 130 or greater?

11. What is the probability that a randomly selected person has an IQ of 90 or below?

12. Calculate the z scores for IQ scores of 55, 110, 103, and 100.

13. What proportion of the population would be expected to have IQs of 110 or greater?
14. If there were 250 first-grade students,
    a. how many would you expect to have IQs of 110 or greater?
    b. how many would you expect to have IQs of less than 110?
    c. how many would you expect to have IQs less than 100?

Finding the Score That Separates the Population into Two Proportions

Instead of starting with an IQ score and calculating proportions, you can also work backward and answer questions about scores if you are given proportions. For example, what IQ score is required to be in the top 10 percent of the population? The problem is pictured in Figure 6.12.

Error Detection

Drawing pictures of normal distributions is the best way to understand these problems and avoid errors.

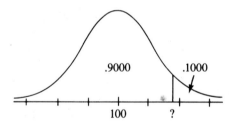

.9000   .1000

100   ?

Figure 6.12   Theoretical distribution of IQ scores divided into an upper 10 percent and a lower 90 percent.

Look at Figure 6.12. You can see that what you need to know is the IQ score associated with the point that divides the distribution into an upper 10 percent and a lower 90 percent. Begin by finding the z score that separates the upper 10 percent of the distribution from the rest. Look in Column C for .1000. It is not there. You have a choice between .0985 and .1003. Since .1003 is closer to the desired .1000, use it.[5] The corresponding z score is 1.28.

Since

$$z = \frac{X - \mu}{\sigma},$$

[5]You might use interpolation to determine the *exact* z score associated with a proportion of .1000. This extra precision (and labor) is unnecessary in this problem because the final result is rounded to the nearest whole number. For IQ scores, the extra precision does not make any difference in the final answer.

multiplying both sides by σ,

$$(z)(\sigma) = X - \mu,$$

and adding μ to both sides,

$$X = \mu + (z)(\sigma).$$

Substituting numbers for the mean, the z score, and the standard deviation,

$$X = 100 + (1.28)(15)$$

$$= 100 + 19.20$$

$$= 119.2$$

$$= 119. \text{ (IQs are usually expressed as whole numbers.)}$$

Therefore, the IQ score required to be in the top 10 percent of the population is 119.

Here is a similar problem. Suppose a mathematics department wants to restrict the remedial math course to those who really need it. It has the scores on the mathematics achievement exam taken by entering freshmen for the past ten years. The scores on this exam are distributed in an approximately normal fashion with $\mu = 58$ and $\sigma = 12$. The department wants to make the remedial course available to those students whose mathematical achievement places them in the bottom third of the freshman class. The question is: what score will divide the lower third from the upper two-thirds? Figure 6.13 should help to clarify the question.

The first step is to look in Column C to find .3333. Again, such a proportion cannot be found. The nearest proportion is .3336, which has a z value of $-.43$. (This time you are dealing with a z score below the mean, where all z scores are negative.) Applying $z = -.43$, find that

$$X = \mu + (z)(\sigma)$$

$$= 58 + (-.43)(12)$$

$$= 58 - 5.16 = 52.84.$$

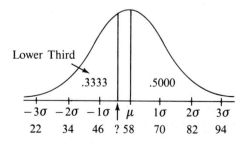

Figure 6.13   Distribution of scores on a mathematics achievement exam.

If the scores on the mathematics achievement exam are always given in whole numbers, a decision must be made either to let persons with scores of 53 or less enroll (which would make the course available to slightly more than one-third of the freshmen), or to require a score of 52 or less (which would make the course available to less than one-third).

Using the theoretical normal curve to establish a cutoff score is efficient. All you need is the mean, the standard deviation, and confidence in your assumption that the scores are distributed normally. The empirical alternative for the math department is to sort physically through all scores for the past ten years, arrange them in a frequency distribution, and calculate the score that separates the bottom one-third.

Problems

15. What IQ score separates the top 25 percent of the population from the lower 75 percent?
16. American females, ages 20–24, have a mean height of 64.0 inches with a standard deviation of 2.4 inches (Stoudt et al., 1960).
    a. What is the shortest height that allows a woman to be in the tallest 5 percent of the population?
    b. What is the tallest height that allows a woman to be in the shortest 5 percent of the population?
17. American males, ages 20–24, have a mean height of 68.7 inches with a standard deviation of 2.6 inches (Stoudt et al., 1960). What proportion of this population would be expected to be over 72 inches in height?
18. The weight of new U.S. pennies is approximately normally distributed with a mean of 3.11 grams and a standard deviation of .05 gram (Youden, 1962).
    a. What proportion of all new pennies would you expect to weigh more than 3.20 grams?
    b. What weights separate the middle 80 percent of the pennies from the lightest 10 percent and the heaviest 10 percent?

Finding the Proportion of the Population between Two Scores

Table C can also be used to determine the proportion of the population between two scores. For example, scores that fall in the range of IQ scores from 90 to 110 are often called "average." What proportion of the population falls in this range? Figure 6.14 is a picture of the problem.

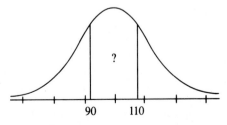

Figure 6.14   The normal distribution showing the IQ scores that define the "average" range.

In this problem, you must add an area on the left of the mean to an area on the right of the mean. First, you need z scores that correspond to the IQ scores of 90 and 110.

$$z = \frac{90 - 100}{15} = \frac{-10}{15} = -.67.$$

$$z = \frac{110 - 100}{15} = \frac{10}{15} = .67.$$

The proportion of the distribution between the mean and $z = .67$ is .2486, and the same proportion is found between the mean and $z = -.67$. Therefore, $(2)(.2486) = .4972$ or 49.72 percent. So, approximately 50 percent of the population is classified as "average," using the "IQ = 90 to 110" definition of "average."

What percent of the population would be expected to have IQs between 90 and 125? Figure 6.15 illustrates this question.

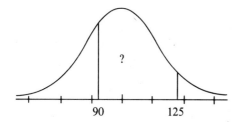

Figure 6.15   The normal distribution showing the area bounded by IQ scores of 90 and 125.

Again, you must add an area that is on the left of the mean to an area on the right of the mean, but this time the areas are unequal. The corresponding z scores are

$$z = \frac{90 - 100}{15} = -.67 \quad \text{and} \quad z = \frac{125 - 100}{15} = 1.67$$

Using Table C, you can find that the proportions between the mean and z scores of $-.67$ and 1.67 are .2486 and .4525, respectively. When these two proportions are added, the result, .7011, is the proportion of the population with IQs between 90 and 125.

Problems

19. The distribution of 800 test scores in an Introduction to Psychology course was approximately normal, with $\mu = 35$ and $\sigma = 6$.
   a. What proportion of the students had scores between 30 and 40?
   b. What is the probability that a randomly selected student would score between 30 and 40?

20. Now that you know the proportion of students with scores between 30 and 40, would you expect to find the same proportion between scores of 20 and 30? If so, why? If not, why not?

21. Calculate the proportion of scores between 20 and 30. Be careful with this one; drawing a picture may prevent an error.

22. How many of the 800 would be expected to have scores between 20 and 30?

Finding the Extreme Scores in a Population

What IQs are so extreme that only one percent of the population could be expected to have them? This is a little different from the earlier questions about the "lower third" and the "upper 25 percent" because this question does not specify which end of the curve is of interest. To answer this type of question, divide the proportion in half, placing each part under one extreme end of the curve. For the current question, the problem is to find the scores that leave $\frac{1}{2}$ of 1 percent (.0050) in each tail of the curve, as seen in Figure 6.16.

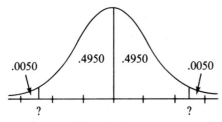

Figure 6.16   Distribution of IQs showing one percent of the scores divided between the two extremes.

Since you are starting with a proportion and looking for a score, look in Column C for .0050. This "1%" problem turns up in other contexts in statistics. By convention, the z score to use is 2.58. You can again use $X = \mu + (z)(\sigma)$ to determine the raw score (IQ). Thus,

$$X = 100 + (2.58)(15) \quad \text{and} \quad X = 100 + (-2.58)(15)$$

$$= 100 + 38.7 \qquad\qquad\qquad = 100 - 38.7$$

$$= 138.7 \text{ or } 139 \qquad\qquad = 61.3 = 61.$$

Thus, IQs of 139 or more and 61 or less are so extreme that they occur in only 1 percent of the population. Ninety-nine percent of the population, of course, falls between these scores.

Clue to the Future   The idea of finding scores and proportions that are extreme in either direction will come up again after Chapter 7. In particular, the extreme 5 percent and the extreme 1 percent will be calculated.

Problems

23. What IQ scores are so extreme that they are achieved by only 5 percent of the population? Set this problem up using the "extreme one percent" example as a model.

24. What is the probability that a randomly selected person has an IQ higher than 129 *or* lower than 71?

25. Look at Figure 6.9 and suppose that the union leadership decided to ask for $.85 per hour as a minimum wage. For those 185,822 workers, the mean was $.99 with a standard deviation of $.17. If $.85 per hour were established as a minimum, how many workers would be affected?

26. Look at Figure 6.8 and suppose that a timber company decided to harvest all trees over 8 inches DBH (Diameter Breast Height) from a 100-acre tract. On a 1-acre tract there were 199 trees with $\mu = 13.68$ and $\sigma = 4.83$. How many trees would be expected to be harvested from 100 acres?

COMPARISON OF THEORETICAL AND EMPIRICAL ANSWERS

You have been using the theoretical normal distribution to find probabilities and to calculate scores and proportions of IQs, diameters, wages, and other measures. Earlier in this chapter, the claim was made that *if* the empirical observations are distributed like a normal curve, accurate predictions can be made. A reasonable question is "How accurate are all these predictions I've just made?" A reasonable answer can be fashioned from a comparison of the predicted proportions (from the theoretical curve) to the actual proportions (computed from empirical data). Figure 6.7 is based on 261 IQ scores of fifth-grade public-school students. You can easily calculate the proportion who had IQs above 120 or below 90 or between 90 and 110. These actual proportions can then be compared with those predicted from the normal distribution. Table 6.2 shows these comparisons.

Table 6.2

Comparison of Predicted and Actual Proportions

IQs	Predicted from Normal curve	Calculated from Actual Data	Difference
Above 120	.0918	.0919	.0001
Below 90	.2514	.2069	.0445
Between 90 and 110	.4972	.5249	.0277

As you can see by examining the "difference" column of Table 6.2, the accuracy of the predictions ranges from good to not so good. Some of this variation can be explained by the fact that the mean IQ of the fifth-grade students was 101 and the standard deviation 13.4. Both the higher mean (101, compared with 100 for the normal curve) and the lower standard deviation (13.4, compared with 15) are due to the systematic exclusion of very-low-IQ children from regular public schools. Thus, the actual proportion of students with IQs less than 90 is less than predicted, which is the result of the fact that our school sample is not representative of all 10- to 11-year-old children. This problem of sampling is the major topic of the next chapter.

Problem

27. An imaginative anthropologist measured the stature of 100 hobbits (using the proper English measure of inches) and found the following:

$$\Sigma X = 3600 \qquad \Sigma X^2 = 130{,}000$$

Assume that the height of hobbits is normally distributed. Find μ and σ and answer the questions below.

a. The Bilbo Baggins Award for Adventure is 32 inches tall. What proportion of the hobbit population is taller than the award?

b. Three hundred hobbits entered a cave that had an entrance 39 inches high. The orcs chased them out. How many could exit without ducking?

c. Gandalf is 44 inches tall. What is the probability that he is a hobbit?

7

SAMPLES AND SAMPLING DISTRIBUTIONS

Objectives for Chapter 7: After studying the text and working the problems in this chapter, you should be able to:

1. define a random sample,
2. obtain a random sample if you are given a population,
3. define and identify biased sampling methods,
4. define a sampling distribution,
5. calculate a standard error of the mean if you are given σ and N or if you are given s and N,
6. use the z-score formula to determine the probability of a certain sample mean being drawn from a population with a certain (known) mean,
7. verbally define the standard error of any statistic, and
8. calculate 95 and 99 percent confidence intervals about a sample mean and explain what they mean.

We want to begin by asking you to be especially attentive to the concept of a sampling distribution. In the topic of inferential statistics, the idea of a **sampling distribution** is clearly the *central concept*. This book on "Tales of Distributions" is a book of tales of sampling distributions. Every chapter after this one is about some kind of sampling distribution; thus, you might consider this paragraph to be a superclue to the future.

An understanding of sampling distributions requires an understanding of samples. A sample, of course, is some part of the whole thing; in statistics the "whole thing" is a population. The population is always the thing of interest; a sample is used only to estimate what the population is like. One obvious problem is to get samples that are representative of the population.

Here is a story illustrating the problems of getting a representative sample and then knowing how accurately it represents the population.

Late one afternoon, two economics students were discussing the average family income of students on their campus.

"Well, the average family income for college students is $28,100, nationwide" (Austin, 1982), asserted the junior, "and I'm sure the mean for this campus is at least that high."

"I don't think so," the sophomore replied. "I know lots of students who have only their own resources or come from pretty poor families. I'll bet you a dollar the mean for students here at State U. is below the national average."

"You're on," grinned the junior.

Together the two went out with a pencil and pad and asked 40 students how much their family income was. Thirty-seven students made estimations, and the mean of these was $24,300.

Now the sophomore grinned, "I told you so; the mean here is $3800 less than the national average of $28,100."

"Curses," cursed the junior. But, after thinking about those 37 interviews, he began to smile. "Actually," he said, "this mean of $24,300 is meaningless because those 37 students are not representative of the entire student body. They are late-afternoon students, and many of them support themselves with temporary jobs while they go to school. Most of the student body is supported from home by parents who have permanent and better-paying jobs. That sample is no good, and I won't pay until we get results for the whole campus or at least from a representative sample."

To get the results for the whole student body, the two went the next day to the Director of Financial Aid, who told them that family incomes for the student body are not public information.

Sampling became a necessity.

With the help of the Office of the Dean of Students, a random sample of 40 students was selected. (A random sample has the best chance of being representative.) The two investigators then got replies from 38 and found the mean family income to be $26,900.

"Pay," demanded the sophomore.

"O.K., O.K., . . . here!"

Later, the junior again began to think about that random sample. He wondered what the mean would be if he polled another random sample. It could be higher. Maybe the mean of $26,900 was just a fluke—a chance event.

Shortly afterward, the junior confronted his co-investigator with his thoughts about repeated sampling. "How do we know if that random sample we got told us the truth about the population? For example, suppose the mean for the entire student body *is* $28,100. Maybe a sample with a mean of $26,900 would occur just by chance fairly often. I wonder just what the chances are of getting a sample mean of $26,900 from a population with a mean of $28,100?"

"To answer that, you would have to know about the sampling distribution of the mean," wisely intoned the sophomore, hurrying from the room.

The two basic points of the story are that samples that are random have the best chance of being representative (the junior paid off only after the results were based on a random sample) and that something called a sampling distribution can tell you how much faith (probability-wise) you can put in results based on a random sample.

Clue to the Future

In the story, the junior introduced a **hypothesis** about the State U. population—namely, that the mean family income for the student body was $28,100.

The basic idea throughout the rest of this book will be to introduce a hypothesis about a population and then, based on the results of a sample, decide either that the hypothesis is reasonable or that it should be rejected.

Before we begin to explain these ideas, we'll expand a little on some terms we introduced in Chapter 1. *Population*, as you know, means all the members of a specified group. Examples of specifications are self-esteem scores of freshmen who were enrolled at State University on October 1, total time of hospitalization of deceased hebephrenic schizophrenics in the state of New York, and weight of the brain of all well-fed albino rats. It is the investigator who defines the population of interest. Sometimes the population is one that could actually be measured, given plenty of time and money, as in the first and second examples. Sometimes, however, such measurements are logically impossible, as in the third example. Many of those rats are already dead, and besides, rats reproduce faster than they can be measured. In the story, the population was the family income of all currently enrolled students at State University. *Inferential statistics* are used when it is not possible or practical to measure an entire population.

So, decisions about unmeasurable populations can be made by using samples and the methods of inferential statistics. Unfortunately, there is some peril in this. Samples are variable, changeable things. Each one produces a different statistic. How can you be sure that the sample you draw will produce a statistic that will lead to a correct decision about the population? Unfortunately, you *cannot* be absolutely sure. **To draw a sample is to agree to accept some uncertainty about the results.** However, in this chapter you will learn how to measure this uncertainty. If a great deal of uncertainty exists, the sensible thing to do is suspend judgment. On the other hand, if there is very little uncertainty, the sensible thing to do is to reach a conclusion, even though there is a small risk of being wrong. *Reread this paragraph. It is important.*

The following problems give you an opportunity to review some concepts.

Problems
1. Define statistic.
2. Define parameter.
3. What is the difference between μ and $\bar{X}$?
4. The quality control engineer at a light-bulb factory was concerned about the number of bulbs on store shelves that would not work (presumably due to rough shipping and handling). A new packaging material was available, but no one was sure it was necessary. The engineer decided to sample from the previous month's production by going to 25 stores in the area and exchanging the bulbs on the shelves for new bulbs. The "recalled" bulbs were tested, and it was found that a proportion, .005 ($\frac{1}{2}$ of 1 percent), would not light. Name the population, the parameter of interest, the sample, and the statistic. What is the numerical value of the statistic?

REPRESENTATIVE AND NONREPRESENTATIVE SAMPLES

Let's begin by assuming that you want to know about an unmeasurable population, a desire common among researchers, both young and old. Unmeasurable populations require you to draw a sample and, naturally, you would like a representative sample. The key to having a representative sample is to use a *method of obtaining samples* that is more likely to produce a representative sample than any other method.[1]

We will name two methods of sampling that are most likely to produce a representative sample, discuss one of them in detail, and then discuss some ways in which nonrepresentative samples are obtained when the sampling method is biased.

Random Samples

A method called *random sampling* is commonly used to obtain a sample that is most likely to be representative of the population. *Random* has a technical meaning in statistics and does not mean haphazard or unplanned. A **random sample** in most research situations is one in which every potential sample of size N has an equal probability of being selected. To obtain a random sample, you must

1. define the population of scores,
2. identify every member of the population, and
3. select scores in such a way that every sample has an equal probability of being chosen.

[1]How well a particular method works can be assessed either mathematically or empirically. For an empirical assessment, start with a population of numbers, the parameter of which can be easily calculated. The particular method of sampling is repeatedly used, and the corresponding statistic is calculated for each sample. The mean of these sample statistics can then be compared with the parameter.

We'll go through these steps with a set of real data—the self-esteem scores of 24 fifth-grade children.[2] We define these 24 scores as our population. From these we will pick a random sample of seven scores.

39   43   40   25   42   42   32   36

36   24   30   31   35   31   45   36

22   31   35   37   46   42   33   36

One method of picking a random sample is to write each self-esteem score on a slip of paper, put the 24 slips in a box, jumble them around, and draw out seven. The scores on the chosen slips become a random sample. This method works fine if the slips are all the same size and there are only a few members of the population. If there are many members, this method is tedious.

Another (easier) method of getting a random sample is to use a table of random numbers, such as Table B in the Appendix. To use the table, you must first assign an identifying number to each of the 24 self-esteem scores, thus:

ID Number	Score	ID Number	Score
01	39	13	35
02	43	14	31
03	40	15	45
04	25	16	36
05	42	17	22
06	42	18	31
07	32	19	35
08	36	20	37
09	36	21	46
10	24	22	42
11	30	23	33
12	31	24	36

Each score has been identified by a two-digit number. Now turn to Table B and pick a row and a column in which to start. Any haphazard method will work; close your eyes and stab a place with your finger. Suppose you started at row 35, columns 70–74. Reading horizontally, the digits are 21105. Since you need only two digits to identify any member of our population, use the first two digits, 21. That identifies one score for the sample—a score of 46. From this point, you can read two-digit numbers in any direction—up, down, or sideways— but the decision should have been made before you looked at the numbers. If you had decided to go down, the next number is 33. No self-esteem score has an identifying number of 33, so skip it and go to 59, which gives you the same problem as 33. In fact, the next five numbers are too large. The sixth number is

[2]The self-esteem scores are based on the Coopersmith Self-esteem Inventory and are from Johnston and Spatz (1973). The Coopersmith inventory consists of 50 items that the student designates as "like me" or "unlike me." Items are similar to "I am often ashamed of myself" and "I am a fairly happy person."

07, which identifies the score of 32 for the random sample. The next usable number is 13, which identifies a score of 35. Continue in this way until you arrive at the bottom. At this point, you can go in any direction. We will skip over two columns to columns 72 and 73 (you were in columns 70 and 71) and start up. The first number is 12, which identifies a score of 31. The next usable numbers are 19, 05, and 10, which give scores of 35, 42, and 24. Thus, the random sample of seven consists of the following scores: 46, 32, 35, 31, 35, 42, and 24. If Table B had produced the same identifying number twice, you would have ignored it the second time.

Problem

5. A random sample is supposed to produce a statistic similar to the parameter. The population μ for the 24 self-esteem scores is 35.375. What is the mean of your random sample of seven?

What is this table of random numbers? In Table B (and in any table of random numbers), the probability of occurrence of any digit from 0 to 9 at any place in the table is the same—.10. Thus, you are just as likely to find 000 as 123 or 381. Incidentally, you cannot generate random numbers out of your head. Certain sequences begin to recur, and (unless warned) you will not include enough repetitions like 666 and 000.

Here are some hints for using a table of random numbers.

1. Make a check beside the identifying number of a score when it is chosen for the sample. This will help prevent duplications.
2. If the population is large (over 100), it is more efficient to get all the identifying numbers from the table first. As you select them, put them in some rough order. This will help prevent duplications. After you have all the identifying numbers, go to the population to select the sample.
3. If the population has exactly 100 members, let 00 be the identifying number for 100. In this way, you can use two-digit identifying numbers, each one of which matches a population score. This same technique can be applied to populations of 10 or 1000 members.

Problems

6. Draw a random sample with $N = 5$ from the population of self-esteem scores on page 133.
7. Draw a random sample of ten from the 24 self-esteem scores. This time, begin at a different place in the table of random numbers. Calculate the mean of the sample.
8. Why should you begin at a different place in the table for your second sample?
9. Draw a random sample with $N = 12$ from the following scores:

76	47	81	70	67	80	64	57	76	81
68	76	79	50	89	42	67	77	80	71
91	72	64	59	76	83	72	63	69	
78	90	46	61	74	74	74	69	83	

Stratified Samples

A method called *stratified sampling* is another way to produce a sample that is very likely to mirror the population. It can be used when an investigator knows the numerical value of some important characteristic of the population. A **stratified sample** is controlled so that it exactly reflects some known characteristic of the population. Thus, in a stratified sample, not everything is left to chance. *not true*

For example, in a public-opinion poll on a sensitive political issue, it is important that the sample reflect the proportions of the population who consider themselves Democrat, Republican, and Independent. Most public-opinion polls show that when a party affiliation is asked for, the answers indicate approximately the following proportions: Democrat, 40 percent; Republican, 20 percent; and Independent, 40 percent. The investigator draws the sample so it will reflect the proportions found in the population. The same may be done for variables such as sex, age, and socioeconomic status. After stratification of the samples has been determined, sampling within each stratum is usually random.[3]

To justify a stratified sample, the investigator must know what variables will affect the results and what the population characteristics are for those variables. Sometimes the investigator has this information (as from census data), but many times such information is just not available (as in most research situations).

Biased Samples

A **biased sample** is one that is drawn using a method that systematically underselects or overselects from certain groups within the population. Thus, in a biased sampling technique, every sample of a given size does *not* have an equal opportunity of being selected. With biased sampling techniques, you are much more likely to get a nonrepresentative sample than you are with random or stratified sampling techniques.

For example, it is reasonable to conclude that some results based on mailed questionnaires are not valid, because the samples are biased. Usually, an investigator defines the population, identifies each member, and mails the questionnaire to a randomly selected sample. Suppose that 70 percent of the recipients respond. Can valid results for the population be based on the questionnaires returned? Probably not. There is good reason to suspect that the 70 percent who responded are different from the 30 percent who did not. Thus, although the population is made up of both kinds of people, the sample reflects only one kind. Therefore, the sample is biased. The probability of bias is particularly high if the questionnaire elicits feelings of pride or despair or disgust or apathy in *some* of the recipients.

Problems   **10.** Suppose a large number of questionnaires about educational accomplishments were mailed out. Do you think that some recipients would be more likely to return the questionnaire than others? Which ones? If the sample is biased, will

[3]A book with a chapter specifically about stratified samples is L. Kish, *Survey Sampling*, New York: John Wiley & Sons, 1965.

it overestimate or underestimate the educational accomplishments of the population?

11. Sometimes newspapers sample opinions of the local population by printing a "ballot" and asking their readers to mark it and mail it in. Evaluate such a sampling technique.

A very famous case of a biased sample occurred in a poll that was to predict the 1936 election for President of the United States. The *Literary Digest* (a popular magazine) mailed ten million "ballots" to registered voters, and two million were returned. The prediction was clear: Alf Landon by a landslide over Franklin Roosevelt. As you may have learned, the actual results were just the opposite; Roosevelt got 61 percent of the vote.

From the ten million who had a chance to express a preference, two million very interested persons had selected themselves. This two million had more than its proportional share of those who were disgruntled with Roosevelt's depression-era programs. The technique was biased and the results were not representative of the population. In fairness to the *Literary Digest*, it should be noted that they knew they had a biased sample and that, in 1932, *using the same biased sampling technique, they had predicted the popular vote within two percentage points.*[4]

With a nice random sample you can predict fairly accurately your chance of being wrong. If it is higher than you would like, you can reduce it by increasing sample size. With a biased sample, however, you do not have any basis for assessing your margin of error and you don't know how much confidence to put in your predictions. You may be right (as the *Literary Digest* was in 1932) or you may be very wrong (the 1936 experience).

Here are some other examples of biased samples. The junior economics major in our story recognized that the sample of passersby was biased toward people taking late afternoon classes (who, perhaps, had lower family incomes). The population of a city cannot be randomly sampled using a telephone book because some persons in the city do not have a telephone and others have an unlisted number. None of these would have a chance of being selected to be in the sample. Intact groups often represent biased samples. An Introduction to Psychology class or a cage of rats may be the handiest sample around, but, for almost any population you can name, they represent biased samples. You may get generalizable results from such samples, but you cannot be sure. The search for biased samples in someone else's research is a popular (and serious) game among researchers.

Problems

12. Explain the difference between a biased and a representative sample.
13. Can a statistic be computed on data from a biased sample?
14. Consider as a population the students at State U. whose names are listed alphabetically in the student directory. Are the following samples biased, random, or stratified?

---

[4]For the postelection analysis, see the *Literary Digest, 122*, November 14, 1936, pp. 7–8.

 a. every fifth name on the list.
 b. every member of the junior class,
 c. 150 names drawn from a box containing all the names in the directory,
 d. one name chosen at random from the directory,
 e. 12 persons chosen at random from the freshman class, 10 chosen at random from the sophomore class, 9 at random from the junior class, and 8 at random from the senior class,
 f. a random sample from those taking the required English course.
15. Was the sample of light bulbs in Problem 4 random, stratified, or biased?

SAMPLING DISTRIBUTIONS

This short section is about sampling distributions in general, and the next section is about a particular sampling distribution—the sampling distribution of the mean. While you study this chapter, begin to build two categories in your thinking: sampling distributions in general and sampling distributions of the mean. Establishing these two categories now will help you as you learn about other sampling distributions in later chapters.

A sampling distribution is a frequency distribution of sample statistics. A sampling distribution could be obtained by drawing many random samples from a population and calculating a statistic on each sample. These statistics would be arranged into a frequency distribution. From such a distribution you could find the probability of obtaining any particular values of the statistic.

Every sampling distribution is for a particular statistic (such as the mean, variance, correlation coefficient, and so forth). In this chapter you will learn only about the sampling distribution of the mean. It will serve as an introduction to sampling distributions in general, some others of which you will find out about in later chapters.

THE SAMPLING DISTRIBUTION OF THE MEAN

Remember the population of 24 self-esteem scores that you sampled from earlier in this chapter? In one of the problems, each student drew a sample ($N = 10$) and calculated the mean. For almost every sample in your class, the mean was different. Each, however, was an *estimate* of the population mean. Think about drawing 200 samples with $N = 10$ from the same population, finding their means, and making a frequency distribution of the means. If it sounds like a tedious task, let us assure you that it is! We worked on it for a while but got so bored that we asked a friend to do the problem for us on a computer.

The computer drew 200 separate random samples, each with $N = 10$, calculated the mean of each, and arranged these 200 $\bar{X}$'s into the frequency polygon seen in Figure 7.1. The mean (parameter) of the 24 self-esteem scores is 35.375. As you can see, most of the statistics (sample means) are fairly good estimates of that parameter. Some $\bar{X}$'s, of course, miss the mark widely, but most

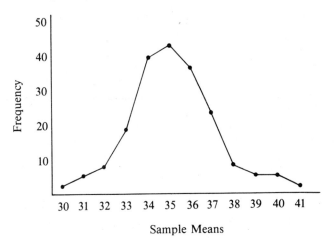

**Figure
7.1**

Empirical sampling distribution of the mean. The population is the self-esteem scores. For each sample, $N = 10$.

are pretty close. Figure 7.1 is an _empirical sampling distribution of the mean_; _that is, a frequency distribution of sample means._

Notice the steps:

1. Every sample is drawn randomly from the same population.
2. The sample size (N) is the same for all samples.
3. The number of samples is very large.

You will never use an empirical sampling distribution of the mean in any of your calculations; you will always use theoretical ones that come from mathematical formulas. An empirical sampling distribution of the mean is easier to understand, though, so we used it for this initial discussion.

We hope that, when you looked at Figure 7.1, you were at least suspicious that it might be the ubiquitous normal curve. It is. This puts you in the position of an educated person: what you learned in the past (how to use the normal curve) can be used to solve new and different problems—finding probabilities about sample means.

Of course, the normal curve is a theoretical curve, and we have presented you with an empirical curve that appears normal. We would like to let you prove for yourself that the form of a sampling distribution of the mean is a normal curve, but, unfortunately, that requires mathematical sophistication beyond that assumed for this course. So, we will resort to a time-honored teaching technique— an appeal to authority. In this case, our authority is mathematical statistics which has proved a theorem called the _Central Limit Theorem_. This important theorem says:

For _any_ population of scores, regardless of form, the sampling distribution of the mean will approach a normal distribution as N (sample size) gets larger. Furthermore, the sampling distribution of the mean will have a mean equal to μ and a standard deviation equal to $\sigma/\sqrt{N}$.

Our appeal to authority resulted in a lot of information. Now you know not only that sampling distributions of the mean are normal curves[5] but also that, if you know the population parameters μ and σ, you can determine the parameters of the sampling distribution. The mean of the sampling distribution of means will be the same as the population mean, μ. The standard deviation of the sampling distribution will be the standard deviation of the population (σ) divided by the square root of the sample size.

We will use this information to calculate the sampling distribution of the mean for the population of self-esteem scores. Figure 7.2 shows the results.

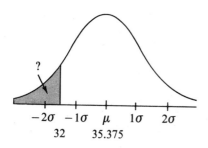

Figure 7.2

Theoretical sampling distribution of the mean, $N = 10$.

Population
mean = 35.375
standard deviation = 6.304

Sampling Distribution of the Mean for $N = 10$
mean = 35.375

standard deviation = $\dfrac{6.304}{\sqrt{10}} = 1.993$.

The most remarkable thing about the Central Limit Theorem is that it works *regardless* of the form of the original population. Thus, the sampling distribution of the mean of scores coming from a rectangular or bimodal population approaches normal if N is large. Figure 7.3 illustrates this and the effects of sample size for two examples you are already familiar with: the distribution of playing cards (Figure 6.1) and the distribution of choices of numbers between 1 and 10 (Figure 6.5).

Getting back to sampling distributions in general, here is further evidence of their importance: the standard deviation and the mean of sampling distributions have special names. The standard deviation of any sampling distribution is called the **standard error**,[6] and the mean is called the **expected value**. In the case of the

---

[5]One qualification is that the sample size (N) must be large. How many does it take to make a large sample? The traditional answer is 30 or more, although, if the population itself is symmetrical, a sampling distribution of the mean will be normal with sample sizes much smaller than 30. If the population is severely skewed, samples with 30 (or more) may be required.

[6]In this context, and in other statistical contexts, the term *error* means deviations or random variation. The word error is left over from the days of Gauss when random variation was referred to as the "normal law of error." Of course, "error" sometimes refers to mistake, so you will have to be alert to the context when this word appears (or you may make an error).

Population of Scores

Playing Cards   Choice of Numbers

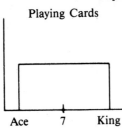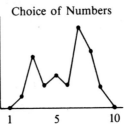

Sampling Distribution $N = 2$

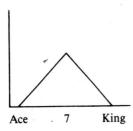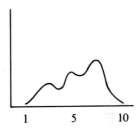

Sampling Distribution $N = 30$

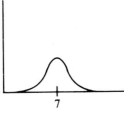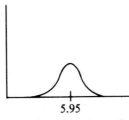

Figure 7.3   Populations of playing cards and number choices, and sampling distributions with $N = 2$ and $N = 30$.

sampling distribution of the mean, we are dealing with a standard error of the mean (symbolized $\sigma_{\bar{x}}$) and the expected value of the mean [symbolized $E(\bar{X})$]. You will not encounter $E(\bar{X})$ again in this book, although you will encounter it if you continue your study of statistics. The standard error, however, will be used many times. Be sure that you understand that it is the standard deviation of the sampling distribution of some statistic. In this chapter, it is the standard deviation of a sampling distribution of the mean.

Use of a Sampling Distribution of the Mean

Since the sampling distribution of the mean is a normal curve, you can apply what you learned in the last chapter about normally distributed *scores* to questions about sample *means*. For example, referring to Figure 7.2, what proportion of all sample means from the population of self-esteem scores would be expected to be 32 or less? You will recall that the z-score formula for an X score is $z = (X - \mu)/\sigma$. The application to a sample mean is similar:

$$z = \frac{\bar{X} - \mu}{\sigma_{\bar{X}}}$$

Thus, for $\bar{X} = 32$,

$$z = \frac{\bar{X} - \mu}{\sigma_{\bar{X}}}$$

$$= \frac{32 - 35.375}{1.993}$$

$$= -1.69.$$

Looking in the normal-curve table (Table C), you can see that, for a z score of 1.69, you would *expect* a proportion of .0455 of the means to be less than 32. We have redrawn Figure 7.2 as Figure 7.4 to summarize these findings.

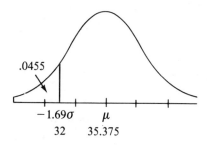

.0455

-1.69σ μ
  32   35.375

Figure 7.4   Proportion of sample means of 32 or less, $N = 10$.

The predicted proportion is .0455. Fortunately, we can check this prediction by determining the proportion of those 200 random samples that had means of 32 or less. By checking the frequency distribution from which Figure 7.1 was drawn, we found the empirical proportion to be .0400. Missing by one-half of 1 percent isn't bad, and once again, you find that a theoretical normal distribution predicts an actual empirical proportion quite nicely.

What effect does sample size have on a sampling distribution? Since $\sigma_{\bar{X}} = \sigma/\sqrt{N}$, $\sigma_{\bar{X}}$ will become smaller as N gets larger. Figure 7.5 shows some sampling distributions of the mean based on the population of 24 self-esteem scores. The sample sizes are 3, 5, 10, and 20. A sample mean of 39 is included in all four figures as a reference point. Notice that, as $\sigma_{\bar{X}}$ becomes smaller, a sample mean of 39 becomes a rarer and rarer event. The good investigator with an experiment to do, will keep in mind what we have just demonstrated about the effect of sample size on sampling distribution and will use reasonably large samples.

Problems   **16.** When the population parameters are known, the standard error of the mean is $\sigma_{\bar{X}} = \sigma/\sqrt{N}$. At each intersection of the table, fill in the value for $\sigma_{\bar{X}}$.

N	σ			
	1	2	4	8
1				
4				
16				
64				

17. On the basis of the table just constructed, write a verbal statement about the relationship between $\sigma_{\bar{X}}$ and N.
18. In order to reduce $\sigma_{\bar{X}}$ to half its size, you must increase N by how much?
19. Under what condition does $\sigma_{\bar{X}} = \sigma$?
20. Back in Chapter 5 you learned how to use a regression equation to predict a Y score if you were given an X score. That Y score is a statistic so it has a sampling distribution with its own standard error. What could you conclude if the standard error were very small? Very large?

CALCULATING A SAMPLING DISTRIBUTION WHEN σ IS NOT AVAILABLE

In calculating the standard error of the mean you have had a population parameter, σ, given to you. As you know, you rarely have population parameters. Fortunately, with a little modification of the formula and no modification of logic, you can estimate σ by calculating s from a random sample. With this estimate of σ, you can calculate an estimate of the standard error of the mean.

When you have only a sample standard deviation with which to estimate the standard error of the mean, the formula is the following:

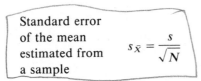

$$\text{Standard error of the mean estimated from a sample} \quad s_{\bar{X}} = \frac{s}{\sqrt{N}}$$

where: $s_{\bar{X}}$ = standard error of the mean,
          s = standard deviation of a sample,
and N = sample size.

The statistic $s_{\bar{X}}$ is an estimate of $\sigma_{\bar{X}}$, and $\sigma_{\bar{X}}$ is required for use of the normal curve. The larger the sample size, the more reliable $s_{\bar{X}}$ is. As a practical matter, $s_{\bar{X}}$ is considered reliable enough if $N \geq 30$. In pure mathematical statistics though, the normal curve is only appropriate when you know μ and σ.

We'll apply this standard error of the mean to a problem. A math educator and an educational psychologist worked out a program in which parents helped their children with arithmetic homework by supplying the answer as soon as the child worked each problem.[7] At the end of the term, the students took a standard-

---

[7] For an actual study on this topic, see J. O. Johnston & N. W. Maertens, "Effects of Parental Involvement in Arithmetic Homework," *School Science and Mathematics*, 1972, *72*, 117–126.

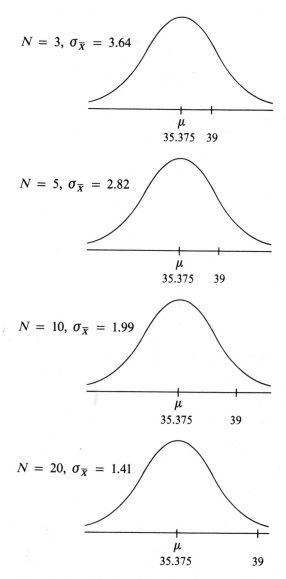

$N = 3, \sigma_{\bar{X}} = 3.64$

μ
35.375 39

$N = 5, \sigma_{\bar{X}} = 2.82$

μ
35.375   39

$N = 10, \sigma_{\bar{X}} = 1.99$

μ
35.375   39

$N = 20, \sigma_{\bar{X}} = 1.41$

μ
35.375   39

**Figure
7.5**   Sampling distributions of the mean for four different sample sizes. All samples are drawn from the same population. Note how a sample mean of 39 becomes rarer and rarer as $\sigma_{\bar{X}}$ becomes smaller.

ized test. Over the years, the mean score on this test has been 46. This year, after the special homework program, the following data were obtained: $\bar{X} = 49.5$, $s = 9.0$, and $N = 36$. We can calculate $s_{\bar{X}}$ and construct the sampling distribution about the assumed parameter of 46. See Figure 7.6.

$$s_{\bar{X}} = \frac{s}{\sqrt{N}} = \frac{9.0}{\sqrt{36}} = 1.50.$$

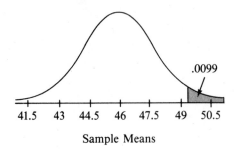

41.5   43   44.5   46   47.5   49   50.5

Sample Means

Figure 7.6   Sampling distribution of the mean for a population with a mean of 46, $N = 36$.

For the sample mean of 49.5, we can find a z score. With the z score, we can determine the probability of a mean of 49.5 or larger being drawn by chance from a population with $\mu = 46$.

$$z = \frac{\bar{X} - \mu}{s_{\bar{X}}} = \frac{49.5 - 46}{1.50} = 2.33.$$

From the normal-curve table, the probability of a z score of 2.33 or larger is .0099. This means that the probability is only about one in a hundred that a sample with a mean of 49.5 or larger would be drawn from a population with a mean of 46, *if chance is the only factor affecting the scores in the sample.*

Besides chance, what other possibilities are there? The investigators would like to conclude that all that immediate feedback given by the parents produced better arithmeticians who then scored higher on the standardized test. We can accept their conclusion if we are sure that all extraneous variables were controlled by the experimenters. As one example, a pretest revealed that the students who participated were not better-than-average arithmeticians before the parental involvement began.

Problems   21. A reporter covering a wildcat teachers' strike wanted to know whether the strike had affected student attendance. Official records would not be available before his deadline. However, he knew that the average daily attendance for classrooms was 25.5, so he walked to the nearest school, counted the children in each of the 12 classrooms with windows next to the sidewalk, and found a mean of 23.3, $s = 3.1$.

a. Which of the following headlines is appropriate?
"SCHOOL ATTENDANCE SIGNIFICANTLY DOWN"
"SCHOOL ATTENDANCE PROBABLY NOT AFFECTED BY TEACHERS' STRIKE"
"REPORTER LAMBASTED BY EDITOR FOR BIASED SAMPLING"

b. Suppose the figures above had been obtained by random sampling. What is the probability that the strike was having *no* effect on school attendance?

22. A social worker thought his clients lacked an important skill—assertiveness. He enrolled 36 of them in an assertiveness training course. Eight weeks later,

the 36 fully trained clients took a test that measured assertiveness, the national mean of which was 30. The clients produced the following statistics.

$$\sum X = 1188 \qquad \sum X^2 = 40,464$$

What is the probability of obtaining a sample with such a mean from a population with a mean of 30? Write a conclusion about the effects of the assertiveness training course.

23. Now you are in a position to return to the story of the two economics students. You should be able to tell the junior how much faith he can have that $26,900 (the sample mean) is the average family income of students at State University. That is, what is the probability of a sample mean of $26,900 being drawn from a population with a mean of $28,100? Remember that $\bar{X} = \$26,900$, $s = \$7500$, and $N = 38$.

If the junior had done his homework as you have just done, and if he had confronted his sophomore friend with the results, the conversation might have gone like this, with the junior speaking:

"... and so, the mean of $26,900 isn't very reliable. It is based on a sample of only 38, and such samples can bounce all over the place. Why, if the real mean (parameter) for this campus is $28,100, we would *expect* about one-sixth of all random samples of 38 students to have means of $26,900 or less."

"Yeah," said the sophomore, "but that *if* is a big one. What if the campus mean is really $26,900? Then the sample we got was right on the nose. After all, the sample mean is the *best estimate* of the population parameter."

"I see your point. And I see mine, too. It seems like either of us could be correct. That leaves me uncertain about the real campus parameter."

"Me too."

Not all statistical stories end with so much uncertainty. However, we said at the beginning that one of the advantages of random sampling was that you could measure the uncertainty. You measured it, and there was a lot. Remember, if you agree to draw a sample, you agree to accept some uncertainty about the results.

Given such uncertainty, there is a practical method that may reduce it: increasing the sample size. You have already worked some problems that showed that a fourfold increase in N will reduce the standard error by half. Applying this principle to our story, suppose the two students got a random sample of 400 students, which again produced a mean of $26,900. Now $s_{\bar{x}}$ becomes

$$\frac{s}{\sqrt{N}} = \frac{7500}{\sqrt{400}} = 375$$

and z becomes

$$z = \frac{26,900 - 28,100}{375} = -3.20$$

compared with -0.99 when $N = 38$. The probability of such a z score is less than .01. Uncertainty has been reduced. With such figures, our heroes would be rather confident that State University students have families with incomes less than the national average.

If you are a typical student, you may still have a nagging doubt about that one chance in a hundred. Hang on to that doubt. It is real and justifiable and is a direct consequence of accepting sampling as a way to find out about parameters.

One other caution about sample size is in order. Increasing sample size will not guarantee a reduction of uncertainty. You know that any new sample will probably produce a new mean. For example, the sample with $N = 400$ might have produced a mean of $27,725.

Problem   **24.** What is the probability that a mean of $27,725 or less would be drawn from a population where $\mu = $28,100. Again, let $s = $7500 and $N = 400$.

CONFIDENCE INTERVALS

Statisticians divide statistical inference into two categories. The first category is called *hypothesis testing,* and the second is called *estimation.* **Hypothesis testing** means to hypothesize a value for a parameter, compare (or test) the parameter with an empirical statistic, and decide whether the parameter is reasonable. Hypothesis testing is just what you have been doing so far in this chapter. Hypothesis testing is the more popular technique of statistical inference among social and behavioral scientists.

The other kind of inferential statistics, estimation, can take two forms—parameter estimation and confidence intervals. **Parameter estimation** means that one particular point is estimated to be the parameter of the population. A **confidence interval** is a range of values bounded by a lower and an upper limit. The interval is expected, with a certain degree of confidence, to contain the parameter. These confidence intervals are based on sampling distributions.

Our purpose in this section is to describe confidence intervals. We will begin with an explanation of the concept of a confidence interval and then describe how to find both the lower and upper limits and the degree of confidence you may have that the limits contain the parameter.

The Concept of a Confidence Interval

A confidence interval is simply a range of values with a lower and an upper limit. With a certain degree of confidence (usually 95 percent or 99 percent), you can state that the two limits contain the parameter. The following example shows how the size of the interval and the degree of confidence are directly related (that is, as one increases, the other increases also).

Suppose you drew a random sample of 36 scores from a population and found a mean of 76 and a standard deviation of 8. The scores ranged from a low of 60 to a high of 91. Since $\bar{X}$ is the best estimate of the parameter, μ, you might just say "I estimate that $\mu = 76$." (This is an example of parameter estimation.) How confident would you be that μ is exactly 76? Not very, we hope. Although 76 is the best estimate, no one would be surprised to get a sample mean of 76 from a population with $\mu = 75$ (or 75.5, 75.8, and so on). Thus, you would not be very confident if you estimated μ with a specific point.

At the other extreme, how confident would you be if you said, "I estimate μ is in the interval between 60 and 91?" Very, we imagine. For most purposes, however, such a wide interval would not be specific enough. The problem, as you may have surmised, is to trade off some of the specificity that goes with a point estimate to get some of the confidence that comes with a wider interval (or conversely, trade off some of the confidence that goes with a wide interval for more specificity).

Fortunately, a sampling distribution can be used to establish both confidence and the interval. The result is a lower and an upper limit for the unknown population parameter.

Here is the rationale for confidence intervals. Suppose you define a population of scores. A random sample is drawn and the mean ($\bar{X}$) is calculated. Using this mean (and the techniques described in the next section), a statistic called a confidence interval is calculated. (We will use a 95 percent confidence interval in this explanation.) Now, suppose that from this population many more random samples are drawn and a 95 percent confidence interval is calculated for each. For most of the samples, $\bar{X}$ will be close to μ and μ will fall within the confidence interval. Occasionally, of course, a sample will produce an $\bar{X}$ far from μ and the confidence interval about $\bar{X}$ will not contain μ. The method is such, however, that the probability of these rare events can be measured and held to an acceptable minimum like 5 percent. The result of all this is a method that produces confidence intervals, 95 percent of which contain μ.

In a real-life situation, you draw *one* sample and calculate *one* interval. You do not know whether or not μ lies between the two limits, but the method you have used makes you 95 percent confident that it does.

Calculating the Limits of a Confidence Interval

Having drawn a random sample and calculated the mean and standard error, the lower limit (LL) of a 95 percent confidence interval may be found by

$$LL = \bar{X} - 1.96 s_{\bar{X}}.$$

In a similar way, the upper limit (UL) may be found by

$$UL = \bar{X} + 1.96 s_{\bar{X}}.$$

The number 1.96 is the z score that leaves $2\frac{1}{2}$ percent of the curve in each tail. For the problem described above, in which a random sample of 36 produced a mean of 76 and a standard deviation of 8, we will calculate a 95 percent confidence interval. To use the formulas for LL and UL, you need $s_{\bar{X}}$.

$$s_{\bar{X}} = \frac{s}{\sqrt{N}} = \frac{8}{\sqrt{36}} = 1.33.$$

To establish the limits of the confidence interval about a mean,

$$\text{LL} = \bar{X} - 1.96 s_{\bar{X}} = 76 - 1.96(1.33) = 73.39.$$

$$\text{UL} = \bar{X} + 1.96 s_{\bar{X}} = 76 + 1.96(1.33) = 78.61.$$

Thus, 73.39 and 78.61 are the lower and upper limits of a 95 percent confidence interval about the sample mean of 76.

A long interpretation is that 73.39 and 78.61 are one set of limits that is calculated by a method that captures μ 95 percent of the time. This leaves us 95 percent confident that 73.39–78.61 contains μ. An abbreviated interpretation is that we have 95 percent confidence that the interval 73.39–78.61 contains μ.

The general formulas for the limits of a confidence interval about a mean are

$$\text{LL} = \bar{X} - z(s_{\bar{X}})$$

and

$$\text{UL} = \bar{X} + z(s_{\bar{X}}),$$

where z is chosen from the normal-curve table according to the confidence level you want for your interval. For a 99 percent confidence interval, a z value of 2.58 is appropriate. To use these formulas, $\bar{X}$ and $s_{\bar{X}}$ should be based on a large (30 or more) random sample from the population.

Here is a real-life problem for which we will calculate a 99 percent confidence interval. We have a friend who was interested in commodity market futures. The commodities in this market are corn, sugar, cattle, frozen pork bellies, and the like. The basic idea of a futures market is to buy a contract that obligates you to pay for several tons of a commodity on a specified future day at a specified price. If things go well, the price others are willing to pay you for this contract will go up before the specified day arrives and the contract can be sold. The risk, of course, is that if the price does not go up, the contract must be sold at a loss (or you must take delivery). The financial pages of many newspapers carry the prices of commodity futures.

Our friend's particular interest was in contracts for frozen pork bellies (from which bacon is produced). He had developed a formula that gave a prediction for the next month's price. (His formula was a version of the regression equation you learned about in Chapter 5.) If the formula predicted higher prices, he would buy now and sell next month. The catch to the scheme was that the prediction was based on sample data, and such predictions are subject to sampling error. So, he calculated a 99 percent confidence interval about his prediction. Now, if the lower limit of the confidence interval were above today's price, he would buy with confidence. If the confidence interval included today's price, he would wait for a better time.

With this background, here is an example of our friend's analysis. Today's price for pork bellies is 65.50 cents per pound. The formula predicts a price of 68.75 cents per pound next month. The standard error of this prediction is 1.75 cents. For a 99 percent confidence interval, lower and upper limits of this predicted $\bar{X}$ are

$$LL = 68.75 - 2.58(1.75) = 64.24.$$

$$UL = 68.75 + 2.58(1.75) = 73.27.$$

Of course, our friend's interest in these numbers is confined to the lower limit. Since 64.24 is below today's price of 65.50, he should not buy but should wait.

We have two things to add to this story, one practical and one statistical. On the practical side, our friend decided that he could do with less than 99 percent confidence, so he bought a contract anyway (which he later sold at a loss). Statistically, our friend's calculation gave him a figure (64.24) for which he could have $99\frac{1}{2}$ percent confidence. That is, he was only concerned about the probability that the sampling distribution of the predicted price was *below* a particular price. In the example, the probability that next month's price will be below 64.24 is $\frac{1}{2}$ of 1 percent. The probability that next month's price may be above 73.27 ($\frac{1}{2}$ of 1 percent) is of no concern to our friend in making a decision to buy.

The concept of confidence intervals has been known to give students trouble. Here is another explanation, with a picture, of the ideas behind confidence intervals.

Figure 7.7 shows a set of 20 confidence intervals calculated from a normal population[8] with a mean, μ (vertical line). The sample size was 25 for each of the random samples. Twenty samples were drawn and for each sample the mean and 95 percent confidence limits were calculated. On each of the 20 horizontal

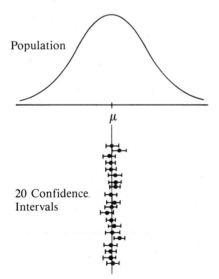

Population

μ

20 Confidence
Intervals

**Figure
7.7**   Twenty 95 percent confidence intervals, each calculated from a random sample of 25 from the normal population.

[8]It is not necessary that the population be normal.

lines, the end points represent the lower and upper limits and the filled circle shows the mean. As you can see, nearly all the confidence intervals have "captured" μ. One has not. Find it. Does it make sense to you that one out of 20 of the 95 percent confidence intervals would not contain μ?

In summary, confidence intervals for means produce a statistic (lower and upper limits) that is destined to contain μ 95 percent of the time (or 99 or 90). Whether or not your particular random sample contains the unknowable μ is uncertain. But, you have control over the degree of that uncertainty.

One more caution about words is in order. It is *not* proper to say that you have "95 percent confidence that μ is in the interval." This implies that μ is variable and you want to do just the opposite: imply that the confidence interval is variable. So, the proper terminology is to say that you have "95 percent confidence that the interval contains μ."

We will end this section with yet another caution about words. The term *confidence level* is used for problems of estimation, such as confidence intervals, and the term *significance level* is used for problems of hypothesis testing.

Problems   **25.** Place 95 percent confidence intervals around $\bar{X}$ for the following samples.

		$\bar{X}$	s	N
	a	36	2	100
Samples	b	36	10	100
	c	36	10	1000

26. Here is one final problem from the data gathered by the economics students. Establish a 99 percent confidence interval about their sample mean of $26,900. Recall that $s = \$7500$ and $N = 38$. Write a statement about the meaning of this confidence interval.

27. Place a 95 percent confidence interval about the mean arithmetic score of 49.5 made by the fifth-grade students, where $s = 9$ and $N = 36$. Tell what this interval means.

28. In Figure 7.7, the 20 lines that represent confidence intervals vary in length. What aspect of the random sample causes this variation?

29. Look at Figure 7.7. Suppose that 20 new 90 percent confidence intervals were calculated, but that the sample size was increased to 100.
  a. How many of the 20 confidence intervals would be expected to capture μ?
  b. Would the new confidence intervals be wider or more narrow? By how much?

OTHER SAMPLING DISTRIBUTIONS

Now you have been introduced to the sampling distribution of the mean. The mean is clearly the most popular statistic among researchers. (Of all statistics, the mean is the mode.) There are times, however, when the statistic necessary to answer a researcher's question is not the mean. For example, to find the degree of relationship between two variables, you need a correlation coefficient. To determine whether a treatment causes more variable responses, you need a standard deviation. Proportions are commonly used statistics. In each of these cases (and indeed, for any statistic), the basic hypothesis testing procedure you have just learned is often used by researchers. That procedure is

1. Hypothesize a population parameter.
2. Draw a random sample and calculate a statistic.
3. Compare the statistic with a sampling distribution of that statistic and decide how likely it is that such a population would produce such a sample statistic.

Where do you find sampling distributions for statistics other than the mean? In the rest of this book you will encounter new statistics and new sampling distributions. Tables D–K in the Appendix represent sampling distributions from which probability figures can be obtained. In addition, some statistics have sampling distributions that are normal, which allows you to use the familiar normal curve. Finally, some statistics and their sampling distributions are covered in other books.

Along with every sampling distribution comes a standard error. Just as every statistic has its sampling distribution, every statistic has its standard error. For example, the standard error of the median is the standard deviation of the sampling distribution of the median. The standard error of the variance is the standard deviation of the sampling distribution of the variance. Worst of all, the standard error of the standard deviation is the standard deviation of the sampling distribution of the standard deviation. If you followed that sentence, you probably understand the concept of standard error quite well.

The main points we want to emphasize are that statistics are variable things, that a picture of that variety is a sampling distribution, and that a sampling distribution can be used to obtain probability figures.

A TASTE OF REALITY

We have used the phrase *random sample(s)* more than 25 times in this chapter. The techniques of inferential statistics that you are learning in this book are based on the assumption that a random sample has been drawn. But how often do you find random samples in actual data analysis? Seldom. However, there are two justifications for the continued use of nonrandom samples.

In the first place, every experiment is an exercise in practicality. Any investigator has a limited amount of time, money, equipment, and personnel to draw upon. Usually, a truly random sample of a large population is just not practical, so the experimenter tries to obtain a representative sample, being careful to balance or eliminate as many sources of bias as possible.

In the second place, the final test of generalizability is empirical—that is, to find out whether the results based on a sample are also true for other samples. This kind of checkup is practiced continually. Usually, the results based on samples that are unsystematic (but not random) are true for other samples from the same population.

Both of these justifications develop a very hollow ring, however, if someone demonstrates that one of your samples is biased and that a representative sample proves your conclusions false.

Problems

30. Give a verbal description of the procedures you would follow to construct an empirical sampling distribution of the median.

31. What is the name of the standard deviation of the distribution described in Problem 30?

8
DIFFERENCES BETWEEN MEANS

Objectives for Chapter 8: After studying the text and working the problems in this chapter, you should be able to:

1. describe the design of a simple experiment,
2. explain the logic of inferential statistics,
3. define with words the null hypothesis,
4. for a sampling distribution of mean differences, describe it in words, calculate its standard error, and use it to determine whether the two sample means came from the same population,
5. define level of significance,
6. define critical value,
7. define Type I and Type II errors,
8. define α and β and describe the relationship between them,
9. analyze the data from a two-group experiment, reject or retain the null hypothesis at some established level of significance, and interpret the results,
10. describe the difference between a one-tailed and a two-tailed test of significance, and
11. describe the factors that affect the rejection of the null hypothesis.

One of the best things about statistics is that it helps you to understand experiments and the experimental method. The experimental method is probably the most powerful method we have of finding out about natural phenomena. Few *ifs*, *ands*, or *buts* or other qualifiers need to be attached to conclusions based on results from a sound experiment.

Besides being powerful, experiments can be interesting. They can answer such questions as:

1. Is there a difference between the racial attitudes of 9th- and 12th-grade students?
2. Is there a difference between the mean IQ of females and males?
3. Does the removal of 20 percent of the cortex of the brain have an effect on the memory of tasks learned before the operation?
4. Can you *reduce* people's ability to solve problems by teaching them other skills?

Our plan in this chapter is to discuss the simplest kind of experiment and then show how the statistical techniques you have learned about sampling distributions can be expanded to answer questions like those just cited.

A SHORT LESSON ON HOW TO DESIGN AN EXPERIMENT

The basic ideas underlying a simple two-group experiment are not very complicated.

The logic of an experiment: Start with two equivalent groups and treat them exactly alike except for one thing. Then measure both groups and attribute any difference between the two to the one way in which they were treated differently.

This summary of an experiment is described more fully in Table 8.1 and in the following explanation.

The fundamental question of the experiment outlined in Table 8.1 is "What is the effect of Treatment A on a person's ability to perform Task Q?" In more formal terms, the question is "For Task Q scores, is the mean of the population of those who have had Treatment A different from the mean of the population of those who have not had Treatment A?" This experiment has an independent variable with two levels (Treatment A or no Treatment A) and a dependent variable (scores on Task Q). A population of subjects is defined and two random samples are drawn.[1] These random samples are both representative of the popula-

---

[1] An equivalent statement is that there are two populations to begin with and that the two population means are equal. One random sample is then drawn from each population. Actually, when two samples are drawn from one population, the correct procedure is to randomly assign each subject to a group immediately after it is drawn from the population. This procedure continues until both groups are filled.

Table
8.1

Summary of a Simple Experiment

Key Words	Tasks for the Experimenter	
Population	1. Define a population	
Sampling	2. Draw Random Sample 1 from the population (experimental group)	Draw Random Sample 2 from the population (control group)
Independent variable	3. Give Treatment A to subjects in experimental group	Withhold Treatment A from subjects in control group
Dependent variable	4. Measure subjects' behavior on Task Q	Measure subjects' behavior on Task Q
	5. Calculate mean score on Task Q, $\bar{X}_I$	Calculate mean score on Task Q, $\bar{X}_C$
Inferential statistics	6. Compare $\bar{X}_E$ and $\bar{X}_C$ and decide whether Treatment A had an effect on Task Q	

tion and (approximately) equivalent to each other. Treatment A is then administered to one group (commonly called the experimental group) but not to the other group (commonly called the control group). Except for Treatment A, both groups are treated in exactly the same way. That is, extraneous variables are held constant or balanced out for the two groups. Both groups perform Task Q and the mean score for each group is calculated. The two sample means almost surely will differ. The question now is whether this observed difference is due to sampling variation (a chance difference) or to Treatment A. You can answer this question by using the techniques of inferential statistics. Table 8.1 summarizes the ideas of this paragraph.

Our generalized example uses procedures familiar to behavioral scientists. (The dependent variable was test scores.) As you know, the experimental method has broad applicability. It has been used to decide how much sugar to use in a cake recipe, what kind of tea tastes the best, whether a drug is useful in treating an illness, the effect of authoritarian parents on the political attitudes of their children, and many other things. One word that we used is recognized by all experimentalists. The word *treatment* refers to different levels of the independent variable. The experiment in Table 8.1 had two treatments.

In some experimental designs, subjects are assigned to treatments by the experimenter; in others, the experimenter uses a group of subjects who have already been "treated" (for example, being males or being children of authoritarian parents). In either of these designs, the methods of inferential statistics are the same, although the interpretation of the first kind of experiment is usually less open to attack.[2]

---

[2]We are bringing up an issue that is beyond the scope of this statistics book. Courses with titles such as "Experimental Methods" or "Research Design" cover the intricacies of interpreting experiments.

Remember that inferential statistics are used to help you decide whether or not a difference between sample means should be attributed to chance. Let's examine closely the logic behind such a decision.

THE LOGIC OF INFERENTIAL STATISTICS

This section presents the logical thought processes and some terms that will be important throughout the remainder of this book. It deserves your careful attention.

A decision must be made about the population of those given Treatment A, but it must be made on the basis of sample data. Accept from the start that because of your decision to use samples, you can never know for sure whether or not Treatment A has an effect. Nothing is ever *proved* through the use of inferential statistics. You can only state probabilities, which are never exactly one or zero. The decision making goes like this. In a well-designed two-group experiment, all the imaginable results can be reduced to two possible outcomes: either Treatment A has an effect or it does not. Make a tentative assumption that Treatment A does *not* have an effect and then, using the results of the experiment for guidance, find out how probable it is that the assumption is correct. If it is not very probable, rule it out and say that Treatment A has an effect. If the assumption is probable, you are back where you began; you have the same two possibilities you started with.[3]

Putting this into the language of an experiment:

1. Begin with two logical possibilities, *a* and *b*.
    a. Treatment A *did not* have an effect. That is, the mean of the population of scores of those who received Treatment A is equal to the mean of the population of scores of those who did not receive Treatment A, and thus the difference between population means is zero. This possibility is symbolized H_0 (pronounced "H sub oh").
    b. Treatment A *did* have an effect. That is, the mean of the population of scores of those who received Treatment A is not equal to the mean of the population of scores of those who did not receive Treatment A. This possibility is symbolized H_1 (pronounced "H sub one").
2. Tentatively assume that Treatment A had no effect (that is, assume H_0). If H_0 is true, the two random samples should be alike except for the usual variations in samples. Thus, the difference in the sample means is tentatively assumed to be due to chance.
3. Determine the sampling distribution for these differences in sample means. This sampling distribution gives you an idea of the differences you can expect if only chance is at work.
4. By subtraction, obtain the actual difference between the experimental-group mean and the control-group mean.

---

[3]In courses in logic this kind of reasoning is called the rule of negative inference (*modus tollens*).

5. Compare the difference obtained to the differences expected (from Step 3) and conclude that the difference obtained was:

   a. expected. Differences of this size are very probable just by chance, and the most reasonable conclusion is that the difference between the experimental group and the control group may be attributed to chance. Thus, retain both possibilities in Step 1.

   b. unexpected. Differences of this size are highly improbable and the most reasonable conclusion is that the difference between the experimental group and the control group is due to something besides chance. Thus, reject H_0 (possibility a in Step 1) and accept H_1 (possibility b); that is, conclude that Treatment A had an effect.

6. Write a conclusion about the experiment using the terms of the experiment. That is, your conclusion should describe the relationship of the dependent variable to the independent variable.

The basic idea is to assume that there is no difference between the two population means and then let the data tell you whether the assumption is reasonable. If the assumption is not reasonable, you are left with only one alternative: the populations have different means.

The assumption of no difference is so common in statistics that it has a name: the **null hypothesis**, symbolized, as you have already learned, H_0. The null hypothesis is often stated in formal terms:

$$H_0: \mu_1 - \mu_2 = 0 \quad \text{or} \quad H_0: \mu_1 = \mu_2.$$

That is, the null hypothesis states that the mean of one population is equal to the mean of a second population.[4] We will frequently use the term *null hypothesis* and the symbol H_0. It will be important that you understand what it is.

H_1 is referred to as an **alternative hypothesis**. Actually, there are an infinite number of alternative hypotheses—that is, the existence of any difference other than zero. In practice, however, it is usual to choose one of three possible alternative hypotheses before the data are gathered:

1. $H_1: \mu_1 \neq \mu_2$. In the example of the simple experiment, this hypothesis states that Treatment A had an effect without stating whether the treatment improves or disrupts performance on Task Q. Most of the problems in this chapter use this H_1 as the alternative to H_0. If you reject H_0 and accept this H_1, *you must examine the means and decide whether Treatment A facilitated or disrupted performance on Task Q.*

2. $H_1: \mu_1 > \mu_2$. The hypothesis states that Treatment A improves performance on Task Q.

3. $H_1: \mu_1 < \mu_2$. The hypothesis states that Treatment A disrupts performance on Task Q.

---

[4]Actually, the concept of the null hypothesis is broader than simply the assumption of no difference, although that is the only version used in this book. Under some circumstances, a difference other than zero might be the hypothesis tested.

The section on one-tailed and two-tailed tests, beginning on page 171 will give you more information on choosing an alternative hypothesis.

A word of caution about terminology: the null hypothesis is proposed and this proposal may meet with one of two fates at the hands of the data. The null hypothesis may either be rejected, which allows you to accept an alternative hypothesis, or it may be retained. If it is retained, it *is not proved as true*; it is simply retained as one among many possibilities.

Perhaps an analogy will help with this distinction about terminology. Suppose a masked man has burglarized a house and stolen all the silver. For purposes of this story suppose there are only two people who could possibly have done the deed, HI and HO. The lawyer for HO tries to establish beyond reasonable doubt that her client was out of state during the time of the robbery. If she can do this, it will exonerate HO (HO will be rejected, leaving only HI as a suspect). However, if she cannot establish this, the situation will revert to its original state: HI or HO could have stolen the silver away, and both are retained as suspects. So the null hypothesis can be *rejected* or *retained*,[5] but it can never be proved with certainty to be true or false by using the methods of inferential statistics. Statisticians are usually very careful with words. That is probably because they are used to mathematical symbols, which are very precise. Regardless of the reason, this distinction between *retained* and *proved*, although subtle, is important. The next section shows you how to do Steps 3 and 5 in the list of steps in an experiment.

Clue to the Future

This section on the logic of inferential statistics describes reasoning that is basic to all the statistical techniques that you will learn to do in the rest of this book. An understanding of this logic now will help you understand this chapter and future chapters, and with statistical arguments and techniques encountered in future articles, books, and discussions.

Problems

1. In your own words, outline the logic of an experiment.
2. In the experiment described in Table 8.1, what method was used to ensure that the two samples were equivalent before treatment?
3. In your own words, outline the logic of inferential statistics.
4. What is H_0 a symbol for?
5. Tell what the null hypothesis is.

[5]Some textbooks use the term accepted instead of retained.

SAMPLING DISTRIBUTION OF A DIFFERENCE BETWEEN MEANS

We have used the term *difference* a great deal in this chapter, including in the title. A difference is simply the answer in a subtraction problem. As explained in the section on the logic of inferential statistics, the difference that is of interest is the difference between two means. You evaluate the obtained difference by comparing it with a sampling distribution of differences between means (often called a sampling distribution of mean differences).

Chapter 7 contained a great deal of discussion about sampling distributions, especially the sampling distribution of the mean. Recall that a sampling distribution is a frequency distribution of sample statistics, all calculated from samples of the same size drawn from the same population; the standard deviation of that frequency distribution is called a standard error. Precisely the same logic holds for a sampling distribution of differences between means.

We can best explain a sampling distribution of differences between means by describing the procedure for generating an *empirical* sampling distribution of mean differences. Find for yourself a single population of scores. Randomly draw *two* samples, calculate the mean of each, and subtract the second mean from the first. Do this many times and then arrange all the differences into a frequency distribution. Such a distribution will consist of a number of scores, each of which is a *difference* between two sample means. Think carefully about the mean of the sampling distribution of mean differences. Stop reading and decide what the numerical value of this mean will be. If you have reasoned out a number, read on. The mean of a sampling distribution of mean differences is zero because positive and negative differences occur equally often, giving you an algebraic sum close to zero. If you were successful on this problem, congratulations! Few students get it right when they first read the chapter.

This sampling distribution of mean differences has a standard deviation called the standard error of a difference between means.

In many experiments, it is obvious that there are *two* populations to begin with. The question, however, is whether they are equal on the dependent variable. To generate a sampling distribution of differences between means in this case, assume that, on the dependent variable, the two populations have the same mean, standard deviation, and form (shape of the distribution). Then draw one sample from each population, calculate the means, and subtract one from the other. Continue this many times. Arrange the differences between sample means into a frequency distribution.

The sampling distributions of differences between means that you will use will be theoretical distributions, not the empirical ones we described in the last two paragraphs. However, a description of the procedures for an empirical distribution, which is what we've just given, is usually easier to understand in the beginning.

Two things about a sampling distribution of mean differences are constant: the mean and the form. The mean is zero, and the form is normal if the sample means are based on large samples. Again, the traditional answer to the question "What is a large sample?" is "30 or more."[6]

We will illustrate these ideas with an experiment your authors did (Johnston and Spatz, 1973). The question we wished to answer was "Are the racial attitudes of 9th graders different from those of 12th graders?" The null hypothesis was that the population means were equal (H_0: $\mu_1 = \mu_2$). The alternative hypothesis was that they were not equal (H_1: $\mu_1 \neq \mu_2$). The subjects in this experiment were 9th- and 12th-grade black and white students who expressed their attitudes about persons of their own sex but different race. Higher scores represent more positive attitudes. Thus, the independent variable was grade and the dependent variable was racial attitude score. Table 8.2 shows the summary data.

Table 8.2   **Data From an Experiment That Compared the Racial Attitudes of 9th- and 12th-Grade Students**

Grade	N	$\bar{X}$	s
9th	200	48.42	12.16
12th	200	52.52	11.83

As you can quickly calculate from Table 8.2, the obtained mean difference between samples of 9th and 12th graders is 4.10. Now a decision must be made. Should this difference in samples be ascribed to chance (retain H_0; there is no difference between the population means)? Or should we say that such a difference is so unlikely that it is due not to chance but to the different characteristics of 9th- and 12th-grade students (reject H_0 and accept H_1; there is a difference between the populations)? Using a sampling distribution of mean differences (Figure 8.1), a decision can be made.

Figure 8.1 shows a sampling distribution of differences between means that is based on the assumption that there are no population differences between 9th and 12th graders—the standard assumption that any experiment would begin with. (You will learn to construct sampling distributions of mean differences in later sections.) In other words, Figure 8.1 assumes H_0.

Figure 8.1 is a normal curve that shows you z scores, possible differences between sample means in the racial-attitudes study, and probabilities associated with the z scores and difference scores.

Study Figure 8.1 carefully and notice the following characteristics:

1. The mean is zero since it is based on the assumption of no difference between means ($H_0 = \mu_1 = \mu_2$).

2. As the size of the difference between means increases, z scores increase.

3. As z scores and differences between means become larger, probabilities become smaller.

4. As before, small probabilities go with small areas of the curve.

_____

[6]The sampling distribution of mean differences will be normal with smaller samples if the population(s) from which they are drawn is (are) symmetrical.

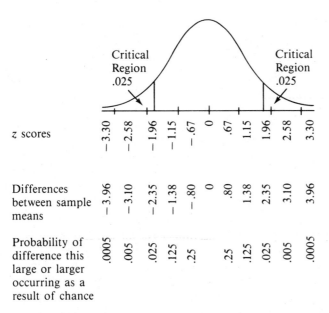

z scores

Differences
between sample
means

Probability of
difference this
large or larger
occurring as a
result of chance

**Figure
8.1**   Sampling distribution from the racial-attitudes study. It is based on chance and shows z scores,
probabilities of those z scores, and differences between sample means.

Our obtained difference, 4.10, is not even shown on the distribution. Such events
are extremely rare when the true difference is zero and only chance is at work.
From Figure 8.1, you can see that a difference of 3.96 or more would be expected
five times in 10,000 (.0005). Since a difference of -3.96 or greater also has a
probability of .0005, we can add the two probabilities together to get .001 (more
about this in the section on one- and two-tailed tests). Since our difference was
4.10 (less probable than 3.96), we can conclude that the probability of a difference
of ±4.10 being due to chance is less than .001. This probability is very small,
indeed, and it seems reasonable to rule out chance; that is, to reject H_0 and, thus,
accept H_1. By examining the means of the two groups in Table 8.2, we can write
a conclusion using the terms in the experiment. "Twelfth graders have more
positive attitudes toward people of their own sex, but different race, than do
ninth graders."

A PROBLEM AND ITS ACCEPTED SOLUTION

The probability that populations of 9th- and 12th-grade attitude scores are
the same was so small ($p < .001$) that it was easy to rule out chance as an
explanation for the difference. But what if that probability had been .01, or .05,
or .25, or .50? How to divide this continuum into a group of events that is "due
to chance" and another that is "not due to chance"—that is the problem.

It may already be clear to you that any solution will be an arbitrary one.
Breaking any continuum into two parts will leave you uncomfortable about the

events close to either side of the break. Nevertheless, a solution does exist, and there are two important concepts involved: *level of significance* and *critical region*.

The generally accepted solution is to say that the .05 level of probability is the cutoff between "due to chance" and "not due to chance." The name of the cutoff point that separates "due to chance" from "not due to chance" is the **level of significance**. If an event has a probability of .05 or less (for example, $p = .03$, $p = .01$, or $p = .001$), H_0 is rejected, and the event is considered **significant** (not due to chance). If an event has a probability of .051 or greater (for example, $p = .06$, $p = .50$, or $p = .99$), H_0 is retained, and the event is considered not significant (may be due to chance). Here, the word *significant* is not synonymous with "important." A significant event in statistics is one that is not ascribed to chance.

The area of the sampling distribution that covers the events that are "not due to chance" is called the **critical region**. If an event falls in the critical region, H_0 is rejected. Figure 8.1 identifies the critical region for the .05 level of significance. As you can see, the difference in means between 9th- and 12th-grade racial attitudes (4.10) falls in the critical region, so H_0 should be rejected.

Although widely adopted, the .05 level of significance is not universal. Some investigators use the .01 level in their research. When the .01 level is used and $H_1: \mu_1 \neq \mu_2$, the critical region consists of .005 in each tail of the sampling distribution. In Figure 8.1, differences greater than -3.10 or 3.10 are required in order to reject H_0 at the .01 level.

In textbooks, a lot of lip service is paid to the .05 level of significance as the cutoff point for decision making. In actual research, the practice is to run the experiment and report any significant differences at the smallest correct probability value. Thus, in the same report, some differences may be reported as significant at the .001 level, some at the .01 level, and some at the .05 level. At present, it is uncommon to report probabilities greater than .05 as significant, although some researchers argue that the .10 or even the .20 level may be justified in certain situations.

Problems

6. What does level of significance mean?
7. What events are included in the critical region?
8. What is the largest probability usually adopted for a level of significance?

HOW TO CONSTRUCT A SAMPLING DISTRIBUTION OF DIFFERENCES BETWEEN MEANS

You already know two important characteristics of a sampling distribution of differences between means. The mean is 0, and the form is normal. When we constructed Figure 8.1, the sampling distribution of differences between the racial attitudes of 9th and 12th graders, we used the normal curve table and a form of the familiar z score[7]

$$z = \frac{\bar{X}_1 - \bar{X}_2}{s_{\bar{X}_1 - \bar{X}_2}}$$

where: $\bar{X}_1$ = mean of one sample,

$\bar{X}_2$ = mean of a second sample,

and $s_{\bar{X}_1 - \bar{X}_2}$ = standard error of a difference.

The only new term in this formula is $s_{\bar{X}_1 - \bar{X}_2}$, which is the standard error of a sampling distribution of mean differences. This longer, more descriptive name is usually shortened to that above, the standard error of a difference. The formula for it is

$$s_{\bar{X}_1 - \bar{X}_2} = \sqrt{s_{\bar{X}_1}^2 + s_{\bar{X}_2}^2},$$

where: $s_{\bar{X}_1 - \bar{X}_2}$ = standard error of a difference between means,

$s_{\bar{X}_1}$ = standard error of the mean of Sample 1,

and $s_{\bar{X}_2}$ = standard error of the mean of Sample 2.

Now you are in a position to follow all the steps used in constructing Figure 8.1. First, we drew a normal curve and labeled the mean 0. Next, we calculated the standard error of a difference. Remember that $s_{\bar{X}} = s/\sqrt{N}$.

$$s_{\bar{X}_1 - \bar{X}_2} = \sqrt{s_{\bar{X}_1}^2 + s_{\bar{X}_2}^2} = \sqrt{\left(\frac{s_1}{\sqrt{N_1}}\right)^2 + \left(\frac{s_2}{\sqrt{N_2}}\right)^2}$$

$$= \sqrt{\left(\frac{12.16}{\sqrt{200}}\right)^2 + \left(\frac{11.83}{\sqrt{200}}\right)^2} = \sqrt{.7393 + .6997} = 1.1996 = 1.20.$$

Although the baseline of Figure 8.1 does not specifically indicate standard error units of 1.20, the hash marks on the X axis represent distances of 1.20.

Our next step was to look in Table C, Column C, for probability figures of .25, .125, .025, .005, and .0005. We found the corresponding z scores and placed both z scores and probability figures at the appropriate places on Figure 8.1.

The final thing we needed for Figure 8.1 was the difference between sample means associated with each z score. Since

$$z = \frac{\bar{X}_1 - \bar{X}_2}{s_{\bar{X}_1 - \bar{X}_2}},$$

---

[7]The formula in the text is the "working model" of the more general formula

$$z = \frac{(\bar{X}_1 - \bar{X}_2) - (\mu_1 - \mu_2)}{s_{\bar{X}_1 - \bar{X}_2}}$$

Since our null hypothesis is that $\mu_1 - \mu_2 = 0$, the term in parentheses on the right is 0, leaving you with the "working model." This more general formula is of a form you have seen before and will see again: the difference between a statistic ($\bar{X}_1 - \bar{X}_2$) and a parameter ($\mu_1 - \mu_2$) divided by the standard error of the statistic.

then

$$\bar{X}_1 - \bar{X}_2 = (z)(s_{\bar{X}_1 - \bar{X}_2}).$$

Solving this formula using each z score gave us the difference between means that we placed between the z score and its probability.

There you have it, the procedures we used in constructing Figure 8.1—a picture of the differences we would expect between samples of 9th and 12th graders if the populations really did have the same attitudes toward race.

A SHORTCUT—TAILS OF DISTRIBUTIONS

Some students suspect that the construction of an entire sampling distribution is unnecessary. You may be such a student, and if so, you are correct. All that is needed is to determine if the difference between sample means falls into the *critical region of the sampling distribution*, and there is a fairly easy way to determine this.

To begin with, adopt a level of significance of .05, which gives you a critical region of .025 in each tail of the distribution. Since all sampling distributions of mean differences (based on large samples) are normal curves, the z scores that mark off the critical regions will always be the same, ±1.96.[8] Thus, any two samples that produce a z score whose *absolute value* is equal to or greater than 1.96 will fall into the critical region and lead to a rejection of the null hypothesis. If the absolute value of the z score is less than 1.96, you retain the null hypothesis. If you will glance back at Figure 8.1 you will see that the critical region is defined by z scores of ±1.96.

This universal z score of 1.96 that defines the .05 critical region for every normal sampling distribution has a name: **critical value**. We mentioned earlier that the .05 level of significance is not universal; some investigators adopt a .01 level. Recall also that most investigators report significant differences at their smallest probability value (whether .05, .01, or .001). Each of these probabilities has its own critical value. Stop reading for a moment and test your understanding by deciding whether the critical values associated with critical regions of .01 and .001 are larger or smaller than 1.96.

The critical value to use with a .01 level of significance is 2.58. With a .001 level, use 3.30. You can check these values for yourself by dividing the significance levels in half (.005 and .0005), finding these values in Column C of Table C, and noting the corresponding z score.

Using the Shortcut

Here is the way we would analyze the data in Table 8.2 if we were starting from scratch. We would begin by finding the z score for the observed mean

---

[8]This is for a two-tailed test. See p. 171 for the z-score value used for a one-tailed test.

difference:

$$z = \frac{\bar{X}_1 - \bar{X}_2}{s_{\bar{X}_1 - \bar{X}_2}} = \frac{\bar{X}_1 - \bar{X}_2}{\sqrt{\left(\frac{s_1}{\sqrt{N_1}}\right)^2 + \left(\frac{s_2}{\sqrt{N_2}}\right)^2}} = \frac{48.42 - 52.52}{\sqrt{\left(\frac{12.16}{\sqrt{200}}\right)^2 + \left(\frac{11.83}{\sqrt{200}}\right)^2}} = -3.42.$$

The obtained z score of -3.42 allows us to reject the null hypothesis at the .001 level of significance. It is unlikely that the two samples of racial attitudes came from the same population. By examining the means, we can write our conclusion, "Twelfth graders have more positive racial attitudes than do ninth graders, $p < .001$."

The "p" in "$p < .001$" refers to the probability of obtaining a difference between samples as large as the one actually observed if the true difference is zero and only chance is at work. This is the conventional way to convey the level of significance to the reader.

Here is another problem. Do males or females have higher IQ scores? The independent variable here is sex and the dependent variable is IQ. Suppose the data in Table 8.3 were obtained.[9]

Table 8.3   **Hypothetical Data on Sex Differences in IQ**

	Males	Females
$\bar{X}$	103.3	101.2
s	15.4	14.3
N	72	57

We will begin by tentatively adopting H_0. In this case, H_0 is the assumption that there is no population difference between the mean IQs of males and females. Next, we will find the standard error of the difference.

$$s_{\bar{X}_1 - \bar{X}_2} = \sqrt{s_{\bar{X}_1}^2 + s_{\bar{X}_2}^2} = \sqrt{\left(\frac{s_1}{\sqrt{N_1}}\right)^2 + \left(\frac{s_2}{\sqrt{N_2}}\right)^2}$$

$$= \sqrt{\left(\frac{15.4}{\sqrt{72}}\right)^2 + \left(\frac{14.3}{\sqrt{57}}\right)^2} = \sqrt{6.8814} = 2.62.$$

Using the z-score formula,

$$z = \frac{\bar{X}_1 - \bar{X}_2}{s_{\bar{X}_1 - \bar{X}_2}} = \frac{103.3 - 101.2}{2.62} = \frac{2.1}{2.62} = .80.$$

[9] For a short discussion of the technical problems besetting anyone who makes this comparison on real data, see J. P. Guilford, *The Nature of Human Intelligence*, New York: McGraw-Hill, 1967, p. 403. The central problem is that an IQ is a composite score, and females do better on some parts and males do better on others. The importance assigned to a particular part can throw the results one way or the other.

Since our obtained z score is less than 1.96, we may not reject H_0. Thus, our data are not conclusive. In statistical terms, we retain H_0, and we have the same number of hypotheses now as we had before we gathered any data—namely, H_0 and H_1. In terms of the experiment, we have no evidence that females and males differ in IQ. A terminology convention, widely adopted by researchers, is to refer to nonsignificant differences as NS.

Problems

9. In an experiment that compared two methods of teaching Spanish, Method A was found to be superior to Method B (based on a test of comprehension). The article reporting the experiment included the phrase "$p < .01$." Write a description of what this phrase means. Be sure to name the event that the probability refers to.

10. What critical value would be appropriate for a .02 level of significance?

11. An educational psychologist was interested in the effect of set (previous experience) on a problem-solving task. The task was to tie together two strings (A and B), which were suspended from the ceiling 12 feet apart. The problem was that the subject could not reach both strings at one time. The subject's plight is illustrated below. The solution to the problem is to tie a weight to

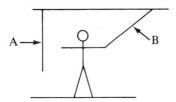

string A, swing it out, and then catch it on the return. The psychologist supplied the weight (an electrical light switch) under one of two conditions. In one condition, subjects had previously wired the switch into an electrical circuit and used it to turn on a light. In the other condition, subjects had no experience with the switch until they were confronted with the two-string problem. The question was whether having used the switch as an electrical device would have any effect on the time required to solve the problem.[10]

Data from an Experiment on Set

	Wired the Switch into Circuit	No Previous Experience with Switch
$\bar{X}$	7.40 minutes	5.05 minutes
s	2.13 minutes	2.31 minutes
N	43	41

a. Identify the independent and dependent variable.
b. State the null hypothesis for these data.

[10]This experiment is based on a study by H. G. Birch & H. S. Rabinowitz, "The Negative Effect of Previous Experience on Productive Thinking," *Journal of Experimental Psychology*, 1951, *41*, 121–125.

c. Perform a z-score test.

d. Interpret the results of the probability you obtain.

12. Examine the following data and tell why you cannot test the difference between the means with a z-score test.

	Experimental Group	Control Group
$\bar{X}$	148	122
s	118	109
N	13	9

13. Karl Lashley (1890–1958) did a number of studies designed to find out how the brain stores memories. In one experiment, rats learned a simple maze that had one blind alley. Afterward, the rats were anesthetized and operated on. For half the rats, 20 percent of the cortex of the brain was removed. For the other half, the same operation was performed except that no brain tissue was removed (a sham operation). After the rats recovered, they were given 20 trials in the simple maze and the number of errors was recorded. Identify the independent and dependent variable and then analyze the data below. Write a conclusion about the effect of a 20 percent loss of cortex on retention by rats of a simple maze task.

	Percent of Cortex Removed	
	0	20
$\sum X$	208	252
$\sum X^2$	1706	2212
N	40	40

14. Estimates of the incidence of asthma run from 2 to 5 percent of the population. In this experiment, a group whose attacks began before age 3 were compared with a group whose attacks began after age 6. The dependent variable was the number of months the asthma persisted. Analyze the difference between means and write a conclusion that tells what you have found.

	Onset of Asthma	
	Before age 3	After age 6
$\bar{X}$	60	36
s	12	6
N	45	52

15. Suppose Somebody Else conducted an experiment on two methods of teaching statistics: the lecture method and the individualized approach. Students knew of the two methods and chose the method they preferred. The lecture class was taught by Professor Y, and the individualized class was taught by Professor Z. At the end of the course, each professor made up his own final exam for this course. The following data were obtained.

	Lecture	Individualized
$\bar{X}$	109	121
s	26	39
N	41	37

Analyze the experiment and write a conclusion.

AN ANALYSIS OF POTENTIAL MISTAKES

At first glance, the idea of adopting a significance level of 5 percent seems preposterous to some students.

"You do an experiment," they say, "and you reject H_0 at the .05 level and you conclude that there is a difference in the populations. But in your heart you are uncertain. Perhaps a rare event happened and the difference really is due to chance."

This line of reasoning is fairly common. Many thoughtful students take the next step.

"I don't have to use the .05 level. I'm going to use a level of significance of one in a million. That way I can reduce the uncertainty."

It is true that adopting the .05 level of significance leaves some room for mistaking a chance difference for a real difference and that lowering the level of significance will reduce the probability of this kind of mistake. However, it increases the probability of another kind. Uncertainty about the conclusion will remain. In this section, we will discuss the two kinds of mistakes that are possible. You will be able to pick up some hints on reducing uncertainty, but to quote a phrase in a famous statistics textbook, "If you agree to draw a sample, you agree to accept some uncertainty about the results."

Look at Table 8.4, which shows the two ways to make a mistake.

Table 8.4   **Type I and Type II Errors***

		The True Situation in the Population	
		H_0 True	H_0 False
The Decision Made on the Basis of Sample Data	Retain H_0	1. Correct Decision	3. Type II Error
	Reject H_0	2. Type I Error	4. Correct Decision

*In this case, error means mistake.

In Table 8.4, cell 1 shows the situation when the null hypothesis is true and you retain it—your sample data led you to a correct decision. However, if H_0 were true, and your sample data led you to reject it (cell 2), you would have made a mistake called a **Type I error**. The probability of a Type I error is symbolized by α (alpha).

On the other hand, suppose H_0 is really false (the second column). If, on the basis of your sample data, you retain H_0 (cell 3), you have made a mistake—a **Type II error**, the probability of which is symbolized by β (beta). If your sample data led you to reject H_0 (cell 4), you made a correct decision.

You are already somewhat familiar with α from your study of level of significance. When the .05 level of significance is adopted, the experimenter concludes that an event with $p \leq .05$ is not due to chance. The experimenter could be wrong; if so, a Type I error has been made. The probability of a Type I error—α—is controlled by the level of significance you adopt.

A proper way to think of α and a Type I error is in terms of "in the long run." Figure 8.1 is a theoretical sampling distribution of mean differences. It is a picture of repeated sampling (that is, the long run). All those differences came from sample means that were drawn from the same population, but some differences were so large that they could be expected to occur only 5 percent of the time. In an experiment, however, you have only one difference, which is based on your two sample means. If this difference is so large that you conclude that there are two populations whose means are not equal, you *may* have made a Type I error. However, the probability of such an error is not more than .05.

The calculation of β is a more complicated matter. For one thing, a Type II error can be committed only when the two populations have different means. Naturally, the farther apart the means are, the more likely you are to detect it, and thus the lower β is. We will discuss other factors that affect β in the last section of this chapter, "How To Reject the Null Hypothesis."

The general relationship between α and β is an inverse one. As α goes down, β goes up. That is, if you insist on a larger difference between means before you call the difference nonchance, you are less likely to detect a real nonchance difference if it is small. Figures 8.2 and 8.3 illustrate this relationship.

Figure 8.2 is a picture of two populations. Since these are populations, the "truth" is that the mean of the experimental group is four points higher than that of the control group.[11] If a sample is drawn from each population, there is only one correct decision: reject H_0. However, will the investigator make the correct decision? Suppose samples are drawn and the mean difference is exactly

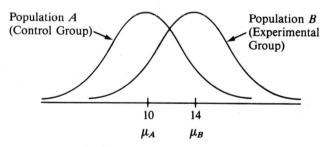

Figure 8.2   Frequency distribution of raw scores when H_0 is false.

[11]Such "truth" is available only in hypothetical examples in textbooks. In the real world of experimentation, you do not know population parameters. This example, however, should help you understand the relation of α to β.

4. Our samples are telling us the exact truth about the populations. But will techniques of inferential statistics lead us to a correct decision?

Our decision will be based on a sampling distribution like the one in Figure 8.3. (We constructed this particular sampling distribution so we could illustrate the following points.)

You can see in Figure 8.3 that if the critical region covers five percent of the curve (all the gray areas), a mean difference of 4 points falls within it. So, if α were set at .05, you would correctly reject the null hypothesis. However, if the critical region covers only one percent of the curve (the dark gray areas), the mean difference of 4 points is not within it. So, if α were .01 you would not reject H_0. Failure to reject H_0 in this case is a Type II error.

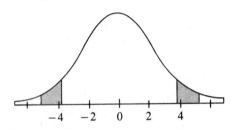

Differences Between Means

**Figure
8.3**   Sampling distribution of differences between means from Populations A and B if H_0 were true.

At this point, we can return to our discussion of setting the significance level. The suggestion was "Why not reduce the significance level to one in a million?" From the analysis of the potential mistakes, you can answer that when you decrease α, you increase β. So protection from one error is traded for liability to another kind of error.

Most persons who use statistics as a tool set α (usually at .05) and let β fall where it may. The actual calculation of β, although important, will not be discussed in this book.[12]

Problems   16. What is a Type I error?
17. What is a Type II error?
18. Suppose the real situation is that the treatment has an effect on scores. Is it possible to make a Type I error?
19. Distinguish among α, Type I error, and level of significance.
20. Suppose an experimenter chose to use $\alpha = .01$ rather than .05. What effect would this have on the probability of Type I and Type II errors?

---

[12]A discussion of the calculation of β can be found in J. P. Guilford & B. Fruchter, *Fundamental Statistics in Psychology and Education* (5th ed.), New York: McGraw-Hill, 1978, pp. 176–182.

ONE-TAILED AND TWO-TAILED TESTS

Earlier, we discussed the fact that in practice it is usual to choose one of three possible alternative hypotheses before the data are gathered.

1. H_1: $\mu_1 \neq \mu_2$. This hypothesis simply says that the population means differ but it makes no statement about the direction of the difference.

2. H_1: $\mu_1 > \mu_2$. Here, the hypothesis is made that the mean of the first population is greater than the mean of the second population.

3. H_1: $\mu_1 < \mu_2$. The mean of the first population is smaller than the mean of the second population.

So far in this chapter, you have been working with the first H_1. You have tested the null hypothesis, $\mu_1 = \mu_2$, against the alternative hypothesis $\mu_1 \neq \mu_2$. The null hypothesis was rejected when you found large positive deviations $(\bar{X}_1 > \bar{X}_2)$ *or* large negative deviations $(\bar{X}_1 < \bar{X}_2)$. When α was set at .05, the .05 was divided into .025 in each tail of the sampling distribution, as seen in Figure 8.4. If the absolute value of the obtained z score is greater than 1.96, you can reject H_0 and accept either of the possible alternative hypotheses, $\mu_1 > \mu_2$ or $\mu_1 < \mu_2$. This is called a **two-tailed test of significance**, for reasons that should be obvious from Figure 8.4.

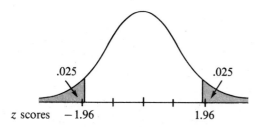

.025   .025

z scores   −1.96   1.96

Figure 8.4   A two-tailed test of significance, with $\alpha = .05$.

Sometimes, however, an investigator is concerned only with deviations in one direction; that is, the alternative hypothesis of interest is either $\mu_1 > \mu_2$ or $\mu_1 < \mu_2$. In either case, a **one-tailed test** is appropriate. For a one-tailed test, H_0 and H_1 become:

$$H_0: \mu_1 \geqslant \mu_2 \qquad\qquad H_0: \mu_1 \leqslant \mu_2$$
$$\text{or}$$
$$H_1: \mu_1 < \mu_2 \qquad\qquad H_1: \mu_1 > \mu_2$$

Figure 8.5 is a picture of the sampling distribution for a one-tailed test, for $\mu_1 > \mu_2$. In a one-tailed test, the critical region is all in one tail of the sampling distribution. The only outcome that allows you to reject H_0 is one in which μ_1 is so much larger than μ_2 that the z score is 1.65 or more. Notice in Figure 8.5 that if you are running a one-tailed test there is *no way* to conclude that μ_1 is less than μ_2, even if $\bar{X}_2$ is many times the size of $\bar{X}_1$. In a one-tailed test, you are interested in only one kind of difference. One-tailed tests are generally used

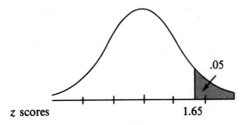

z scores   1.65

**Figure
8.5**

A one-tailed test of significance, with $\alpha = .05$.

when an investigator knows a great deal about the particular research area or when *practical* reasons dictate an interest in establishing $\mu_1 > \mu_2$ but not $\mu_1 < \mu_2$.

Here are two examples that illustrate when one- or two-tailed tests should be used. The first is an example that calls for a two-tailed test. Jojoba (ho-ho-ba) and guayule (wy-òo-lee) are plants that grow well in semiarid regions such as the Sonoran Desert of Arizona and California (Maugh, 1977). The seeds of the jojoba can be processed into a liquid wax that has many industrial applications. At present, these needs are being met by expensive synthetics. The other plant, guayule, can be processed into rubber that is virtually identical to rubber that is now imported. Given a particular area for cultivation, an experiment could be run. The question to be answered is "Which of the two plants is more productive in this particular area?" To answer this question, you need a two-tailed test because you want to be able to show that $\mu_1 < \mu_2$ *or* that $\mu_1 > \mu_2$.

The second example, also from agronomy, calls for a one-tailed test. Suppose a newly developed variety of rice is being compared with an old, established variety. For the established variety, much is already known: yield, resistance to disease, optimal schedules for flooding, and the like. Research on the new variety will continue only if the new variety shows promise of a greater yield. Now the question is "Is the new variety more productive?" and the appropriate statistical test for an experiment is one-tailed. If the new variety is equal to the old standard, or if it is worse, research will stop. The experimenter is only interested in being able to detect a difference such as $\mu_1 > \mu_2$.

There is some controversy among behavioral scientists about the use of one- or two-tailed tests.[13] When in doubt, most behavioral scientists use a two-tailed test. The decision to use a one-tailed or a two-tailed test should be made before the data are gathered. The answers to the problems in this text are based on two-tailed tests, unless a one-tailed test is specified.

Problems

21. Two sisters, who attended different colleges, began a friendly discussion of the relative intellectual capacities of their fellow students. Each was sure that she had to compete with brighter students than her sister. Though both were majoring in literature, each had an appreciation for quantitative thinking so they decided

[13]See Jones, 1952, 1954; Burke, 1953; Kimmel, 1957; Goldfield, 1959—all of which are reprinted in Kirk (1972).

to settle their discussion with admission scores made by entering freshmen at the two schools. Suppose they bring you the following data and ask for your analysis. The N's represent all the freshmen for one year.

	The U.	State U.
$\bar{X}$	21.4	21.5
s	3	3
N	8000	8000

Be sure to carry several decimal places in your work.

The sisters agreed to set $\alpha = .05$. Begin by deciding whether to run a one-tailed or a two-tailed test. Run the test and interpret your results.

22. What z score should be used for a one-tailed test with $\alpha = .01$?

23. A computer salesman is promoting Brand Z by telling a researcher that his brand is much faster than Brand Y, the computer on the researcher's desk. The researcher works some problems on both computers, and records the amount of time required (seconds).

a. Should a one-tailed or a two-tailed test be used?

b. What advice would you give?

	Time Scores		
	ΣX	ΣX^2	N
Brand Z	1216	30830	54
Brand Y	984	19390	54

24. A class of nursing students, interested in preventive medicine, set up a blood-pressure monitoring station in a gym where many students and faculty engaged in a noontime aerobics program. Some of those who had their blood pressure measured were veterans of the aerobics program and some were newcomers to it. Identify the independent and dependent variables, analyze the following data, and write a sentence about the effect of participating in the aerobics program.

	Veterans	Newcomers
$\bar{X}$	116	125
s	15	19
N	36	36

SIGNIFICANT RESULTS AND IMPORTANT RESULTS

Once, when one of us was a fledgling instructor, we did some research with an undergraduate student, David Cervone, who was interested in hypnosis. We got permission to use two rooms in the library to conduct the experiment. When the results were analyzed, two of the groups were significantly different. One day, the librarian asked how the experiment had come out.

"Oh, we got some significant results," I said.

"I imagine David thought they were more significant than you did," was the reply.

At first I was confused. Why would David think they were more significant than I would? Point oh-five was point oh-five. Then, I realized that the librarian and I were using the word significant in two quite different ways. He meant "important" and I meant "not due to chance."

The word "significant" has a precise technical meaning in statistics and other meanings in other contexts.

Let's take a cue from the librarian and ask about the *importance* of a difference. A significant difference may not be an important difference. When you worked Problem 21, you found that the difference in the college admissions scores between two schools was significant at the .05 level. However, the difference was only one-tenth of a point, which doesn't seem important at all.

When you worked Problem 14, you found that the amount of time a person was subject to asthma attacks depended on whether the first attack occurred before age 3 or after age 6. Such a difference would be important if you wanted to tell parents what to expect.

In an excellent article on the importance of good experimental designs in research in clinical psychology, Garfield (1978) wrote about the distinction between statistical significance and clinical importance (pp. 600–602). Several examples are given in his well-written article.

The basic point we are making is that a study that has statistically significant results may or may not have important results. You have to decide about the importance without the help of inferential statistics.

HOW TO REJECT THE NULL HYPOTHESIS—THE TOPIC OF POWER

From your reading of this text and other material about experiments, you may have the impression that researchers are excited when they reject H_0 and unhappy when they don't. That impression is correct. It is also reasonable. To reject H_0 is to be left with only one alternative, H_1, from which a conclusion can be drawn. To retain H_0 is to be left up in the air. You don't know whether the null hypothesis is really true or whether it is false and you just failed to detect it. So, if you are going to design and run an experiment, you should maximize your chances of rejecting H_0.[14] There are three factors to consider: actual difference, standard error, and α.

In order to get this discussion out of the realm of the hypothetical and into the realm of the practical, consider the following problem. Suppose you must do

---

[14]Some of you may object to this philosophy of deliberately setting out to reject H_0. You might state the position that a scientist should try to discover the "true situation," which certainly may be that H_0 is true. That position is OK, theoretically, but the methods of statistics cannot be used to establish the null hypothesis.

an experiment for a class. You are free to choose any project, and you want to reject H_0. You decide to try to show that widgets are different from controls. Accept for a moment the idea that widgets *are* different—that H_0 *should* be rejected. What are the factors that determine whether you will conclude from your experiment that widgets are different?

1. *Actual difference.* The larger the actual difference between widgets and controls, the more likely you are to reject H_0. There is a practical limit, though. If the difference is too large, other people, including your instructor, will call your experiment trivial, saying that it demonstrates the obvious and that anyone can see that widgets are different. On the other hand, small differences can be difficult to detect. Pre-experiment estimations of actual differences are usually based on your own experience.[15]

2. *The standard error of a difference.* Look at the formula for z on page 163. You can see that as $s_{\bar{X}_1-\bar{X}_2}$ gets smaller, z gets larger, and you are more likely to reject H_0. This is true, of course, *only if widgets are really different from controls.* Here are two ways you can reduce the size of $s_{\bar{X}_1-\bar{X}_2}$.

    a. *Sample size.* The larger the sample, the smaller the standard error of the difference. Figure 7.5 shows that the larger the sample size, the smaller the standard error of the mean. The same relationship is true for the standard error of a difference.

        Some texts (for example, Guilford & Fruchter, 1978, pp. 182–186) show you how to calculate the sample size required to reject H_0. In order to do this calculation, you must make assumptions about the size of the actual difference. Many times, the size of the sample is dictated by practical consideration—time, money, or the availability of widgets.

    b. *Sample variability.* Reducing the variability in the sample will produce a smaller $s_{\bar{X}_1-\bar{X}_2}$. You can reduce variability by using reliable measuring instruments, recording data correctly, and, in short, reducing the "noise" or random error in your experiment.

3. *Alpha.* The larger α is, the more likely you are to reject H_0. The limit to this factor is your colleagues' sneer when you report that widgets are "significantly different at the .40 level." Everyone believes that such differences should be attributed to chance. Sometimes practical considerations may permit the use of $\alpha = .10$. If both widgets and controls could be used to treat a deadly illness and both have the same side effects, but "widgets are significantly better at the .10 level," then widgets will be used. (Also, more data will then be gathered [sample size increased] to see whether the difference between widgets and controls is reliable.)

---

[15]There is some danger in trusting your experience. Otto Loewi received a Nobel Prize for demonstrating the chemical nature of nerve transmission. The idea for the experiment came to him during the night. He got out of bed, went to his laboratory, and performed an experiment that involved isolated frog hearts and some Ringer solution. Later he wrote "If I had carefully considered it in the daytime, I would undoubtedly have rejected the kind of experiment I performed." Loewi goes on to explain that it seems improbable that the actual difference between the Ringer solution with the chemical and that without could be detected by the method he used. "It was good fortune that at the moment of the hunch I did not think but acted" (Loewi, 1963).

These three factors are discussed by statisticians under the topic *power*.[16] The power of a statistical test is defined as $1 - \beta$. The more powerful the test, the more likely it is to detect any actual difference between widgets and controls.

We will close this section on power by asking you to imagine that you are a researcher directing a project that could make a Big Difference. (Since you are imagining this, the difference can be in anything you would like to imagine—the health of millions, the destiny of nations, your bank account, whatever.) Now, suppose that the success or failure of the project hinges on one final statistical test. One of your assistants comes to you with the question, "How much power do you want for this last test?"

"All I can get," you answer.

If you will examine the list of factors that influence power you will find that there is only one item that you have some control over, and that is the standard error of a difference. Of the factors affecting the size of the standard error of a difference, the one that most researchers can best control is N.

Our advice to you is to allocate plenty of power to important statistical tests—use large Ns.

Problems   25. In your own words, distinguish between a significant difference and an important difference.

26. List the four factors that determine the power of an experiment. (If you can recall them without looking at the text, you will know that you have been reading actively and effectively.)

---

[16]Two texts that devote a separate chapter to power are Loftus and Loftus (1982), Chapter 8 and Minium and Clark (1982), Chapter 15.

TRANSITION PAGE

The techniques you have learned so far require the use of the normal distribution to assess probabilities. These probabilities will be accurate if you have used σ in your calculations or if N is so large that s is a reliable estimate of σ. In the next chapter, you will learn about a distribution that will give you accurate probabilities when you do not know σ and N is not large. The logic you have used, however, will be used again. That is, you assume the null hypothesis, draw random samples, introduce the independent variable, and calculate a mean difference on the dependent variable. If these differences cannot be attributed to chance, reject the null hypothesis and interpret the results.

At this point in their studies, most students suspect that the normal curve is an indispensable part of modern statistical living. Up until now, in this book, it has been. However, in the next five chapters you will encounter several sampling distributions, none of which is normal, but all of which can be used to determine the probability that a particular event occurred by chance. Deciding which distribution to use is not a difficult task but it does require some practice. Remember that a theoretical distribution is accurate if the assumptions on which it is based are true for the data from the experiment. By knowing the assumptions a distribution requires and the nature of your data, you can pick an appropriate distribution.

9
THE *t* DISTRIBUTION
AND THE *t* TEST

O bjectives for Chapter 9: After studying the text and working the problems in this chapter, you should be able to:

1. contrast the *t* distribution with the normal distribution,
2. use the *t* distribution to determine whether a sample mean came from a particular population,
3. distinguish an independent-samples design from a correlated-samples design and calculate and interpret a *t* value for both,
4. calculate and interpret a confidence interval for the difference between two sample means,
5. list the assumptions for the *t* test, and
6. use the *t* distribution to determine whether a Pearson product-moment correlation coefficient is significantly different from .00.

This chapter is about a theoretical distribution called the ***t* distribution**. The *t* is a lowercase one; capital *T* has entirely different meanings, one of which you will learn about in Chapter 13. The *t* distribution is used when σ is not known and sample sizes are too small to ensure that *s* is a reliable estimate of σ.

In this chapter, the *t* distribution will be used to find answers to the four kinds of problems listed next.

1. Did a sample with a mean $\bar{X}$ come from a population with a mean μ?
2. Did two samples, with means $\bar{X}_1$ and $\bar{X}_2$ come from the same population?
3. What is the confidence interval about the difference between two sample means?
4. Did a Pearson product-moment correlation coefficient, based on sample data, come from a population with a true correlation of .00 for the two variables?

Problems 1, 2, and 4 are problems of hypothesis testing. Problem 3 requires the establishment of a confidence interval. Although the *t* distribution is new to you and requires explanation, the logic of hypothesis testing and confidence intervals is not.

The story of the man who invented (discovered?) the *t* distribution is an interesting one that tells something about the motivation of statisticians. W. S. Gosset (1876–1937)[1] was educated at Oxford in chemistry and mathematics. In 1899, he went to work for Arthur Guinness, Son & Company, a brewery in Dublin, Ireland, where the management had just begun a policy of employing scientists to make tests and recommendations. Gosset was confronted with the problem of gathering, in a limited amount of time, data about the brewing process. He recognized that the sample sizes were so small that *s* was not an accurate estimate of σ and thus the normal-curve model was not appropriate. After working out the mathematics of distributions based on *s*, which is a statistic and, therefore, variable, rather than on σ, which is a parameter and, therefore, constant, Gosset found that the theoretical distribution depended upon sample size, a different distribution for each *N*. These distributions make up a family of curves that have come to be called the *t* distribution.

In Gosset's work, you again see how a practical question forced the development of a statistical tool. (Remember that Francis Galton invented the concept of the correlation coefficient in order to assess the degree to which characteristics of fathers are found in their sons.) In Gosset's case, an example of a practical question was "Will this new strain of barley, developed by the botanical scientists, have a greater yield than our old standard?" Such questions were answered with data from experiments carried out on the ten farms maintained by the Guinness Company in the principal barley-growing regions of Ireland. A typical experiment might involve two one-acre plots (one planted with the old barley, one with the new) on each of the ten farms. Gosset then was confronted with ten one-acre yields for the old barley and ten for the new. Was the difference in yields due to sampling fluctuation, or was it a reliable difference between the two strains? He made the decision using his newly derived *t* distribution.

[1]For more information, see "Gosset, W. S.," in *Dictionary of National Biography, 1931–40*, London: Oxford University Press, 1949, or L. McMullen & E. S. Pearson, "William Sealy Gosset, 1876–1937," *Biometrika*, 1939, 205–253.

Gosset wanted to publish his work on the t distribution in *Biometrika*, a journal founded in 1901 by Francis Galton. However, at the Guinness Company there was a rule that employees could not publish (the rule there, apparently, was publish *and* perish). Because the rule was designed to keep brewing secrets from escaping, there was no particular ferment within the company when Gosset, in 1908, published his new mathematical statistics under the pseudonym "Student." The t distribution, then, came to be known as "Student's t." (No one seems to know why the letter t was chosen. E. S. Pearson surmises that t was simply a "free letter"—that is, no one had yet used t to designate a statistic.) Since he worked for the Guinness Company all his life, Gosset continued to use the pseudonym "Student" for his publications in mathematical statistics. Gosset was very devoted to his company, working hard and rising through the ranks. He was appointed head brewer a few months before his death in 1937.

We will describe some characteristics of the t distribution and then compare t with the normal distribution. The following two sections are on hypothesis testing: one section on samples that are independent of each other and one on samples that are correlated. Next, you will use the t distribution to establish confidence intervals about a mean difference. Then you will learn the assumptions that are required if you choose to use a t test to analyze your data. Finally, you will learn how to determine whether a correlation coefficient is statistically significant. Problems 1–4, mentioned before, will be dealt with in order.

THE t DISTRIBUTION

Rather than just one t distribution, there are many t distributions. In fact, there is a t distribution for each sample size from 1 to ∞. These different t distributions are described as having different **degrees of freedom**, and there is a different t distribution for each degree of freedom. *Degrees of freedom* is abbreviated *df*. For now we will define *degrees of freedom* as sample size minus 1, and then provide you with a more thorough explanation later. Thus $df = N - 1$. If the sample consists of 12 members, $df = 11$.

Figure 9.1 is a picture of four of these t distributions, each based on a different number of degrees of freedom. You can see that, as the degrees of freedom become fewer, a larger proportion of the curve is contained in the tails.

You know from your work with the normal curve that a theoretical distribution is used to determine a probability and that, on the basis of the probability, the null hypothesis is retained or rejected. You will be glad to learn that the logic of using the t distribution to make a decision is just like the logic of using the normal distribution. Recall that

$$z = \frac{\bar{X} - \mu}{\sigma_{\bar{X}}}$$

and that z is normally distributed. You probably also recall that at the critical z value of ± 1.96, the chances are only 5 in 100 that the mean $\bar{X}$ came from the population with mean μ.

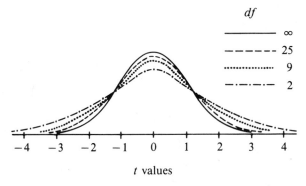

$$df$$

————— ∞

—————— 25

.... 9

—·—·—·— 2

-4 -3 -2 -1 0 1 2 3 4

t values

**Figure
9.1**

Four different *t* distributions. The difference between $df = 25$ and $df = \infty$ is exaggerated here so that the curves will print distinctively.

In a similar way, if the samples are small, you can calculate a *t* value from the formula

$$t = \frac{\bar{X} - \mu}{s_{\bar{X}}}.$$

The number of degrees of freedom (*df*) determines which *t* distribution is appropriate, and from it you can find a *t* value that would be expected to occur by chance 5 times in 100. Figure 9.2 separates the *t* distributions of Figure 9.1. The critical values in Figure 9.2 are those associated with the interval that contains 95 percent of the cases, leaving 2.5 percent in each tail. Look at each of the four curves.

If you looked at Figure 9.2 carefully, you may have been suspicious that the *t* distribution for $df = \infty$ is a normal curve. It is. As *df* approaches ∞, the *t* distribution approaches the normal distribution. When $df = 30$, the *t* distribution is almost normal. Now you understand why we repeatedly cautioned, in chapters that used the normal curve, the *N* must be at least 30 (unless you know σ or that the distribution of the population is symmetrical). Even when $N = 30$, the *t* distribution is more accurate than the normal distribution for assessing probabilities and so, in most research studies (that use samples), *t* is used rather than *z*.

A reasonable question now is "Where did those critical values of ±4.30, ±2.26, ±2.06, and ±1.96 come from?" The answer is Table D. Table D is really a condensed version of 34 *t* distributions. Turn to Table D and note that there are 34 different degrees of freedom in the left-hand column. Across the top under "α Levels for Two-Tailed Test" you will see six selected probability figures, .20, .10, .05, .02, .01, and .001. Follow the .05 column down to $df = 2, 9, 25$, and ∞ and you will find critical values of 4.30, 2.26, 2.06, and 1.96.

Table D differs in several ways from the normal-curve table. In the normal-curve table, the *z* scores are on the margin of the table and the probability figures are in two other columns. In the *t*-distribution table, the opposite is true; the *t* values are in the columns and the probability figures are on the top and bottom margins. Also, in the normal-curve table, you can find the exact probability of

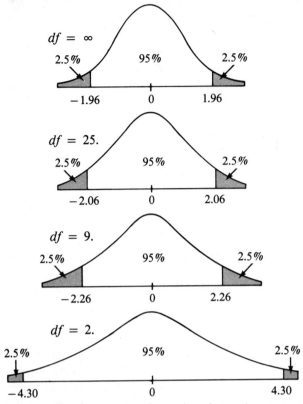

$df = \infty$

2.5%   95%   2.5%

−1.96   0   1.96

$df = 25.$

2.5%   95%   2.5%

−2.06   0   2.06

$df = 9.$

2.5%   95%   2.5%

−2.26   0   2.26

$df = 2.$

2.5%   95%   2.5%

−4.30   0   4.30

Figure 9.2   Four *t* distributions showing the *t* values that enclose 95 percent of the cases.

any *z* score; in Table D, the exact probability is given for only six *t* values. These six are commonly chosen as α levels by experimenters. Finally, if you wish to conduct a one-tailed test, use the probability figures shown under that heading at the bottom of Table D. Note that the probability figures are one-half those for a two-tailed test. You might draw a *t* distribution, put in values for a two-tailed test, and see for yourself that reducing the probability figure by one-half is appropriate for a one-tailed test.

As a general rule, researchers run two-tailed tests. If a one-tailed test is used, a justification is usually given. In this text we will routinely use two-tailed tests.

We'll use student's *t* distribution to decide whether a particular sample mean came from a particular population.

A Belgian, Adolphe Quételet (`Ka-tle) (1796–1874), is regarded as the first person to recognize that social and biological measurements may be distributed according to the "normal law of error" (the normal distribution). Quételet made this discovery while developing actuarial (life expectancy) tables for a Brussels life insurance company. Later, he began making anthropometric (body) measurements and, in 1836, he developed Quételet's Index (QI), a ratio in which weight in grams was divided by height in centimeters. This index was supposed to permit evaluation of a person's nutritional status: very large numbers indicated obesity and very small numbers indicated starvation.

Suppose a present-day anthropologist read that Quételet had found a mean QI value of 375 on the entire population of French army conscripts. No standard deviation was given because it had not yet been invented. Our anthropologist, wondering if there has been a change during the last hundred years, obtains a random sample of 20 present-day Frenchmen who have just been inducted into the Army. He finds a mean of 400 and a standard deviation of 60. One now familiar question remains, "Should this mean increase of 25 QI points be attributed to chance or not?" To answer this question, we will perform a *t* test. As usual, we will require $p \leq .05$ to reject chance as an explanation.

$$t = \frac{\bar{X} - \mu}{s_{\bar{X}}}$$

Recall that

$$s_{\bar{X}} = \frac{s}{\sqrt{N}} = \frac{60}{\sqrt{20}} = 13.42.$$

Thus,

$$t = \frac{400 - 375}{13.42} = 1.86,$$

and

$$df = N - 1 = 20 - 1 = 19.$$

Upon looking in Table D under the column for a two-tailed test with $\alpha = .05$ at the row for 19 *df*, you'll find a critical value of 2.09. Our anthropologist's *t* is less than 2.09, so the null hypothesis should be retained. The difference between present-day soldiers and those of old may be due to chance.

Quételet's Index is not currently used by anthropologists. There were several later attempts to develop a more reliable index of nutrition and most of those attempts were more successful than Quételet's. Some of Quételet's ideas are still around, though. For example, it was from Quételet, it seems, that Francis Galton got the idea that the phenomenon of genius could be treated mathematically, an idea that led to the concept of correlation. (Galton seems to turn up in many stories about important concepts.)

Problems

1. Fill in the blanks in the statements below with "*t*" or "normal."
   a. There is just one _____ distribution, but there are many _____ distributions.
   b. Given a value of 2.00 on the horizontal axis, the _____ distribution has a greater proportion of the curve to the right than does the _____ distribution.
   c. The _____ distribution must be used if the samples are small.

2. Would the standard deviation of a t distribution with $df = 3$ be larger than, smaller than, or equal to the standard deviation of the normal curve?

3. Determine the df for the following samples.
   a. $N = 25$.
   b. $N = 4$.
   c. $N = 42$.

4. Who invented the t distribution? Why?

5. Find the t value for the following means and standard errors of the mean, given that $\mu = 29.00$.
   a. $\bar{X} = 31.50$; $s_{\bar{x}} = 2.50$.
   b. $\bar{X} = 25$; $s_{\bar{x}} = 12.00$.
   c. $\bar{X} = 29.60$; $s_{\bar{x}} = .15$.
   d. $\bar{X} = 38.00$; $s_{\bar{x}} = 18.00$.
   e. $\bar{X} = 23.00$; $s_{\bar{x}} = 1.33$.

6. In the following problems, the investigator thought that μ might be 81.00. Your task is to calculate t values for each of the five problems, find the appropriate critical values in Table D, and assign each sample to one of the following categories: $p > .05$, $p = .05$, $.01 < p < .05$, $p = .01$, $p < .01$. This table should help you.

---

if $t_{ob} < t_{.05}$, then $p > .05$

if $t_{ob} = t_{.05}$, then $p = .05$

if $t_{.01} > t_{ob} > t_{.05}$, then $.01 < p < .05$

if $t_{ob} > t_{.01}$, then $p < .01$

where t_{ob} is the value calculated from the data (observations), and $t_{.05}$ and $t_{.01}$ are t values found in Table D.

---

   a. $\bar{X} = 84.92$; $s = 7.75$, $N = 15$
   b. $\bar{X} = 76.59$; $s = 5.56$, $N = 7$
   c. $\bar{X} = 76.59$; $s = 10.29$, $N = 24$
   d. $\bar{X} = 132.21$; $s = 82.49$, $N = 21$
   e. $\bar{X} = 80.014$; $s = .45$, $N = 5$

7. You just determined a number of probabilities. What event do these probabilities refer to?

DEGREES OF FREEDOM

You have been determining "degrees of freedom" by a rule-of-thumb technique: $N - 1$. Now it is time for us to explain the concept more thoroughly, in order to prepare you for statistical techniques in which $df \neq N - 1$.

It is somewhat difficult to obtain an intuitive understanding of the concept of degrees of freedom without the use of mathematics. If the following explanation leaves you scratching your head, you might read Helen Walker's excellent article in the *Journal of Educational Psychology* (Walker, 1940).

The *freedom* in *degrees of freedom* refers to freedom of a number to have any possible value. If you were asked to pick two numbers, and there were no restrictions, both numbers would be free to vary (take any value) and you would have two degrees of freedom. If, however, a restriction is imposed—namely, that $\sum X = 0$—one degree of freedom is lost because of that restriction. That is, when you now pick the two numbers, only one of them is free to vary. As an example, if you choose 3 for the first number, the second number *must be* -3. The second number is not free to vary, because of the restriction that $\sum X = 0$. In a similar way, if you were to pick five numbers, with a restriction $\sum X = 0$, you would have four degrees of freedom. Once four numbers are chosen (say, -5, 3, 16, and 8), the last number (-22) is determined.

The restriction that $\sum X = 0$ may seem to you to be an "out-of-the-blue" example and unrelated to your earlier work in statistics, but some of the statistics you have calculated have had such a restriction built in. For example, when you found $s_{\bar{X}}$, as required in the formula for *t*, you used some algebraic version of

$$s_{\bar{X}} = \frac{s}{\sqrt{N}} = \frac{\sqrt{\dfrac{\sum (X - \bar{X})^2}{N - 1}}}{\sqrt{N}}.$$

The restriction that is built in is that $\sum (X - \bar{X})$ is always zero and, in order to meet that requirement, one of the X's is determined. All X's are free to vary except one, and the degrees of freedom for $s_{\bar{X}}$ is $N - 1$. Thus, for the problem of using the *t* distribution to determine whether a sample came from a population with a mean μ, $df = N - 1$. Walker (1940) summarizes the reasoning above by stating: "A universal rule holds: The number of degrees of freedom is always equal to the number of observations minus the number of necessary relations obtaining among these observations." A necessary relationship for $s_{\bar{X}}$ is that $\sum (X - \bar{X}) = 0$. Another way of stating this rule is that the number of degrees of freedom is equal to the number of original observations minus the number of parameters estimated from the observations. In the case of $s_{\bar{X}}$, one degree of freedom is subtracted because $\bar{X}$ is used as an estimate of μ.

INDEPENDENT-SAMPLES AND CORRELATED-SAMPLES DESIGNS

Now we switch from the question of whether a sample came from a population with a mean, μ, to the more common question of whether two samples came from populations with identical means. That is, the mean of one group is compared with the mean of another group, and the difference is attributed to chance (null hypothesis retained) or to a treatment (null hypothesis rejected).

However, there are two kinds of two-group designs. With an **independent-samples design**, the subjects serve in only one of the two groups, and there is no reason to believe that there is any correlation between the scores of the two groups. With a **correlated-samples design** subjects may serve in both groups or in just one group, but in either case, there is a correlational relationship between the scores of the two groups. The difference between these designs is important because the calculation of the *t* value for independent samples is different from the calculation of the *t* value for correlated samples. You may not be able to tell which design has been used just by looking at the numbers or by identifying the independent and dependent variable; instead, you must examine the description of the procedures of the experiment.

Although the *t* formulas appear to be different, the purpose of both designs is the same: to determine the probability that two such samples could have a common population mean.

**Clue
to the
Future**

Most of the rest of this chapter is organized around independent-samples and correlated-samples designs. Three-fourths of Chapter 13 is also organized around these two designs. In Chapters 10 and 11 the procedures you will learn are appropriate only for independent samples.

Correlated-Samples Designs

In a correlated-samples[2] design, there is a logical reason to pair a score in one group with one particular score in the second group. Correlated-samples designs always consist of *pairs* of scores.

A correlated-samples design may come about in a number of ways. Fortunately, the actual arithmetic in calculating a *t* value is the same for any of three correlated-samples designs. The three types of designs are **natural pairs**, **matched pairs**, and **repeated measures**.

Natural pairs. In a natural-pairs investigation, the experimenter does not assign the subjects to one group or the other—the pairing occurs prior to the investigation. Table 9.1 identifies one way in which natural pairs may occur—father and son. In such an investigation you might ask whether fathers are shorter than their sons (or more religious, or more racially prejudiced, or whatever). Notice, though, that it is easy to decide that these are correlated-samples data; there is a logical way to pair up the scores in the two groups.

**Table
9.1**

Illustration of a Correlated-Samples Design

Fathers	Height		Sons	Height
Mark Smith	5′ 9″	his son	Mark, Jr.	5′ 10″
Kenneth Johnson	6′ 1″	his son	Kenneth, Jr.	6′ 2″
Bruce Brown	5′ 7″	his son	Bruce, Jr.	5′ 8″
Paul Williams	5′ 10″	his son	Paul, Jr.	5′ 9″
$\bar{X}_{\text{fathers}}$			$\bar{X}_{\text{sons}}$	

---

[2]Some texts use the term *dependent samples* instead of *correlated samples*.

Did you notice that Table 9.1 is the same as Table 5.1, which outlined the basic requirements for the calculation of a correlation coefficient? As you will soon see, that correlation coefficient is a part of determining a *t* value.

Matched pairs. In some situations, the experimenter has control over the ways pairs are formed and a match can be arranged. One method is for two subjects to be paired on the basis of similar scores on a pretest that is related to the dependent variable. For example, a hypnotic susceptibility test might be given to a group of subjects. Subjects with similar scores could be paired and then one member of each pair randomly assigned to either the experimental group or control group. The result is two groups equivalent in hypnotizability.

Another variation of matched pairs is the split-litter technique used with non-human animals. A pair from a litter is split and one is put in each group. In this way, the genetics of one group is matched with that of the other. The same technique has been used in human experiments with twins or siblings. Student's barley experiments are examples of starting with two similar subjects (adjacent plots of ground) and assigning them at random to one of two treatments.

Still another example of the matched-pairs technique is the treatment of each member of the control group according to what happens to its paired member in the experimental group. Because of the forced correspondence, this is called a *yoked-control* design.

The difference between the matched-pairs design and a natural-pairs design is that, with the matched pairs, the investigator can randomly assign one member of the pair to a treatment. In the natural-pairs design, the investigator has no control over assignment. Although the statistics are the same, the natural pairs design is usually open to more interpretations than the matched-pairs design.

Repeated measures. A third kind of correlated-samples design is called a repeated-measures design because more than one measure is taken on each subject. This design often takes the form of a before-and-after experiment. A pretest is given, some treatment is administered, and a post-test is given. The mean of the scores on the post-test is compared with the mean of the scores on the pretest to determine the effectiveness of the treatment. Clearly, there are two scores that should be paired: the pretest and the post-test scores of each subject. In such an experiment, each person is said to serve as his or her own control.

All three of these methods of forming groups have one thing in common: a meaningful correlation may be calculated for the data. The name *correlated samples* comes from this fact. With a correlated-samples design, one variable is designated X, the other Y.

Independent-Samples Design

In an independent-samples design,[3] there is no reason to pair a score in one group with a particular score in the second group. Often in this design, the whole pool of subjects is assigned in a random fashion to the groups.

---

[3]Some texts use the terms *noncorrelated* or *uncorrelated* for this design.

The designs you analyzed in Chapter 8 were all independent-samples designs. This design is outlined in Table 8.1, which you might want to review now.

One caution is in order. Both of these designs utilize random sampling, but with an <u>independent-samples design</u>, the subjects are randomly selected from a population of *individuals*. In a correlated-samples design, *pairs* are randomly selected from a population of pairs.

Finally, by way of comparison, both these designs have an independent variable that has two levels. Both have a dependent variable on which every subject has a score. The basic difference is that with the correlated-samples design, there is a logical reason to pair up scores from the two groups and in an independent-samples design there is not.

Problems   Identify each of the following experiments as an independent-samples design or a correlated-samples design. Work them all before checking your answers.

8. An investigator gathered as many case histories as she could of situations in which identical twins were raised apart—one in a "good" environment, and one in a "bad" environment. The group raised in the "good" environment was compared with that raised in the "bad" environment on attitude toward education.

9. A researcher interested in voluntary control of bodily functions attached a sensitive thermometer to both forefingers of a subject. The subject's task was to increase the flow of blood into the left forefinger (thereby raising the temperature). At the end of ten minutes, the temperature of both forefingers was recorded. The mean temperature of the left forefinger was compared with that of the right forefinger. Data were gathered on 38 subjects. (See Maslach, Marshall, & Zimbardo, 1971.)

10. A researcher counted the number of aggressive encounters among children who were playing with parts of toys. Then he lifted a screen to reveal other children playing with whole toys. The children with the parts of toys watched for a while, and then the researcher lowered the screen. The children resumed playing, and the researcher counted the number of aggressive encounters. The procedure was completed for six groups of children, and the investigator compared the number of aggressive encounters within the groups before and after the screen was lifted. (For a similar experiment, see Barker, Dembo, & Lewin, 1941.)

11. This is an experiment on the effects of vigilance on gastric secretions. One monkey must pay attention to a stimulus light. When the light goes on, the monkey has five seconds to press a lever, or he will be shocked. A control monkey in the next cage has no control over the shocks. If the first monkey gets shocked, so does the second. If the first monkey avoids the shock, so does the second. The experiment consists of nine monkeys who must be vigilant and press the lever to avert the shock (the "executive monkeys") and nine who do not. (The monkeys who are not in control of the shocks are called a "yoked control group.") The dependent variable is the gastric secretions for each monkey during a three-hour period of the light-shock sequence. (See Brady, Porter, Conrad, & Mason, 1958.)

12. A college dean faced a problem that suggested an experiment. Thirty-two freshmen had applied for the sophomore honors course. Only 16 could be accepted, so the dean flipped a coin for each applicant, with the result that 16

were selected. At graduation, she compared the mean grade-point average of the "winners" (those who had taken the course) with that of the "losers" to see if the sophomore honors course had any effect on G.P.A.

13. A social psychologist was interested in the effect of birth order on IQ. He obtained data from many families with two children and then compared the IQs of the firstborn with those of the second born. (See Zajonc, 1975.)

USING THE *t* DISTRIBUTION FOR INDEPENDENT SAMPLES

Using the *t* distribution to test a hypothesis is very similar to using the normal distribution. The null hypothesis is that the two populations have the same mean, and thus any difference between the two sample means is due to chance. The *t* distribution tells you the probability that the difference you observe is due to chance if the null hypothesis is true. You simply establish an α level, and if your observed difference is less probable than α, reject the null hypothesis and conclude that the two means came from populations with different means. If your observed difference is more probable than α, retain the null hypothesis. Does this sound familiar? We hope so.

The way to find the probability of the observed difference is to use a *t* test. The probability of the resulting *t* value can be found in Table D. For an independent-samples design, the formula for the *t* test is

$$t = \frac{\bar{X}_1 - \bar{X}_2}{s_{\bar{X}_1 - \bar{X}_2}}.$$

The *t* test, like many other statistical tests, is a ratio of a statistic over a measure of variability. $\bar{X}_1 - \bar{X}_2$ is a statistic and, of course, $s_{\bar{X}_1 - \bar{X}_2}$ is a measure of variability. You have seen this basic form before and you will see it again.

Table 9.2 shows two formulas for calculating $s_{\bar{X}_1 - \bar{X}_2}$. Use the formula in the top half of the table when the two samples have an unequal number of scores. In the special situation where $N_1 = N_2$, the formula simplifies to that shown in the bottom half of Table 9.2.

Table 9.2 **Formulas for $s_{\bar{X}_1 - \bar{X}_2}$ for Independent-Samples *t* Tests**

If $N_1 \neq N_2$

$$s_{\bar{X}_1 - \bar{X}_2} = \sqrt{\left(\frac{\sum X_1^2 - \dfrac{(\sum X_1)^2}{N_1} + \sum X_2^2 - \dfrac{(\sum X_2)^2}{N_2}}{N_1 + N_2 - 2}\right)\left(\frac{1}{N_1} + \frac{1}{N_2}\right)}.$$

If $N_1 = N_2$

$$s_{\bar{X}_1 - \bar{X}_2} = \sqrt{s_{\bar{X}_1}^2 + s_{\bar{X}_2}^2} = \sqrt{\frac{\sum X_1^2 - \dfrac{(\sum X_1)^2}{N} + \sum X_2^2 - \dfrac{(\sum X_2)^2}{N}}{N(N-1)}}.$$

The formula for degrees of freedom for independent samples is $df = N_1 + N_2 - 2$. The reasoning is as follows. For each sample, the number of degrees of freedom is $N - 1$, since, for each sample, a mean has been calculated with the restriction that $\sum (X - \bar{X}) = 0$. Thus, the total degrees of freedom is $(N_1 - 1) + (N_2 - 1) = N_1 + N_2 - 2$.

Here is an example of an experiment in which the results were analyzed with an independent-samples t test. Thirteen monkeys[4] were randomly assigned to either an experimental group (drug) or a control group (placebo). The experimental group ($N = 7$) was given the drug for eight days, while the control group ($N = 6$) was given a placebo (an inert substance). After eight days of injections, training began on a complex problem-solving task. Training and shots were continued for six days, after which the number of errors was tabulated. The number of errors each animal made and the t test are presented in Table 9.3. The null hypothesis is that the drug made no difference—that the difference obtained was due just to chance. Since the N's are unequal for the two samples,

Table 9.3   Number of Errors for Each Monkey during Six Days of Training

	Experimental Group (X_1)	Control Group (X_2)
	34	39
	52	57
	26	68
	47	74
	42	49
	37	
	40	
	$\sum X = 278$	344
	$\sum X^2 = 11478$	20520
	$N = 7$	6
	$\bar{X} = 39.71$	57.33

Applying the t test,

$$t = \frac{\bar{X}_1 - \bar{X}_2}{s_{\bar{X}_1 - \bar{X}_2}} = \frac{\bar{X}_1 - \bar{X}_2}{\sqrt{\dfrac{\sum X_1^2 - \dfrac{(\sum X_1)^2}{N_1} + \sum X_2^2 - \dfrac{(\sum X_2)^2}{N_2}}{N_1 + N_2 - 2}\left(\dfrac{1}{N_1} + \dfrac{1}{N_2}\right)}}$$

$$= \frac{39.71 - 57.33}{\sqrt{\left(\dfrac{11478 - \dfrac{(278)^2}{7} + 20520 - \dfrac{(344)^2}{6}}{7 + 6 - 2}\right)\left(\dfrac{1}{7} + \dfrac{1}{6}\right)}}$$

$$= \frac{-17.62}{\sqrt{\left(\dfrac{437.43 + 797.33}{11}\right)(.31)}} = \frac{-17.62}{5.90} = -2.99$$

$df = N_1 + N_2 - 2 = 7 + 6 - 2 = 11.$

[4]Monkey research is very expensive, so experiments are carried out with small N's. Thus, small-sample statistical techniques are a must.

the longer formula for the standard error must be used. Consulting Table D for 11 *df*, you'll find that a *t* = 2.20 is required in order to reject the null hypothesis with α = .05. Since the obtained *t* = −2.99, reject the null hypothesis. The final (and perhaps most important) step is to interpret the results. Since the experimental group, on the average, made fewer errors (39.71 vs. 57.33), we may conclude that the drug treatment *facilitated* learning. We will often express tabled *t* values as $t_{.05}$ (11 *df*) = 2.20. This gives you the critical value of *t* (2.20) for a particular *df* (11) and level of significance (α = .05).

Notice that the absolute value of the obtained *t* (|*t*| = |− 2.99| = 2.99) is *larger* than the tabled *t* (2.20). In order to reject the null hypothesis, the absolute value of the obtained *t* must be as great as, or greater than, the tabled *t*. The larger the obtained |*t*|, the smaller the probability that the difference between means occurred by chance. Figure 9.3 should help you see why this is so. Notice in Figure 9.3 that, as the values of |*t*| become larger, less and less of the area of the curve remains in the tails of the distribution. Remember that the area under the curve is a probability.

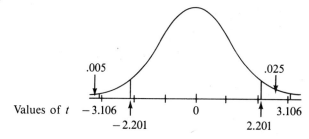

Figure 9.3   Values of *t* at the .05 and .01 levels of significance when *df* = 11.

Recall that we have been conducting a two-tailed test. That is, the probability figure for a particular *t* value is the probability of +*t* or larger plus the probability of −*t* or smaller. In Figure 9.3, $t_{.05}$ (11 *df*) = 2.201. This means that, if the null hypothesis is true, a *t* value of +2.201 would occur $2\frac{1}{2}$ percent of the time and a *t* value of −2.201 would occur $2\frac{1}{2}$ percent of the time.

Problems   For each problem in this section, decide whether a one-tailed or two-tailed test is appropriate. (For review see p. 171.)

    14. A French teacher was interested in the effect of "immediate, concrete experience" on learning vocabulary. He decided to conduct an experiment using volunteers from his beginning French class. The college was 30 miles from a city that students often visited. As he drove to the city one day, he made a tape describing in French the terrain, signs, distances, and so forth. ("L'auto est sur le pont. Carrefour prochain est dangereux.") Some students listened to the trip tape while driving (immediate, concrete experience) and some listened to the trip tape in the language laboratory. Students then took a vocabulary test and the number of errors was recorded. Analyze the data below with an independent-samples *t* test. Write a conclusion about "immediate, concrete experience."

Listened to Tape While in Laboratory	Listened to Tape While in Car
14	6
9	8
18	5
11	2
	9

15. Here are the data from the hypothetical experiment described in Problem 12. Reread Problem 12, analyze the summary data below, and write a conclusion.

	"Winners"	"Losers"
$\sum X$	55.36	56.32
$\sum X^2$	193.38	199.60
N	16	16

16. A purchasing agent must buy 50 new desk calculators. Suppose she decides to test a new brand against the ones her company already has. Both calculator companies loan her ten new calculators, and she arranges training for the calculator operators. Then, during a two-week test period, 20 operators are randomly assigned—ten to the new brand of machine and ten to the old brand. Three of the operators are not at work for part of the two weeks, and they are not included in the results. The final data consist of the mean number of minutes required to work a problem during the test period. Analyze the data for the agent and draw a conclusion.

New Brand	Old Brand
4.3	6.3
4.7	5.8
6.4	7.2
5.2	8.1
3.9	6.3
5.8	7.5
5.2	6.0
5.5	5.6
4.9	

USING THE t DISTRIBUTION FOR CORRELATED SAMPLES

The formula for t when the data come from correlated samples has a familiar theme: a difference between means divided by the standard error of the difference. The standard error of the difference between means of correlated samples is symbolized $s_{\bar{D}}$.

One formula for a *t* test between correlated samples is

$$t = \frac{\bar{X} - \bar{Y}}{s_{\bar{D}}},$$

where $s_{\bar{D}} = \sqrt{s_{\bar{X}}^2 + s_{\bar{Y}}^2 - 2r_{XY}(s_{\bar{X}})(s_{\bar{Y}})}.$

$df = N - 1$, where $N = $ *the number of pairs.*

The number of degrees of freedom in a correlated-samples case is the number of *pairs* minus one. Although each pair has two values, once one value is determined, the other is restricted to a similar value. (After all, they are called *correlated* samples.) In addition, another degree of freedom is subtracted when $s_{\bar{D}}$ is calculated. This loss is similar to the loss of 1 *df* when $s_{\bar{X}}$ is calculated.

As you can see by comparing the denominator of the correlated-samples *t* test with that of the *t* test on page 189 for independent samples (when $N_1 = N_2$), the difference lies in the term $2r_{XY}(s_{\bar{X}})(s_{\bar{Y}})$. Of course, when $r_{XY} = 0$, this term drops out of the formula, and the standard error is the same as for independent samples.

Also notice what happens to the standard-error term in the correlated-samples case where $r > 0$: the standard error is reduced. Such a reduction will increase the size of *t*. Whether this reduction will increase the likelihood of rejecting the null hypothesis depends on how much *t* is increased, since the degrees of freedom in a correlated-samples design are fewer than in the independent-samples design.

The formula $s_{\bar{D}} = \sqrt{s_{\bar{X}}^2 + s_{\bar{Y}}^2 - 2r_{XY}(s_{\bar{X}})(s_{\bar{Y}})}$ is used only for illustration purposes. There is an algebraically equivalent but arithmetically easier calculation called the *direct-difference method*, which does not require you to calculate *r*. To find the $s_{\bar{D}}$ by the direct-difference method, find the difference between each pair of scores, calculate the standard deviation of these difference scores, and divide the standard deviation by the square root of the number of pairs. Thus,

$$t = \frac{\bar{X} - \bar{Y}}{s_{\bar{D}}} = \frac{\bar{X} - \bar{Y}}{s_D / \sqrt{N}}$$

where

$$s_D = \sqrt{\frac{\sum D^2 - \frac{(\sum D)^2}{N}}{N - 1}}$$

where $D = X - Y,$
and $N = $ the number of *pairs* of scores.

Here is an example of a correlated-samples design and a *t*-test analysis. Suppose you were interested in the effects of interracial contact on racial attitudes. You have a fairly reliable test of racial attitudes, in which high scores indicate more positive attitudes. You administer the test one Monday morning to a biracial group of fourteen 12-year-old boys who do not know each other but who have signed up for a week-long community day camp. The campers then spend the next week taking nature walks, playing ball, eating lunch, swimming, and doing the kinds of things that camp directors dream up to keep 12-year-old boys busy. On Saturday morning, the boys are again given the racial-attitude test. Thus, the data consist of 14 pairs of before-and-after scores. The null hypothesis is that the mean of the population of "before" scores is equal to the mean of the population of "after" scores or, in terms of the specific experiment, that a week of interracial contact has no effect on racial attitudes.

Suppose the data in Table 9.4 were obtained. We will set $\alpha = .01$ and perform the analysis. Using the sum of the D and D^2 columns in Table 9.4, we can find s_D.

Table 9.4 **Hypothetical Data from a Racial Attitudes Study**

| | Racial-Attitude Scores | | | |
	Before Day Camp X	After Day Camp Y	D	D^2
Lonn	34	38	−4	16
Nicholas	22	19	3	9
Kenneth	25	36	−11	121
Bruce	31	40	−9	81
Dan	27	36	−9	81
Paul	32	31	1	1
Joshua	38	43	−5	25
Nathan	37	36	1	1
Darren	30	30	0	0
Butch	26	31	−5	25
Phillip	16	34	−18	324
Mark	24	31	−7	49
David	26	36	−10	100
Harton	29	37	−8	64
Sum	397	478	−81	897
Mean	28.36	34.14		

$$s_D = \sqrt{\frac{\sum D^2 - \dfrac{(\sum D)^2}{N}}{N-1}} = \sqrt{\frac{897 - \dfrac{(-81)^2}{14}}{13}} = \sqrt{32.95} = 5.74.$$

$$s_{\bar{D}} = \frac{s_D}{\sqrt{N}} = \frac{5.74}{\sqrt{14}} = 1.53.$$

Thus,

$$t = \frac{\bar{X} - \bar{Y}}{s_{\bar{D}}} = \frac{28.36 - 34.14}{1.53} = \frac{-5.78}{1.53} = -3.78.$$

$$df = N - 1 = 14 - 1 = 13.$$

Since $t_{.01}(13\ df) = 3.01$, this difference is significant beyond the .01 level. That is, $p < .01$. The "after" mean was larger than the "before" mean; therefore, we may conclude that, after the week of camp, racial attitudes were significantly more positive than before.

You might note that $\bar{X} - \bar{Y} = \bar{D}$, the mean of the difference scores. In this problem, $\sum D = -81$ and $N = 14$, so $\bar{D} = \sum D/N = -81/14 = -5.78$.

Gosset preferred the correlated-samples design. In his agriculture experiments, there were significant correlations between the yields of the old barley and the new barley grown on adjacent plots. This correlation reduced the standard-error term in the denominator of the *t* test, making the correlated-samples design more sensitive than the independent-samples design for detecting a difference between means.

Problems

17. Give a formula and definition for each of the following symbols.
    a. s_D
    b. D
    c. $s_{\bar{D}}$
    d. *t* (verbal definition)
    e. $\bar{Y}$

18. Here are data based on the experiment described in Problem 10. Reread Problem 10 and analyze the scores below. Write a conclusion.

Before	After
16	18
10	11
17	19
4	6
9	10
12	14

19. A researcher—a high-drive, worrying type—set up a behavior-modification program for high-drive, worrying persons who were concerned about their hearts. He measured anxiety behavior before (X) and after (Y) a ten-week program, with the following results.

$$\sum X = 192. \qquad \sum Y = 157. \qquad N = 13.$$

$$\sum X^2 = 2990. \qquad \sum Y^2 = 1985. \qquad \sum XY = 2385.$$

Set $\alpha = .05$ and test the hypothesis that the behavior-modification program had no effect on the anxiety-behavior scores. Note: You may have to do some thinking before you can set this problem up. Start by looking at the data you have.

20. Decide whether the following study is an independent or correlated-samples experiment and analyze the data. Choose your own α level. An experimental-methods class was randomly divided into two groups. One group found their reaction time (RT) to an auditory "go" signal, and the other group found their RT to a visual "go" signal. The groups then switched places and found their RT to the other stimulus. Each person thus found his or her RT for both stimuli, and these paired scores are arranged in rows for the 11 members of the class.

Auditory RT (Seconds)	Visual RT (Seconds)
.16	.17
.19	.23
.23	.20
.14	.19
.19	.19
.18	.22
.21	.23
.18	.16
.17	.21
.16	.19
.17	.18

21. Two groups were chosen randomly from a large sociology class for a study on the effects of primacy versus recency. Both groups were given a two-page description of a person at work. This description had a paragraph that told how the person had been particularly helpful to a new employee. Half of the descriptions were arranged so that the "helping" paragraph was in the first half (primacy), and the other half of the descriptions were arranged with the "helping" paragraph in the second half (recency). One group read the primacy description and the other the recency description. The subjects were then asked to write a page summary of what the worker would do during his or her leisure time. The dependent variable was the number of positive adjectives (words like *good, exciting, cheerful*) in the page summary of leisure activities. Below are the number of positive adjectives. Test for a difference at the .05 level.

Recency	Primacy
13	10
14	17
4	9
8	11
11	12
6	5
6	10
10	16
14	14

USING THE *t* DISTRIBUTION TO ESTABLISH A CONFIDENCE INTERVAL ABOUT A MEAN DIFFERENCE

As you probably recall from Chapter 7, a confidence interval is a range of values within which a parameter is expected to be. A confidence interval is established for a specified degree of confidence, usually 95 percent or 99 percent.

In this section, you will learn how to establish a confidence interval about a mean difference. The problems here are similar to those dealt with in Chapter 7, except that

1. probabilities will be established with the *t* distribution rather than with the normal distribution, and
2. the parameter of interest is a *difference* between two population means rather than a population mean.

The first point can be dispensed with rather quickly. You have already practiced using the *t* distribution to establish probabilities; you will use Table D in this section, too.

The second point will require a little more explanation. The questions you have been answering so far in this chapter have been *hypothesis-testing* questions, of the form "Does $\mu_1 - \mu_2 = 0$?" You answered each question by drawing two samples, calculating the means, and finding the difference. If the probability of the difference was very small, the hypothesis $H_0: \mu_1 - \mu_2 = 0$ was rejected. Suppose you have rejected the null hypothesis, but someone wants more information and asks "What is the real difference between μ_1 and μ_2?" The person recognizes that the real difference is not zero and wonders what it is. You are being asked to make an estimate of $\mu_1 - \mu_2$. If you establish a confidence interval about the difference between $\bar{X}_1$ and $\bar{X}_2$ or $\bar{X}$ and $\bar{Y}$, you can state with a specified degree of confidence that $\mu_1 - \mu_2$ lies within the interval.

Confidence Intervals for Independent Samples

The sampling distribution of $\bar{X}_1 - \bar{X}_2$ is a *t* distribution with $N_1 + N_2 - 2$ degrees of freedom. The lower and upper limits of the confidence interval about a mean difference are found with the following formulas:

$$LL = (\bar{X}_1 - \bar{X}_2) - t_\alpha(s_{\bar{X}_1 - \bar{X}_2}).$$
$$UL = (\bar{X}_1 - \bar{X}_2) + t_\alpha(s_{\bar{X}_1 - \bar{X}_2}).$$

For a 95 percent confidence interval, use the *t* value in Table D associated with $\alpha = .05$. For 99 percent confidence, change α to .01.

For an example, we will use the calculations you worked up in Problem 16 on the time required to do problems on the two different brands of desk calculators. We will establish a 95 percent confidence interval about the difference found.

As your calculations revealed,

$$\bar{X}_1 - \bar{X}_2 = 6.6 - 5.1 = 1.5 \text{ minutes.} \qquad s_{\bar{X}_1 - \bar{X}_2} = .40.$$

$$N_1 = 8. \qquad N_2 = 9. \qquad df = 8 + 9 - 2 = 15.$$

From Table D, $t_{.05}$ (15 df) = 2.13.

$$LL = (\bar{X}_1 - \bar{X}_2) - t_\alpha(s_{\bar{X}_1 - \bar{X}_2}) = 1.5 - 2.13(.40) = .65.$$

$$UL = (\bar{X}_1 - \bar{X}_2) + t_\alpha(s_{\bar{X}_1 - \bar{X}_2}) = 1.5 + 2.13(.40) = 2.35.$$

Thus, .65 and 2.35 are the lower and upper limits of a 95 percent confidence interval for the mean difference between the two kinds of calculators.

One of the benefits of establishing a confidence interval about a mean difference is that you also test the null hypothesis, $\mu_1 - \mu_2 = 0$, in the process (see Natrella, 1960). If 0 is outside the confidence interval, then the null hypothesis would be rejected using hypothesis-testing procedures. In the example we just worked, the confidence interval was .65 to 2.35 minutes; a value of 0 falls outside this interval. Thus, we can reject H_0: $\mu_1 - \mu_2 = 0$ at the .05 level. (You confirmed this in working Problem 16.)

Sometimes, hypothesis testing is not sufficient, and the extra information of confidence intervals is desirable. Here is one example of how this "extra information" on confidence intervals might be put to work in this calculator-purchasing problem. Suppose the new brand is faster, but also more expensive. Is it still a better buy? Through cost-benefit-analysis procedures, the purchasing agent can show that, given a machine life of five years, a reduction of time per problem of .50 minute justifies the increased cost. If she has the confidence interval you just worked out, she can see immediately that such a difference in machines (.50 minute) is below the lower limit of the confidence interval. The new machines are the better buy.

Confidence Intervals for Correlated Samples

The sampling distribution of $\bar{X} - \bar{Y}$ is also a t distribution. The number of degrees of freedom is $N - 1$. As in the section on hypothesis testing of correlated samples, N is the number of *pairs* of scores. The lower and upper limits of the confidence interval about a mean difference between correlated samples are

$$LL = (\bar{X} - \bar{Y}) - t_\alpha(s_{\bar{D}}) \qquad and \qquad UL = (\bar{X} - \bar{Y}) + t_\alpha(s_{\bar{D}}).$$

The interpretation of a confidence interval about a difference between means is very similar to the interpretation you made of confidence intervals about a sample mean. Again, the method is such that repeated sampling from two populations will produce a series of confidence intervals, 95 (or 99) percent of which will contain the true difference between the population means. You have sampled only once so the proper interpretation is that you are 95 (or 99) percent confident that the true difference lies between your lower and upper limits. It would probably be helpful for you to reread the material on interpreting a confidence interval about a mean, which is found on pages 146–150.

Error Detection

For confidence intervals for either independent or correlated samples, use a t value from Table D, not one calculated from the data.

Problems

22. Suppose $s_{\bar{X}_1 - \bar{X}_2} = .04$ for the problem on calculators. Establish the 95 percent confidence interval about the mean difference.

23. Based on the results of Problem 22, what decision should the purchasing agent make about which calculator to buy if a difference of 1.7 minutes is required to justify the extra expense?

24. Suppose a difference in time per problem of .25 minute is required to justify buying the more expensive machine. What decision would the purchasing agent make?

25. Establish a 99 percent confidence interval about the difference you found in Problem 20.

26. For an experiment on the effects of sleep on memory, eight volunteers were randomly divided into two groups. Everyone learned a list of ten nonsense syllables (now called consonant-vowel-consonants—CVCs). One group then slept for four hours, while the other group engaged in daytime activities. Then each subject recalled the CVCs. Two days later, everyone learned a new list. The group that had slept now engaged in daytime activities, and the group that had been active slept. After four hours, each person recalled as many CVCs as he or she could. The scores below represent the number of CVCs recalled by each person under the two conditions. Establish a 90 percent confidence interval about the mean difference. (This problem is modeled after a 1924 study by Jenkins and Dallenbach.)

Awake	Asleep
2	3
0	2
4	4
0	3
1	2
2	5
1	2
0	4

27. Interpret the confidence interval established in Problem 26.

28. This problem requires you to establish a 95 percent confidence interval about a mean rather than a mean difference. This is just what you did in Chapter 7, but this time use the *t* distribution rather than the normal distribution.

A researcher had a graph that showed a generally upward trend, but with a dip in the middle. In order to illustrate that the dip was not a chance phenomenon, he calculated 95 percent confidence intervals about each mean. Given the data shown, calculate the confidence interval about each mean. Redraw the graph and place a vertical line the length of the confidence interval on each mean. This will give you an idea of the sampling variation you would expect for each mean. By looking at the variation around the means for Trials 3, 4, and 5, you can determine whether the dip at Trial 4 can be attributed to chance.

	Trials					
	1	2	3	4	5	6
$\bar{X}$	21	27	30	25	30	36
s	3.3	2.7	3.0	3.6	5.1	4.5
N	17	17	17	17	17	17

(*continued*)

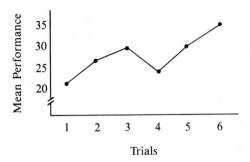

ASSUMPTIONS FOR USING THE t DISTRIBUTION

You can perform a t test on the difference between means on any two-group data you have or any that you can beg, borrow, buy, or steal. No doubt about it, you can easily come up with a t value using

$$t = \frac{\bar{X}_1 - \bar{X}_2}{s_{\bar{X}_1 - \bar{X}_2}}.$$

You can then attach a probability figure to your t value by deciding that the t distribution is an appropriate model of your empirical situation.

In a similar way, you can calculate a confidence interval about the difference between means in any two-group experiment. By deciding that the t distribution is an accurate model, you can claim you are "99 percent confident that the true difference between the population means is between thus and so."

But should you decide to use the t distribution? When is it an accurate reflection of the empirical probabilities?

The t distribution will give correct results when the assumptions it is based on are true for the populations being analyzed. The t distribution, like the normal curve, is a theoretical distribution. In deriving the t distribution, mathematical statisticians make three assumptions.

1. The dependent-variable scores for both populations are normally distributed.
2. The variance of the dependent-variable scores for the two populations are equal.
3. The scores on the dependent variable are random samples from the population.

Assumption 3 requires three explanations. First, in a correlated-samples design, the *pairs* of scores should be random samples from the population you are interested in.

Second, Assumption 3 ensures that any sampling errors will fall equally into both groups and that you may generalize from sample to population. Many times it is a physical impossibility to sample randomly from the population. In these cases, you should randomly assign the available subjects to one of the two groups. This will randomize errors, but your generalization to the population will be on less secure grounds than if you had obtained a truly random sample.

Third, Assumption 3 ensures the *independence* of the scores. That is, knowing one score within a group does not help you predict other scores in that same group. Either random sampling from the population or random assignment of subjects to groups will serve to achieve this independence.

Now we can return to the major question of this section: "When will the *t* distribution produce accurate probabilities?" The answer is "When random samples are obtained from populations that are normally distributed and have equal variances."

This may appear to be a tall order. It is, and in practice no one is able to demonstrate these characteristics exactly. The next question becomes "Suppose I am not sure that my data have these characteristics. Am I likely to reach the wrong conclusion if I use Table D?"

We don't have a simple answer to your reasonable question. In the past, several studies have suggested that the answer is, "No, because the *t* test is a robust test." (Robust means that a test gives you fairly accurate probabilities even when the data do not meet the assumptions on which it is based.) However, Bradley (1978) points out that robustness has not been defined quantitatively and further, that several studies (largely ignored by textbooks) show sizable departures from accuracy in a variety of situations. In actual practice, many researchers routinely use a *t* test unless one or more of these assumptions is clearly not justified.

USING THE *t* DISTRIBUTION TO TEST THE SIGNIFICANCE OF A CORRELATION COEFFICIENT

In Chapter 5, you learned to calculate Pearson product-moment correlation coefficients. This section is on testing the statistical significance of these coefficients. The question is whether an obtained *r*, based on a sample, could have come from a population of pairs of scores for which the parameter correlation is .00. The answer to this question is based on the size of a *t* value that is calculated from the correlation coefficient. The *t* value is found using the formula[5]

$$t = (r) \sqrt{\frac{N - 2}{1 - r^2}}$$

$df = N - 2$, where N = number of pairs.

The null hypothesis is that the population correlation is .00. Samples are drawn and an *r* is calculated. The *t* distribution is then used to determine whether the obtained *r* is significantly different from .00.

As an example, suppose you had obtained an $r = .40$ with 22 pairs of scores. Does such a correlation indicate a significant relationship between the two variables, or should it be attributed to chance?

---

[5]This formula is an algebraic manipulation of $t = r/s_r$, where $s_r = \sqrt{(1 - r^2)/(N - 2)}$. The form $t = r/s_r$ should be familiar to you now: a statistic divided by the standard error of the statistic.

$$t = (r)\sqrt{\frac{N-2}{1-r^2}} = (.40)\sqrt{\frac{22-2}{1-(.40)^2}} = 1.95.$$

$$df = N - 2 = 22 - 2 = 20.$$

Table D shows that, for 20 df, a t value of 2.09 is required to reject the null hypothesis. The obtained t for $r = .40$, where $N = 22$, is less than the tabled t, so the null hypothesis is retained. That is, a coefficient of .40 would be expected by chance alone more than 5 times in 100.

In fact, for $N = 22$, an $r = .43$ is required for significance at the .05 level and an $r = .54$ for the .01 level. As you can see, even medium-sized correlations can be expected by chance alone for samples as small as 22. Most researchers strive for N's of 30 or more for correlation problems.

Sometimes you may wish to determine whether the *difference* between two correlations is statistically significant. Several texts discuss this test (Ferguson, 1981, p. 196 and Guilford & Fruchter, 1978, p. 163).

Problems

29. A novice researcher developed a questionnaire on attitudes toward "consumerism." Responses were obtained from a random sample of males and females. A t test showed no significant differences between sexes. The researcher then went back to the same random sample, administered the questionnaire a second time, pooled the data with those obtained earlier, and calculated a new t value, which was significant. Which of the assumptions of the t test was violated?

30. Determine whether the following Pearson product-moment correlation coefficients are significantly different from .00. Set $\alpha = .05$.
   a. $r = .62$; $N = 10$.
   b. $r = -.19$; $N = 122$.
   c. $r = .50$; $N = 15$.
   d. $r = -.34$; $N = 64$.

31. This is a think problem. Suppose you had a summer job in an industrial research laboratory testing the statistical significance of hundreds of correlation coefficients based on varying sample sizes. An α level of .05 had been adopted by the management and, for each coefficient, your task was to test whether it was "significant" or "not significant" (N.S.). How could you construct a table of your own, using Table D and the t formula for testing the significance of a correlation coefficient, that would allow you to label each coefficient without working out a t value for each?

The table you mentally designed in working Problem 31 already exists. One version is reproduced in Table A in the Appendix of this book. α values of .10, .05, .02, .01, and .001 are included there. Use Table A in the future to determine if a Pearson product-moment correlation coefficient is significantly different from .00, but remember that this table is based on the t distribution.

In summary, you have learned about a new sampling distribution in this chapter. You can now use it to assess the probability that a sample came from a population with a specific mean, μ. You can determine the probability that two samples (either independent or correlated) have a common population mean or that a sample correlation coefficient came from a population with a correlation of .00. Finally, you learned that a t distribution can be used to establish a confidence interval about a mean or a mean difference.

TRANSITION PAGE

The t test, at which you are now skilled, is a very efficient method of testing the significance of the difference between two sample means. Its limitation is that it is inappropriate for dealing with more than two means at once—something experimenters often want to do. If an experimental question can be answered using two treatment conditions, the t test is the method to use; but what if you need to use three or four or more treatment conditions?

The answer to this problem is a technique called the **analysis of variance** (ANOVA for short, pronounced uh-nóve-uh). ANOVA was invented by Sir Ronald Fisher, an Englishman, and it is appropriate for both small and large samples, just as t is. In fact, it is a close relative of t.

So the transition this time is from a t test and its sampling distribution, the t distribution, to ANOVA, and its sampling distribution, the F distribution. Chapter 10 will show you how to use ANOVA to make comparisons among two or more groups. Chapter 11 will show you that the ANOVA technique can be extended to the analysis of experiments in which there are *two* independent variables, each of which may have two or more levels of treatment.

The analysis of variance is one of the most widely used statistical techniques, and Chapters 10 and 11 are devoted to an introduction to its more elementary forms. Many advanced books are available that explain more sophisticated (and complicated) analysis-of-variance designs.

10 ANALYSIS OF VARIANCE: ONE-WAY CLASSIFICATION

Objectives for Chapter 10: After studying the text and working the problems in this chapter, you should be able to:

1. identify the independent and dependent variables in a one-way ANOVA,
2. explain the rationale of ANOVA,
3. define F and explain its relationship to t,
4. compute sums of squares, mean squares, degrees of freedom, and F for an ANOVA,
5. interpret an F value obtained in an experiment,
6. construct a summary table of ANOVA results,
7. distinguish between *a priori* and *a posteriori* tests,
8. use orthogonal comparisons to make *a priori* comparisons among means following ANOVA,
9. use Scheffé tests to make *a posteriori* comparisons among means following an ANOVA, and
10. list and explain the assumptions of ANOVA.

In this chapter, you will become acquainted with the most simple of analysis-of-variance designs. You will learn to use ANOVA to examine the effects of two or more treatment levels in a single experiment. Such experiments are common in all disciplines that use statistics. Here are some examples:

1. Samples of lower-, middle-, and upper-class persons were compared on attitudes toward religion.
2. An experimenter determined the effect of 10, 20, 40, and 80 grams of reinforcement on the rate of response of four groups of rats.
3. Three methods of teaching Spanish were compared on their effectiveness with fourth graders.
4. Five species of honeybees were observed to determine which would produce the greatest number of kilograms of honey.

These experiments are similar to those whose results you analyzed with the *independent-samples t test* in Chapter 9. Again, there is an independent variable and a dependent variable. Again, the subjects in each group are independent of subjects in the other groups. Again, the null hypothesis is that the population mean is the same for all samples. The only difference is that, instead of only two levels of the independent variable, there are two or more. The name of this design is *one-way ANOVA* because there is only one independent variable.[1]

In Example 1, the independent variable is social class, and it has three levels. The dependent variable is attitudes toward religion. The null hypothesis is that the religious attitudes are the same in all three populations of social classes; that is, $H_0: \mu_{lower} = \mu_{middle} = \mu_{upper}$.

Problem

1. For Examples 2, 3, and 4, identify the independent variable, the number of levels of the independent variable, the dependent variable, and the null hypothesis.

A common reaction when confronted with three or more means is to run *t* tests on all possible combinations. For three means, three *t* tests would be required, for four means, six tests, and so on.[2] *This will not work.* The reason is that, if you perform more than one *t* test involving a particular mean, you will increase the chance of making a Type I error. That is, if you run several tests, each with $\alpha = .05$, the *overall* probability of making a Type I error is greater than .05. If you had an experiment with 15 groups, 105 *t* tests would be required in order to compare each group with every other group. If all 15 groups came from populations with the same mean, and you set $\alpha = .05$ for each test, you would *expect* five *t* tests to be significant just by chance.[3] If you then pulled those five tests out and claimed they were significant, you would be violating the spirit of

---

[1] Some writers prefer to call this a *completely randomized design.*
[2] The formula for the number of combinations of *n* things taken two at a time is $n(n-1)/2$.
[3] Remember that, if $\alpha = .05$, and the null hypothesis is true, then five times in a hundred you will wrongly reject the null hypothesis on the basis of sample data.

inferential statistics. What is needed in the case of more than two groups is a sampling distribution that gives the probability that the several means could have come from identical populations. This is exactly what Sir Ronald Fisher produced with the analysis of variance.

Fisher (1890–1962) was an Englishman whose important contributions in genetics are overshadowed by his fundamental work in statistics. (See biography in Greene, 1966.) In genetics, he explained how a recessive gene produced by mutation can become established in a population. For these experiments, he chose wild jungle fowl and their domesticated descendants, poultry.

In statistics, Fisher developed the techniques you will be studying in this chapter and the next one, discovered the exact sampling distribution of r, and developed a way to find the exact probability of results from a particular small-sample design. His *Statistical Methods for Research Workers*, first published in 1925, went into a 14th edition in 1973. Before getting into genetics and statistics in such a big way, Fisher worked for an investment company for two years and taught in a public school for four years.

RATIONALE OF ANOVA

The question to be answered by ANOVA is whether the samples all came from populations with the same mean μ or whether at least one of the samples came from a population with a different mean. The assumption is made that if more than one population is involved, the variances in the populations are equal.

Fisher, who was a friend of Gosset, is said to have looked at Student's t and realized that it used a principle that was also applicable to experiments having more than two groups. *The principle is that of dividing one estimate of the population variability by another.*[4]

The sampling distribution that Fisher derived is the **F distribution**. As will be shown, F values that make up the F distribution are obtained by dividing one estimate of the population variance by a second estimate. Thus,

$$F = \frac{\text{estimate of } \sigma^2}{\text{estimate of } \sigma^2}.$$

These two estimates of σ^2 are obtained by different methods. The numerator is obtained by a method that accurately estimates σ^2 *only when H_0 is true*. If H_0 is false, the estimate of σ^2 in the numerator will be too large.

The denominator is obtained by a method that is unaffected by the truth or falsity of H_0. Thus, when the null hypothesis is true, the expected value of F is about 1.00, since both methods are good estimators of σ^2, and $\sigma^2/\sigma^2 \cong 1.00$. Values somewhat larger and smaller than 1.00 are to be expected because of

---

[4]In the case of t,

$$t = \frac{\bar{X}_1 - \bar{X}_2}{s_{\bar{X}_1 - \bar{X}_2}} = \frac{\text{a range}}{\text{standard error of the difference between means}} = \frac{\text{a measure of variability}}{\text{a measure of variability}}.$$

sampling fluctuation, but, if an F value is too *large*, there is cause to suspect that H_0 is false.

We'll take these two estimates of σ^2 one at a time and discuss them. The estimate of σ^2 in the numerator is obtained from the two or more *sample means*. The conceptual steps follow. (Computational steps will come later.)

1. Find the standard deviation of the two or more sample means. This standard deviation of sample means is an old friend of yours, the standard error of the mean, $s_{\bar{x}}$.
2. Since $s_{\bar{x}} = s/\sqrt{N}$, squaring both sides gives $s_{\bar{x}}^2 = s^2/N$. Multiplying both sides by N and rearranging, $s^2 = N s_{\bar{x}}^2$.
3. s^2 is, of course, an estimate of σ^2.

Thus, to find this s^2, you need to multiply the sample size (N) by the variance of the sample means, both of which you can calculate. This estimate of σ^2 is called the *between-means estimate* (or the between-groups estimate). Notice that this between-means estimate of σ^2 is accurate only if the sample means are all drawn from the same population. If one or more means come from a population with a larger or smaller mean, the variance of the sample means will be larger.

The other estimate of σ^2 (the denominator of the F ratio) is obtained from the variability within each of the samples. Each sample variance is an independent estimate of σ^2, so, by averaging them, an even better estimate can be made. This estimate is called the *within-groups estimate*, and it is an unbiased estimate even if the null hypothesis is false. Once calculated, the two estimates can be compared. If the between-means estimate is much larger than the within-groups estimate, the null hypothesis is rejected.

We'll express these same ideas with pictures. Figure 10.1 illustrates the situation when the null hypothesis is true. Four samples have been drawn from identical populations and a mean calculated for each sample. As the projection of the four sample means on the vertical axis shows, the means are fairly close, and, therefore, the variability of these four means (the between-means estimate) will be small.

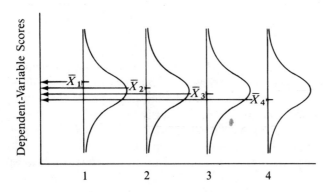

Levels of the Independent Variable

Figure 10.1 H_0 is true. The normal curves are the populations from which the four samples are drawn. The sample means are all estimates of the common population mean, μ.

Figure 10.2 illustrates one situation in which the null hypothesis is false (one group comes from a population with a larger μ). The projection of the means this time shows that $\bar{X}_4$ will greatly increase the variability of the four means.

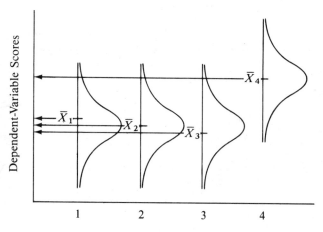

Levels of the Independent Variable

**Figure
10.2**
H_0 is false. Three of the samples are drawn from populations with the same mean, μ. The fourth sample is drawn from a population with a different mean.

Study Figures 10.1 and 10.2. They illustrate how the between-means estimate is larger when the null hypothesis is false. So, if you have a small amount of variability between means, retain H_0. If you have a large amount of variability between means, reject H_0. Small and large, however, are relative terms and, in this case, they are relative to the population variance. A comparison of Figures 10.3 and 10.1 illustrates how the amount of between-means variability *depends* upon the population variance. In both of these figures, the null hypothesis is true, but notice the projection of the sample means on the vertical axis. There is more variability among the means that come from populations with greater variability. Figure 10.3, then, shows a large between-means estimate that is the result of large population variances and not the result of a false null hypothesis.

So, in order to decide whether a large between-means estimate is due to a false null hypothesis or to a population variance, you need another estimate of the population variance. The best such estimate is the average of the sample variances.

All of this discussion brings us back to the principle that Fisher found Gosset to be using in the *t* test: dividing one estimate of the population variability by another. In the case of ANOVA, if the null hypothesis is true, the two estimates should be very similar, and dividing one by the other should produce a value close to 1.0. If the null hypothesis is false, dividing the between-means estimate by the within-groups estimate will produce a value greater than 1.0.

Be sure you understand this rationale of ANOVA. It is the basic rationale underlying the procedures to be explained in this chapter and the next.

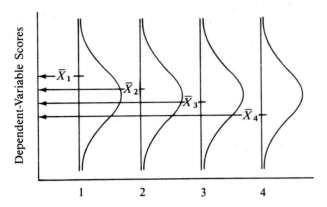

Levels of the Independent Variable

**Figure
10.3** H_0 is true. The normal curves are the populations from which the four samples are drawn.
The populations have more variability than those in Figure 10.1.

Sir Ronald Fisher developed a mathematical way to express the reasoning
we have just outlined. He worked out a sampling distribution that was later
named F in his honor.

As is the case with t, there is more than one F distribution. There is, in
fact, a different distribution for every possible combination of degrees of freedom
for the two variance estimates. All F distributions are positively skewed. The
fewer the degrees of freedom, the greater the skew. When the numbers of degrees
of freedom for both variance estimates are very large, the distribution approaches
the shape of the normal distribution. Figure 10.4 demonstrates the shape of one
F distribution (when one variance estimate has 9 degrees of freedom and the
other estimate has 15).

Table F in the Appendix is a table developed from sampling distributions
of different F ratios. The table was first compiled in 1934 by George W. Snedecor,
a gentleman to whom users of statistics owe a debt of gratitude. The formulation

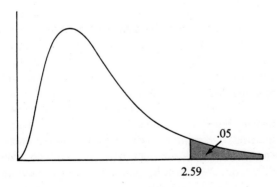

.05

2.59

Values of F

**Figure
10.4**   Form of the F distribution for $df_1 = 9$ and $df_2 = 15$. Adapted from Kirk (1982, p. 54), with
permission.

of the F table required Snedecor to set up sampling distributions of ratios of variances for all combinations of degrees of freedom in the table—a monumental task in precomputer times. The existence of the F table permits experimenters to simply compare the F value obtained in an experiment with those listed in the table at the appropriate degrees of freedom to determine significance at the .05 and .01 levels. *If the obtained F value is as large or larger than the tabled value, the null hypothesis can be rejected.* If the F value is not that large, the null hypothesis may not be rejected.

The F distribution and the t distribution are closely related, both mathematically and conceptually. The mathematical relation is

$$t^2 = F$$

for a two-group experiment. Theoretically, ANOVA made the t test obsolete; but t continues to be widely used for comparing two groups.

This section completes our explanation of the rationale of ANOVA. Soon you will learn how to actually compute an F value and interpret it. First, however, work the following problems and then learn the terms that are used in ANOVA.

Problems

  2. Tell what the two designs below have in common and why the ANOVA technique described in this chapter is *not* appropriate.

    a. An experimenter studied the effect of 0, 2, 4, and 8 ounces of alcohol on hand steadiness. Nine subjects participated for four days, each taking a steadiness test after consuming 0, 2, 4, or 8 ounces on one day. Thus, the mean for each level of the independent variable was based on the same nine subjects.

    b. An experimenter set up a design to study the effect of three types of propaganda on opinion change. In order to make sure the family background was equivalent for the three groups, the subjects were sets of three siblings from the same family. One sibling was assigned to each of the three groups.

  3. If three means come from a population with $\mu = 100$ and a fourth mean from a population with $\mu = 50$, will the between-means estimate of variability be larger or smaller than the between-means estimate of variability of four means drawn from a population with $\mu = 100$?

  4. How can estimates of a population variance be obtained?

  5. Define F.

  6. Interpret the meaning of the F values given below for the experiments described in Examples 1–4 on page 206.

    a. a very large F value for Example 1

    b. a very small F value for Example 2

    c. a very large F value for Example 3

    d. a very small F value for Example 4

MORE NEW TERMS

Sum of squares. In the "Clue to the Future" on page 75, we pointed out that, in the computation of the standard deviation, certain values were obtained that would be important in future chapters. That future is now. The term $\sum x^2$ (the numerator of the basic formula for the standard deviation) is called the **sum of squares** (abbreviated *SS*). So, $SS = \sum x^2 = \sum (X - \bar{X})^2.$ A more descriptive name for sum of squares is "sum of the squared deviations."

Mean square.   **Mean square** (*MS*) is the ANOVA term for a variance s^2. The mean square is a sum of squares divided by its degrees of freedom.

Grand mean.   The **grand mean** is the mean of all scores; it is computed without regard for the fact that the scores come from different groups (samples).

tot.   The subscript *tot* after a symbol makes the symbol stand for all such numbers in the experiment; for example, $\sum X_{tot}$ is the sum of all scores.

g.   The subscript *g* after a symbol means that the symbol applies to a group; for example, $\sum (\sum X_g)^2$ tells you to sum the scores in each group, square each sum, and then sum these squared values.

K.   *K* is the number of groups in the experiment. This is the same as the number of levels of the independent variable.

SUMS OF SQUARES

Analysis of variance is based on the fact that the variability of all the scores in an experiment can be attributed to two or more sources. In the case of simple analysis of variance, just two sources contribute all the variability to the scores. One source is the variability between groups, and the other source is variability within each group. The sum of these two sources is equal to the total variability.

We will illustrate this, using the sum of squares as a measure of variability. Calculations will be made on the data in Table 10.1, which shows fictional data from an experiment in which there were three groups of mental patients. The independent variable is drug therapy. Three levels of the independent variable were used—Drug A, Drug B, and Drug C. The dependent variable is the number of psychotic episodes observed for each patient during the therapy period. The experimenter believed that the use of Drug C would reduce the number of psychotic episodes more than would Drug A or B, which were in common use.

This experimenter's belief is an alternative hypothesis (sometimes called a private hypothesis or experimenter's hypothesis), but it is not the alternative hypothesis being tested in ANOVA. The alternative hypothesis in ANOVA is that there are one or more differences among the population means. No direction of differences is specified. The null hypothesis is that $\mu_1 = \mu_2 = \mu_3$ or, in the terms of the experiment, that any differences among the psychotic episodes of the three groups are due to sampling fluctuation and not to differences caused by the three drugs.

First, we'll focus on the total variability as measured by the total sum of squares. Actually, as you will see, you are already familiar with the total sum

Table 10.1   **Computation of Sums of Squares of Fictional Data from a Drug Study**

	Drug A (1)		Drug B (2)		Drug C (3)	
	X_1	X_1^2	X_2	X_2^2	X_3	X_3^2
	9	81	9	81	4	16
	8	64	7	49	3	9
	7	49	6	36	1	1
	5	25	5	25	1	1
Σ	29	219	27	191	9	27
$\bar{X}$	7.25		6.75		2.25	

$\Sigma X_{tot} = 29 + 27 + 9 = 65$

$\Sigma X_{tot}^2 = 219 + 191 + 27 = 437$

$\bar{X}_{tot} = 5.42.$

$$SS_{tot} = \Sigma X_{tot}^2 - \frac{(\Sigma X_{tot})^2}{N_{tot}} = 437 - \frac{65^2}{12} = 84.92.$$

$$SS_{drugs} = \Sigma \left[\frac{(\Sigma X_g)^2}{N_g} \right] - \frac{(\Sigma X_{tot})^2}{N_{tot}}$$

$$= \frac{29^2}{4} + \frac{27^2}{4} + \frac{9^2}{4} - \frac{65^2}{12}$$

$$= 210.25 + 182.25 + 20.25 - 352.08 = 60.67.$$

$$SS_{wg} = \Sigma \left[\Sigma X_g^2 - \frac{(\Sigma X_g)^2}{N_g} \right] = \left(219 - \frac{29^2}{4} \right) + \left(191 - \frac{27^2}{4} \right) + \left(27 - \frac{9^2}{4} \right)$$

$$= 8.75 + 8.75 + 6.75 = 24.25.$$

Check: $SS_{tot} = SS_{bg} + SS_{wg} = 84.92 = 60.67 + 24.25.$

of squares (SS_{tot}). To find SS_{tot}, subtract the grand mean from each score. Square these deviation scores and sum them up.

$$SS_{tot} = \Sigma (X - \bar{X}_{tot})^2.$$

Computationally, SS_{tot} is more readily (and accurately) obtained using the raw-score formula

$$SS_{tot} = \Sigma X_{tot}^2 - \frac{(\Sigma X_{tot})^2}{N_{tot}},$$

which you may recognize as the numerator of the raw-score formula for s. This formula is equivalent to Σx^2. Its computation, as illustrated in Table 10.1, requires you to square each score and sum the squared values to obtain ΣX_{tot}^2. Next, the scores are summed and the sum squared. That squared value is divided by the total number of scores to obtain $(\Sigma X_{tot})^2 / N_{tot}$. Subtraction of $(\Sigma X_{tot})^2 / N_{tot}$ from ΣX_{tot}^2 yields the total sum of squares. For the data in Table 10.1,

$$SS_{tot} = \Sigma X_{tot}^2 - \frac{(\Sigma X_{tot})^2}{N_{tot}} = 437 - \frac{65^2}{12} = 84.92.$$

Thus, the total variability of all the scores in Table 10.1 is 84.92 when measured by the sum of squares. This total comes from two sources: the between-groups sum of squares and the within-groups sum of squares. Each of these can be computed separately.

The *between-groups sum of squares*, SS_{bg}, is the variability of the group means from the grand mean of the experiment, weighted by the size of the group:

$$SS_{bg} = \sum [N_g(\bar{X}_g - \bar{X}_{tot})^2].$$

SS_{bg} is more easily computed by the raw-score formula:

$$SS_{bg} = \sum \left[\frac{(\sum X_g)^2}{N_g} \right] - \frac{(\sum X_{tot})^2}{N_{tot}}.$$

This formula tells you to sum the scores for each group and then square the sum. Each squared sum is then divided by the number of scores in that group. These values (one for each group) are then summed, giving you $\sum [(\sum X_g)^2 / N_g]$. From this sum is subtracted the value $(\sum X_{tot})^2 / N_{tot}$, which was obtained in the computation of SS_{tot}.

When describing experiments in general, the term SS_{bg} is used. In a specific analysis, *between groups* is changed to a summary word for the independent variable. For the experiment in Table 10.1, that word is *drugs*. Thus,

$$SS_{drugs} = \sum \left[\frac{(\sum X_g)^2}{N_g} \right] - \frac{(\sum X_{tot})^2}{N_{tot}}$$

$$= \frac{29^2}{4} + \frac{27^2}{4} + \frac{9^2}{4} - \frac{65^2}{12} = 60.67.$$

Finally, we will focus on the *within-groups sum of squares* (SS_{wg}), which is the sum of the variability in each of the groups. SS_{wg} is defined as

$$SS_{wg} = \sum (X_1 - \bar{X}_1)^2 + \sum (X_2 - \bar{X}_2)^2 + \cdots + \sum (X_K - \bar{X}_K)^2$$

or the sum of the squared deviations of each score from the mean of its group added to the sum of the squared deviations from all other groups for the experiment. As with the other SS's, there is an arrangement of the arithmetic that is easiest. For SS_{wg},

$$SS_{wg} = \sum \left[\sum X_g^2 - \frac{(\sum X_g)^2}{N_g} \right].$$

This formula tells you to square each score in a group and sum them $(\sum X_g^2)$. Subtract from this a value that you obtain by summing the scores, squaring the sum, and dividing by the number of scores in the group: $(\sum X_g)^2 / N_g$. For each group, a value is calculated, and these values are summed to get SS_{wg}. For the data in Table 10.1,

$$SS_{wg} = \sum \left[\sum X_g{}^2 - \frac{(\sum X_g)^2}{N_g} \right]$$

$$= \left(219 - \frac{29^2}{4} \right) + \left(191 - \frac{27^2}{4} \right) + \left(27 - \frac{9^2}{4} \right) = 24.25.$$

We mentioned that, if you work with SS as a measure of variability, the total variability is the sum of the variability of the parts. Thus,

$$SS_{tot} = SS_{bg} + SS_{wg}.$$

$$84.92 = 60.67 + 24.25.$$

Error Detection

$SS_{tot} = SS_{bg} + SS_{wg}$. All sums of squares are always zero or positive, never negative. This check will not catch errors made in summing the scores ($\sum X$) or in summing squared scores ($\sum X^2$).

Problems

7. Show that $SS_{tot} = SS_{bg} + SS_{wg}$ for the data below.

X_1	X_2	X_3	X_4
9	2	5	7
5	6	5	6
3	7	9	7
6	8	2	3
5	3	4	2

8. In this problem, the N's per group are not equal. Find SS_{tot}, SS_{bg}, and SS_{wg}.

X_1	X_2	X_3
8	9	4
9	5	7
6	1	7
	3	2
	6	

9. Thirty undergraduate students enrolled in an introductory statistics course were randomly assigned to three groups. Group 1 students were taught by the traditional lecture method. Group 2 students were also taught by the lecture method but, in addition, had a two-hour laboratory each week in which they worked on statistics problems. Group 3 students used a book designed for individualized teaching. They attended no lectures but worked problems on their own and took short tests when they felt they were ready. At the end of the term, all students took the same comprehensive statistics test. The number of errors made by each student is recorded below. Compute SS_{tot}, SS_{bg}, and SS_{wg}.

Group 1 Lecture	Group 2 Lecture and Lab	Group 3 Individualized
19	13	12
18	12	11
17	10	11
15	10	10
15	9	9
14	9	8
12	8	7
11	6	5
9	6	5
7	5	4

MEAN SQUARES AND DEGREES OF FREEDOM

The next step in an analysis of variance is to find the mean squares. A mean square is simply a sum of squares divided by its degrees of freedom. It is an estimate of the population variance, σ^2.

Each sum of squares has a particular number of degrees of freedom associated with it. Thus, in a one-way classification, the df are df_{tot}, df_{bg}, and df_{wg}. Also, like SS,

$$df_{tot} = df_{bg} + df_{wg}.$$

The df_{tot} is $N_{tot} - 1$. The df_{bg} is the number of groups minus one ($K - 1$). The df_{wg} is the sum of the degrees of freedom for each group $[(N_1 - 1) + (N_2 - 1) + \cdots + (N_K - 1)]$. If there are equal numbers of scores in the K groups, the formula for df_{wg} reduces to $K(N_g - 1)$. A little algebra will reduce this still further.

$$df_{wg} = K(N_g - 1) = KN_g - K.$$

But $KN_g = N_{tot}$,

so $df_{wg} = N_{tot} - K.$

This formula for df_{wg} works whether the number in each group is the same or not.

Error Detection

$df_{tot} = df_{bg} + df_{wg}$. Degrees of freedom are always positive.

Mean squares, then, can be found using the following formulas.[5]

$$MS_{bg} = \frac{SS_{bg}}{df_{bg}}, \qquad \text{where } df_{bg} = K - 1.$$

$$MS_{wg} = \frac{SS_{wg}}{df_{wg}}, \qquad \text{where } df_{wg} = N_{tot} - K.$$

For the data in Table 10.1,

$$df_{drugs} = K - 1 = 3 - 1 = 2.$$
$$df_{wg} = N_{tot} - K = 12 - 3 = 9.$$

Check: $df_{tot} = df_{bg} + df_{wg} = 2 + 9 = 11.$

$$MS_{drugs} = \frac{SS_{drugs}}{df_{bg}} = \frac{60.67}{2} = 30.34.$$

$$MS_{wg} = \frac{SS_{wg}}{df_{wg}} = \frac{24.25}{9} = 2.69.$$

Notice that, although $SS_{tot} = SS_{bg} + SS_{wg}$ and $df_{tot} = df_{bg} + df_{wg}$, mean squares are *not* additive.

CALCULATION AND INTERPRETATION OF F VALUES USING THE F DISTRIBUTION

We said earlier that F is a ratio of two estimates of the population variance. MS_{bg} is an estimate based on the variability between means. MS_{wg} is an estimate based on the sample variances. An **F test** consists of dividing MS_{bg} by MS_{wg} to obtain an F value.

$$F = \frac{MS_{bg}}{MS_{wg}},$$

and the expected value of F is about 1.00 when H_0 is true. Of course, sampling error will produce some F's greater than 1.00 and some less than 1.00.

There are two different degrees of freedom associated with any F value. They are the df associated with $MS_{bg}(K - 1)$ and the df associated with $MS_{wg}(N_{tot} - K)$. For the data in Table 10.1,

$$F = \frac{MS_{drugs}}{MS_{wg}} = \frac{30.34}{2.69} = 11.26; \qquad df = 2, 9.$$

---

[5] MS_{tot} is not used in ANOVA; only MS_{bg} and MS_{wg} are calculated.

The next question is "What is the probability of obtaining an $F = 11.26$ if all three samples come from populations with the same mean?" As before, if the probability is less than α, reject the null hypothesis.

Turn now to Table F in the Appendix, which gives the critical values of F when $\alpha = .05$ and when $\alpha = .01$. Across the top of the table are degrees of freedom associated with the numerator or MS_{bg}. For Table 10.1, $df_{bg} = 2$, so 2 is the column you want. Along the left side of the table are degrees of freedom associated with the denominator or MS_{wg}. In this case, $df_{wg} = 9$, so look for 9 along the side. The tabled value for 2 and 9 df is 4.26 at the .05 level (lightface type) and 8.02 at the .01 level (boldface type). Our obtained F is 11.26. Therefore, if the three samples were drawn from populations with the same mean (the null hypothesis), an $F = 11.26$ would occur less than 1 percent of the time. Thus, reject the null hypothesis and conclude that the three samples do not have a common population mean. At least two drugs had different effects on the number of psychotic episodes.

At this point, your interpretation must stop. An ANOVA does not tell you which of the population means is greater than or less than the others. Such an interpretation requires a further statistical analysis, which is the topic of the last part of this chapter.

It is customary to summarize the results of an ANOVA in a *summary table*. Table 10.2 is an example. The values in the table are those we have calculated. Look at the right side of the table under p (for probability). The notation "$p < .01$" is shorthand for "the probability of an $F = 11.26$ or greater occurring as a result of chance fluctuations is less than one in a hundred."

Table 10.2

Summary Table of ANOVA for Data in Table 10.1

Source	df	SS	MS	F	p
Between drugs	2	60.67	30.34	11.26	$<.01$
Within groups	9	24.25	2.69		
Total	11	84.92			

$F_{.01}(2,9\ df) = 8.02$.

Sometimes Table F does not contain an F value for the df in your problem. For example, an F with 2,35 df or 4,90 df is not tabled. When this happens, be conservative; use the F value that is given for *fewer df* than you have. Thus, the proper F values for those two examples would be based on 2,34 df and 4,80 df, respectively.[6]

[6]The F distribution may also be used to test hypotheses about variances rather than hypotheses about means. To determine the probability that two sample variances came from the same population (or from populations with equal variances), form a ratio with the larger sample variance in the numerator. The resulting F value can be interpreted with Table F. The proper df are $N_1 - 1$ and $N_2 - 1$ for the numerator and denominator, respectively. For more information, see Ferguson (1981, p. 189).

A LEARNING EXPERIMENT

Thus far, we have demonstrated the ANOVA technique using a small hypothetical study. At this point, we will present a four-group experiment with its analysis. Table 10.3 shows the data and the ANOVA. Read the following explanation of the study carefully.

To learn any new task or skill, you must know whether your responses are correct or incorrect. Progress in the task depends on this knowledge of results (KR). In the following experiment, a psychologist was interested in two things: (1) the effect of delaying KR after the response and (2) the effect of requiring the subject to make an irrelevant response before receiving KR.

Sixty subjects were randomly assigned making four groups of 15. The task was to learn which of two nonsense syllables was the "correct" answer to a stimulus syllable. The correct syllable was chosen arbitrarily by the experimenter, so, on the first trial, the subject was just guessing. Twelve such problems were presented to each subject. Four treatment conditions from the study are illustrated here. Participants in Group 1 received immediate KR after each response. In Group 2, participants had to wait five seconds before receiving KR. Group 3 participants waited ten seconds for their KR. Participants in Group 4 waited four seconds and then viewed an irrelevant syllable for two seconds, after which there was another four-second delay before KR was provided. The number of trials required to reach a criterion of two perfect performances in a row was the dependent variable.

The complete ANOVA is presented at the bottom of Table 10.3. A summary table is presented as Table 10.4 that shows an F value of 3.73. Using the tabled F value for 3,55 df (2.78), we conclude that an $F = 3.73$ has a probability less than .05.

"So what?" you may justifiably ask. "What does this $F = 3.73$, which is significant beyond the .05 level, tell me about the effect of knowledge of results on learning?" It tells you this: somewhere among the four treatment means there are significant differences. That is, one or more of the means came from a population with a different mean. It does not tell you which means differ. To discover this requires more sleuthing through the data. Methods for carrying out this detective work are covered in the next section; first, here are a few problems.

Problems

10. Calculate the F value for the data in Problem 7 and compose a summary table. Determine the critical value of F and retain or reject the null hypothesis. Tell what the analysis has shown.
11. Same as Problem 10 but use the data in Problem 8.
12. Same as Problem 10 but use the data in Problem 9.
13. Suppose you obtained an $F = 2.56$ with 5,75 df. How many groups are in the experiment? What are the critical values for the .05 and .01 levels? What conclusion should be reached?

Table 10.3   **Analysis of Variance of a Knowledge-of-Results Study**

Group 1		Group 2		Group 3		Group 4	
X_1	X_1^2	X_2	X_2^2	X_3	X_3^2	X_4	X_4^2
10	100	11	121	15	225	15	225
9	81	8	64	9	81	15	225
9	81	6	36	8	64	14	196
6	36	6	36	7	49	12	144
6	36	6	36	7	49	12	144
6	36	5	25	6	36	10	100
6	36	5	25	6	36	8	64
5	25	5	25	6	36	7	49
5	25	5	25	5	25	7	49
4	16	4	16	5	25	6	36
4	16	4	16	5	25	5	25
3	9	4	16	5	25	5	25
3	9	4	16	4	16	4	16
3	9	4	16	4	16	4	16
3	9	2	4	1	1	4	16

$\sum X_1 = 82.$ $\sum X_2 = 79.$ $\sum X_3 = 93.$ $\sum X_4 = 128.$

$\sum X_1^2 = 524.$ $\sum X_2^2 = 477.$ $\sum X_3^2 = 709.$ $\sum X_4^2 = 1330.$

$\bar{X}_1 = 5.4667.$ $\bar{X}_2 = 5.2667.$ $\bar{X}_3 = 6.20.$ $\bar{X}_4 = 8.5333.$

$\sum X_{tot} = 82 + 79 + 93 + 128 = 382.$

$\sum X_{tot}^2 = 524 + 477 + 709 + 1330 = 3040.$

$$SS_{tot} = \sum X_{tot}^2 - \frac{(\sum X_{tot})^2}{N_{tot}} = 3040 - \frac{(382)^2}{60} = 607.93.$$

$$SS_{conditions} = \sum \left[\frac{(\sum X_g)^2}{N_g} \right] - \frac{(\sum X_{tot})^2}{N_{tot}} = \frac{(82)^2}{15} + \frac{(79)^2}{15} + \frac{(93)^2}{15} + \frac{(128)^2}{15} - \frac{(382)^2}{60}$$

$$= 101.13.$$

$$SS_{wg} = \sum \left[\sum X_g^2 - \frac{(\sum X_g)^2}{N_g} \right]$$

$$= \left[524 - \frac{(82)^2}{15} \right] + \left[477 - \frac{(79)^2}{15} \right] + \left[709 - \frac{(93)^2}{15} \right] + \left[1330 - \frac{(128)^2}{15} \right]$$

$$= 506.80.$$

$$SS_{wg} = SS_{tot} - SS_{bg} = 607.93 - 101.13 = 506.80.$$

$$MS_{conditions} = \frac{SS_{conditions}}{K - 1} = \frac{101.13}{3} = 33.71.$$

$$MS_{wg} = \frac{SS_{wg}}{N_{tot} - K} = \frac{506.80}{56} = 9.05.$$

$$F = \frac{MS_{conditions}}{MS_{wg}} = \frac{33.71}{9.05} = 3.73. \qquad df = 3,56.$$

COMPARISONS AMONG MEANS

The information obtained thus far in our knowledge of results study doesn't answer our original question. Our significant F value has told us only that the

Table
10.4

Summary Table of the ANOVA Analysis of the KR Study

Source	df	SS	MS	F	p
Between conditions	3	101.13	33.71	3.73	<.05
Within groups	56	506.80	9.05		
Total	59	607.93			

method of providing KR does, indeed, have an effect on the number of trials required to reach the criterion. We still don't know which methods are significantly different. To discover this, we must make further comparisons among the means.

The problem of making several comparisons following an F test has been troublesome for statisticians. There are several solutions, each with its advantages and disadvantages. Roger E. Kirk (1982), Geoffery Keppel (1982) and B. F. Winer (1971) have excellent summaries of several of these methods. We will present two methods here. Our first step is to make a distinction between two circumstances that experimenters find themselves in.

In a study with, say, four levels of the independent variable, there are many specific comparisons that an experimenter might make (Group 1 vs. Group 3; Group 2 vs. the mean of Groups 3 and 4, and so forth). Nearly always, the experimenter is especially interested in one or two of these outcomes (the first circumstance). Tests for the significance of these results are planned *in advance of* the data collection. Such comparisons are called *planned* or *a priori* (ah pre òre ee) comparisons. In the KR study, for example, the experimenter expected that the participants who viewed an irrelevant syllable before receiving KR would require more learning trials than participants who did not. A test for the significance of this difference was planned when the experiment was first designed. One restriction of *a priori* methods is that they limit you to just a few comparisons.

After the experiment has been completed and the means have been calculated, experimenters usually indulge in an activity called "data snooping" (the second circumstance). When you look at the differences between means, you find some that are large. Although you hadn't thought of it before, such differences are meaningful, so you would like to test the significance of these differences. Such comparisons are called *post hoc*, or *a posteriori* (ah post 'tear ee óre ee) comparisons. Since the comparisons were not planned in advance, but were selected from many possible comparisons only because the mean differences were large, the probability is high that the differences are the result of chance. (Remember the example on p. 206 about multiple t tests.) For this reason, *a posteriori* tests must pay the penalty of a much more conservative test of significance.

In this section we will explain two methods of comparing means after an ANOVA. The first is an *a priori* method called *orthogonal comparisons.* The second is an *a posteriori* method called the *Scheffé* test.

We chose these two methods because:

1. They are among those favored by leading statisticians.
2. They can be used when N's are unequal.
3. If there is a more advanced statistics course in your future, you are nearly certain to encounter them there.

Orthogonal Comparisons, An *A Priori* Test

The **orthogonal comparisons method** permits you to make a few preselected comparisons among the means. The number and kind of comparisons are limited. These comparisons result in a t value that is interpreted using the t distribution you used in Chapter 9.

The basic idea is to partition the total variability among the means into independent (uncorrelated) components. Such uncorrelated components of variability are called orthogonal components. Once the variability has been partitioned in this way, comparisons (sometimes called contrasts) can be made among them. These comparisons will test for the significance of the differences, which was the initial goal. The number of orthogonal comparisons that can be made is $K - 1$.

Table 10.5 will be used to illustrate how to determine orthogonality for any three-group experiment. The body of this table consists of *coefficients*, which are weights that are assigned to the means. If a mean is not part of a comparison, its weight is zero. The other coefficients could be any numbers as long as they satisfy the two requirements presented next. We selected the smallest integers possible as coefficients in Table 10.5 and Table 10.6.

Table 10.5   **Examples of Orthogonal and Nonorthogonal Coefficients for any Three-Group Experiment**

		Comparison	Group 1	Group 2	Group 3	Σ
	Set A	1 vs. 2	1	-1	0	0
		3 vs. 1 and 2	-1	-1	2	0
		Products	-1	1	0	0
Orthogonal sets	Set B	2 vs. 3	0	1	-1	0
		1 vs. 2 and 3	2	-1	-1	0
		Products	0	-1	1	0
	Set C	1 vs. 3	1	0	-1	0
		2 vs. 1 and 3	-1	2	-1	0
		Products	-1	0	1	0
Nonorthogonal set	Set D	1 vs. 2	1	-1	0	0
		2 vs. 3	0	1	-1	0
		Products	0	-1	0	-1

Set A in Table 10.5 shows two comparisons that can be made. The first, 1 vs. 2, tests the null hypothesis $H_0: \mu_1 - \mu_2 = 0$. The second, 3 vs. 1 and 2, tests the null hypothesis $H_0: \mu_3 - \dfrac{\mu_1 + \mu_2}{2} = 0$; that is, the difference between the mean of Group 3 and the mean of Groups 1 and 2 is zero.

Two requirements must be met for comparisons to be orthogonal.

1. For each comparison the sum of the coefficients is zero. Thus, in Set A, $(1) + (-1) + (0) = 0$ and $(-1) + (-1) + (2) = 0$.

2. The sum of the products of the coefficients is zero. Thus, in Set A, $(1)(-1) + (-1)(-1) + (0)(2) = 0$.

Therefore, the two comparisons in Set A are orthogonal and might be used in an *a priori* test, subsequent to ANOVA.

Sets B and C in Table 10.5 show two other ways to analyze data from a three-group experiment. You could choose A, B, or C, depending on which set made the most sense for your particular experiment. Note that you can use only one of the three possible sets.

Set D in Table 10.5 is an example of coefficients that are *not* orthogonal. Notice that the sum of the products of corresponding coefficients does not equal zero.

$$(1)(0) + (-1)(1) + (0)(-1) = -1.$$

This nonzero sum of products means that these two comparisons involve overlapping information and are, therefore, not independent. They should not be used as *a priori* comparisons. You could choose to do either one of the comparisons in Set D, but not both.

Table 10.6 gives examples of orthogonal coefficients that may be used in any four-group experiment (of which our KR study is one example). In four-group experiments, $K - 1 = 3$. Three *a priori* comparisons, then, can be made in four-group experiments. The orthogonality requirement for three or more comparisons is referred to as *mutual* orthogonality. This means that every comparison must be orthogonal with every other comparison. Look at Set A of Table 10.6.

Table 10.6   **Examples of Orthogonal Coefficients for Comparisons in any Four-Group Experiment**

	Comparison	1	2	Group 3	4	Σ
Set A	1 vs. 2	1	−1	0	0	0
	3 vs. 4	0	0	1	−1	0
	1 and 2 vs. 3 and 4	1	1	−1	−1	0
Set B	1 vs. 3	1	0	−1	0	0
	2 vs. 4	0	1	0	−1	0
	1 and 3 vs. 2 and 4	1	−1	1	−1	0
Set C	1 vs. 2, 3, and 4	3	−1	−1	−1	0
	3 vs. 4	0	0	1	−1	0
	2 vs. 3 and 4	0	2	−1	−1	0

When the coefficients for the first two comparisons are multiplied and summed across the four groups, the sum is zero.

$$(1)(0) + (-1)(0) + (0)(1) + (0)(-1) = 0.$$

For the first and third comparisons,

$$(1)(1) + (-1)(1) + (0)(-1) + (0)(-1) = 0.$$

For the second and third comparisons,

$$(0)(1) + (0)(1) + (1)(-1) + (-1)(-1) = 0.$$

These three comparisons, then, are mutually orthogonal.

Further information about orthogonal comparisons may be found in Kirk (1982), p. 95; Winer (1971), p. 172; and Edwards (1972), p. 136.

Problems

14. Show the orthogonality of Sets B and C of Table 10.5.

15. Show the mutual orthogonality of Sets B and C of Table 10.6.

Now we can take up where we left off on the KR study and answer the "so what?" question. Remember that there were four groups, all treated differently, and that the overall F was significant beyond the .05 level. We know, then, that some group or groups differed from others by more than could be expected by chance. The question to be answered now is "Which ones differ from which?"

The experimenter hypothesized that a short time delay would not disrupt learning. He also hypothesized that having to make an irrelevant response (view an irrelevant syllable) before receipt of KR *would* disrupt learning. To test these hypotheses, two comparisons were planned before the experiment was carried out.

Groups 1 and 2 were treated alike, except that Group 1 received immediate KR and Group 2 had a 5-second delay between the response and KR. A test of the effect of time delay could be made by testing the null hypothesis

$$H_0: \mu_1 - \mu_2 = 0.$$

A test of the effect of the irrelevant response could be made by comparing the means of Groups 3 and 4, since the time delay was practically the same for both groups, but Group 4 saw an irrelevant syllable while Group 3 did not. The null hypothesis for this test is

$$H_0: \mu_3 - \mu_4 = 0.$$

Remember that $K - 1$ planned orthogonal comparisons can be made. Since four treatments were used in the KR experiment, $K - 1 = 3$. Three mutually orthogonal comparisons could be made; however, the only other comparison that could be made with the two comparisons selected and still satisfy the mutual orthogonality requirement is a comparison of the average of means 1 and 2 with the average of means 3 and 4. (Prove this for yourself.) In this case, such a comparison would be meaningless. Therefore, only two planned comparisons were made.

Table 10.7 shows the formulas and computation for carrying out these comparisons. At the top of Table 10.7 are the orthogonal coefficients for the two

Table 10.7

Planned Orthogonal Comparisons for the Knowledge of Results Experiment

	Orthogonal Coefficients			
		Group		
Hypothesis	1	2	3	4
$\mu_1 - \mu_2 = 0$	1	-1	0	0
$\mu_3 - \mu_4 = 0$	0	0	1	-1

Hypothesis	Computations

$H_0: \mu_1 - \mu_2 = 0$

$$t = \frac{c_1 \bar{X}_1 + c_2 \bar{X}_2}{\sqrt{MS_{wg} \left[\frac{(c_1)^2}{N_1} + \frac{(c_2)^2}{N_2} \right]}}$$

$$= \frac{(1)(5.4667) + (-1)(5.2667)}{\sqrt{9.05 \left[\frac{(1)^2}{15} + \frac{(-1)^2}{15} \right]}}$$

$$= \frac{.20}{1.0985} = <1.00$$

$H_0: \mu_3 - \mu_4 = 0$

$$t = \frac{(1)(6.20) + (-1)(8.5333)}{\sqrt{9.05 \left[\frac{(1)^2}{15} + \frac{(-1)^2}{15} \right]}}$$

$$= \frac{-2.3333}{1.0985} = -2.1241$$

$\bar{X}_1 = 5.4667 \qquad \bar{X}_2 = 5.2667 \qquad \bar{X}_3 = 6.20 \qquad \bar{X}_4 = 8.5333$

$MS_{wg} = 9.05$

hypotheses. They are presented again to illustrate the orthogonality of the comparisons.[7]

Problem

16. Show that the coefficients in Table 10.7 are orthogonal.

The general formula for making orthogonal comparisons is

$$t = \frac{c_1 \bar{X}_1 + c_2 \bar{X}_2 + \cdots + c_k \bar{X}_k}{\sqrt{MS_{wg} \left[\frac{(c_1)^2}{N_1} + \frac{(c_2)^2}{N_2} + \cdots + \frac{(c_k)^2}{N_k} \right]}}$$

[7]Though the comparisons between means have to be uncorrelated, the t tests do not. Since the same MS_{wg} is used in the denominator of all the t tests, they are, in fact, correlated.

where: c_1, c_2, and c_k are the coefficients assigned to the means,

$\bar{X}_1$, $\bar{X}_2$, and $\bar{X}_k$ are the means to be compared,

MS_{wg} is the obtained value from the ANOVA,

and N_1, N_2, and N_k are the N's for the groups to be compared.

For the hypothesis $\mu_1 - \mu_2 = 0$, the t value is less than one and is therefore not significant. The experimenter's hypothesis that a short time delay would not affect learning is supported. The t value for the hypothesis $\mu_3 - \mu_4 = 0$ is -2.1211. To evaluate this t value, go back to the familiar t table you used in Chapter 9 (Table D). Degrees of freedom for this test are $N - K = 60 - 4 = 56$. The critical value from Table D at the .05 level is between 2.000 (for 60 df) and 2.021 (for 40 df). Our obtained t value, then, is significant beyond the .05 level. Since the mean of those who saw the irrelevant syllable was larger (8.53) and since more trials to criterion means poorer performance, you may conclude that the experimenter's hypothesis is supported: having to view the irrelevant syllable did disrupt learning.

The Scheffé Test, An *A Posteriori* Test

The method we will present for making *a posteriori* comparisons was devised by Scheffé (1953). The **Scheffé Test** allows you to make all possible comparisons among K groups. You can compare each group with every other, and you can compare each group with the mean of two or more groups. You can even compare a mean of two or more groups with the mean of two or more other groups. You can make all these comparisons and still be sure your α level is not above .05. In fact, this Scheffé test has been criticized as being too conservative. That is, it errs in the direction of too many Type II errors.

Keep in mind that this test is appropriate *only* if the overall F test produced a rejection of the null hypothesis.

In the Scheffé test, two statistics called F' are computed and compared. F'_{ob} is based on the data (observed) and F'_α is a critical value computed from a value found in the F table.

If F'_{ob} is larger than F'_α, the null hypothesis is rejected. To find the critical values,

$$F'_{.05} = (K - 1)F_{.05}$$

$$F'_{.01} = (K - 1)F_{.01}$$

where K = the number of groups in the original ANOVA

and $F_{.05}$ and $F_{.01}$ = critical values for F for the original ANOVA.

Now, back to the KR study. Remember that you are data snooping now. You have looked at the data, and by eyeballing the means, you see some differences that you suspect may be significant and meaningful. This process is also referred to as "milking the data."

Look again at the means in Table 10.3. The 5-second delay in KR ($\bar{X}_2$) resulted in a mean almost identical to that for no delay ($\bar{X}_1$). However, the 10-second delay resulted in a higher mean. Might the difference between means

1 and 3 be significant, indicating that a delay of 10 seconds retarded learning? It's worth finding out. To do so, test the hypothesis that

$$H_0: \mu_1 - \mu_3 = 0.$$

The formula for this test uses orthogonal coefficients, but the only requirement is that they sum to zero for each comparison.

$$F'_{ob} = \frac{(c_1 \bar{X}_1 + c_3 \bar{X}_3)^2}{MS_{wg}\left[\dfrac{(c_1)^2}{N_1} + \dfrac{(c_3)^2}{N_3}\right]}$$

$$= \frac{[(1)(5.4667) + (-1)(6.20)]^2}{9.05\left[\dfrac{(1)^2}{15} + \dfrac{(-1)^2}{15}\right]} = \frac{.5377}{1.2067} = <1.00.$$

F values less than 1.00 are never significant; therefore you may conclude that a 10-second delay is not significantly different from no delay. This supports the experimenter's hypothesis that short-time delays between the response and KR do not affect learning.

Let's have another look at Table 10.3. Since the first three treatments seem not to differ, a good test of the effect of the irrelevant syllable would be a comparison of the average of results of the first three treatments with the results of the fourth treatment. You can test the null hypothesis

$$H_0: \frac{\mu_1 + \mu_2 + \mu_3}{3} - \mu_4 = 0.$$

For this test, the formula becomes

$$F'_{ob} = \frac{(c_1 \bar{X}_1 + c_2 \bar{X}_2 + c_3 \bar{X}_3 + c_4 \bar{X}_4)^2}{MS_{wg}\left[\dfrac{(c_1)^2}{N_1} + \dfrac{(c_2)^2}{N_2} + \dfrac{(c_3)^2}{N_3} + \dfrac{(c_4)^2}{N_4}\right]}$$

$$= \frac{[(-1)(5.4667) + (-1)(5.2667) + (-1)(6.20) + (3)(8.5333)]^2}{9.05\left[\dfrac{(-1)^2}{15} + \dfrac{(-1)^2}{15} + \dfrac{(-1)^2}{15} + \dfrac{(3)^2}{15}\right]}$$

$$= \frac{75.1082}{7.24} = 10.3741.$$

Let's check the significance of this F'_{ob} value by calculating a critical value. The F values from Table F are the same as those used in the ANOVA—those based on 3 and 56 df. $K - 1 = 4 - 1 = 3$. Thus,

$$F'_{.05} = (2.78)(3) = 8.34$$

$$F'_{.01} = (4.16)(3) = 12.48.$$

Our $F_{ob} = 10.37$ is larger than $F'_{.05}$, so you can reject the null hypothesis with some confidence. Since the mean of those who viewed the irrelevant syllable (8.53) is larger than that of those who did not (5.64, the mean of the three means), you may conclude that the irrelevant syllable does foul up the learning process.

Problems

17. In Problem 9, how many planned comparisons are legitimate?
18. Assume that the experimenter for Problem 9 planned the following comparisons before data collection

    1 vs. 2
    1 and 2 vs. 3

    a. Show that these comparisons are orthogonal.
    b. State the null hypothesis for these comparisons.
    c. Test the hypotheses. (Pertinent data are repeated below.)

    $$\bar{X}_1 = 13.70 \qquad \bar{X}_2 = 8.80 \qquad \bar{X}_3 = 8.20$$
    $$MS_{wg} = 10.12 \qquad N = 10 \text{ per group}$$

    d. Interpret your results.
    e. Test the *a posteriori* hypothesis that $H_0: \mu_2 - \mu_3 = 0$. Interpret the results.
19. The ANOVA for the drug study (Table 10.1) yielded an F value of 11.26, $p < .01$. The conclusion was that the drugs differed in their effects on the reduction of psychotic episodes. Some summary data from that study are given below.

    Drug A: $\bar{X} = 7.25$ $MS_{wg} = 2.69$
    Drug B: $\bar{X} = 6.75$ $N = 4$ per group
    Drug C: $\bar{X} = 2.25$

    Remember that the experimenter believed that Drug C would be more effective than Drugs A and B.
    a. What is the maximum number of *a priori* comparisons the experimenter can make?
    b. Choose the comparisons the experimenter would likely have planned for this study before seeing the data and state the null hypotheses for them.
    c. Show that those comparisons meet orthogonality requirements.
    d. Make the *a priori* tests for the comparisons you chose and interpret the results.
    e. Make an *a posteriori* test of the hypothesis $H_0: \mu_B - \mu_C = 0$.

ASSUMPTIONS OF THE ANALYSIS OF VARIANCE

For the analysis of variance and the F test to be appropriate, three characteristics of the data must be assumed to be true. To the extent that the data fail to meet these requirements, conclusions from the analysis will be subject to doubt.

1. **Normality.** It is assumed that the dependent variable is normally distributed in the populations from which samples are drawn. It is often difficult or impossible to demonstrate normality or lack of normality in the parent popula-

tions. Such a demonstration usually occurs only with very large samples. On the other hand, because of extensive research, some populations are known to be skewed, and researchers in those fields may decide that ANOVA is not appropriate for their data analysis. Unless there is a reason to suspect that populations depart severely from normality, the inferences made from the F test will probably not be affected. ANOVA is "robust." (It results in correct probabilities even when the populations are not exactly normal.) Where there is suspicion of a severe departure from normality, however, use the nonparametric method explained in Chapter 13.

2. **Homogeneity of variance.** This means that the two or more population variances are equal. In ANOVA, the variances of the dependent-variable scores for each of the populations sampled are assumed to be equal. Figures 10.1 and 10.2, which we used to illustrate the rationale of ANOVA, show populations with equal variances. Several methods for testing this assumption are presented in advanced texts, such as Kirk (1982), Keppel (1982), and Winer (1971). An alternative is to use a nonparametric method for comparing all pairs. This will be discussed in Chapter 13.

3. **Random sampling.** Every care should be taken to assure that sampling is random and that assignment to groups is also random, so that the measurements are all independent of one another.

We hope these assumptions have a familiar ring to you. They are the same as those you learned for the t distribution. This makes sense; t is a special case of F. For a reminder of how to avoid violating Assumption 3, see Problem 29 in Chapter 9.

Problems   Here is the final set of problems on one-way ANOVA. Careful work here will facilitate your understanding of the more complex experiments in the next chapter.

20. List the three assumptions of the analysis of variance.

21. For the data in Problem 8, use a Scheffé test to compare Group 1 with the mean of Groups 2 and 3.

22. A researcher interested in the effects of stimulants on memory gave three groups of rats 20 trials of training to avoid a shock. Immediately after learning, one group of rats was injected with a small amount of strychnine (a stimulant), a second group was injected with saline as a control for the injection, and a third group was not injected. Seven days later, the researcher tested the rats' memory by training them until they reached a criterion of avoiding the shock five trials in a row. The number of trials to reach this criterion was recorded for each rat. (See McGaugh & Petrinovich, 1965.)

a. Perform an ANOVA with $\alpha = .05$ and write a sentence summary.

b. Compare the strychnine group with the mean of the other two groups (a comparison planned by the experimenter). Set $\alpha = .05$, and write a sentence summary.

	Strychnine	Saline	No Injection
$\sum X$	121	138	152
$\sum X^2$	1608	2464	3000
N	10	8	8

23. In a followup study, the researcher trained rats for 20 trials and then gave injections of strychnine immediately, 30 minutes, one hour, or two hours later. Again, the rats were retrained seven days later.

a. Perform an ANOVA.

b. Inspect the means, perform a Scheffé test on the most interesting difference you find, and write a conclusion.

	Delay in Strychnine (Minutes)			
	0	30	60	120
$\sum X$	128	131	168	171
$\sum X^2$	1838	1820	2932	3124
N	10	10	10	10

24. An agricultural experimenter wanted to know which of three varieties of soybeans would produce the highest yield in a particular type of soil. He divided a field with that type of soil into 15 plots of equal size, then randomly assigned the three varieties to five plots each. The dependent variable was the number of bushels harvested from each plot.

Variety of Soybean		
X_1	X_2	X_3
9	13	11
7	12	10
7	12	9
6	10	9
6	9	8

Set $\alpha = .05$ and perform an ANOVA. Construct a summary table. Test for a significant difference between varieties 1 and 2 and between varieties 2 and 3 with Scheffé tests. Write a summary of the results.

11
ANALYSIS OF VARIANCE: FACTORIAL DESIGN

Objectives for Chapter 11: After studying the text and working the problems in this chapter, you should be able to:

1. define the terms *factor*, *level*, and *cell*,
2. identify the sources of variance in a factorial design,
3. compute sums of squares in a factorial design,
4. compute mean squares in a factorial design,
5. compute F values and test their significance in a factorial design,
6. interpret main effects and interactions,
7. make planned comparisons among means with an orthogonal comparisons test, and
8. make comparisons among means with a Scheffé test.

Chapter 10 provided you with the basic concepts and procedures for carrying out a simple analysis of variance. In this chapter, we will extend those concepts and procedures to a somewhat more complex (and more useful) design.

In this chapter, we will use the term *factor*. **Factor** is just another word for independent variable. The different treatments of the independent variable are called **levels** of the factor. Thus, an experiment might use the factor food deprivation, with the levels 12, 24, and 36 hours, or the factor professional training, with the levels of lawyer, teacher, theologian, and physician. In Chapter 10, you analyzed data from an experiment in which three methods of teaching statistics were compared. The factor was teaching method, and there were three levels.

In Chapter 9, you learned to analyze data from a two-group experiment, schematically shown in Table 11.1. In Table 11.1, there is *one independent variable*, Factor A, with data from two levels, A_1 and A_2. For the *t* test, the samples could be independent or correlated.

Table 11.1   **Illustration of a Two-Group Design That Can Be Analyzed with a *t* Test**

Factor A	
A_1	A_2
Scores on the Dependent Variable	Scores on the Dependent Variable

In Chapter 10, you learned to analyze data from a two-or-more-group experiment, schematically shown in Table 11.2 for a four-group design. In Table 11.2, there is *one independent variable*, Factor A, with data from four levels, A_1, A_2, A_3, and A_4. In the ANOVA of Chapter 10, the samples for the levels were independent of one another.

Table 11.2   **Illustration of a Four-Group Design That Can Be Analyzed with an *F* Test**

Factor A			
A_1	A_2	A_3	A_4
Scores on the Dependent Variable	Scores on the Dependent Variable	Scores on the Dependent Variable	Scores on the Dependent Variable

FACTORIAL DESIGN AND INTERACTION

In this chapter, you will learn to analyze data from a design in which there are *two independent variables* (factors), each of which may have two or more levels. Table 11.3 illustrates an example of this design with one factor (Factor

Table 11.3   **Illustration of a 2 × 3 Factorial Design That Can Be Analyzed with F Tests**

		Factor A		
		A_1	A_2	A_3
Factor B	B_1	$A_1 B_1$ Scores on the Dependent Variable	$A_2 B_1$ Scores on the Dependent Variable	$A_3 B_1$ Scores on the Dependent Variable
	B_2	$A_1 B_2$ Scores on the Dependent Variable	$A_2 B_2$ Scores on the Dependent Variable	$A_3 B_2$ Scores on the Dependent Variable

A) having three levels (A_1, A_2, and A_3) and another factor (Factor B) having two levels (B_1 and B_2). Such a design is called a **factorial design**.[1]

Factorial designs are identified with a shorthand notation such as "2 × 3" or "3 × 5." The general term is $R \times C$ (Rows × Columns). The first number tells you the number of levels of one factor; the second number tells you the number of levels of the other factor. The design in Table 11.3 is a 2 × 3 design. Assignment of a factor to a row or column is arbitrary; we could just as well have made Table 11.3 a 3 × 2 table.

In Table 11.3, there are six cells. Each **cell** represents a different way to treat subjects. A subject in the upper left cell is given treatment A_1 *and* treatment B_1. That cell is, therefore, identified as Cell $A_1 B_1$. Subjects in the lower right cell are given treatment A_3 and treatment B_2, and that cell is called Cell $A_3 B_2$.

A factorial ANOVA can be compared to two separate one-way ANOVAs.[2] The principal advantage of a factorial ANOVA is that it provides a test of the interaction of the two independent variables. Look again at Table 11.3. A factorial ANOVA will help you decide whether treatments A_1, A_2, and A_3 produced significantly different scores (a one-way ANOVA with three groups). It will also help you to decide whether treatments B_1 and B_2 produced significantly different scores (a second one-way ANOVA). The interaction test helps you to decide whether the difference in scores between treatments B_1 and B_2 is dependent upon which level of A is being administered.

Perhaps a couple of examples of interactions will help at this point. Suppose a group of friends were sitting in a dormitory lounge one Monday discussing the weather of the previous weekend. What would you need to know to predict each person's rating of the weather? The first thing you probably want to know is what the weather was actually like. A second important variable is the activity each person had planned for the weekend. For purposes of this little illustration, suppose that weather comes in one of two varieties, snow or no snow and that our subjects could plan only one of two activities, camping or skiing. Now we

---

[1]A factorial design is one that has two or more independent variables. In this chapter, you will learn to analyze a two-factor design. Intermediate- and advanced-level textbooks discuss the analysis of three-or-more-factor designs. See Kirk (1982), Keppel (1982), Edwards (1972, Chapter 12), or Winer (1971).

[2]See Kirk (1982) pp. 422–423 for both the advantages and disadvantages of a factorial ANOVA.

have the ingredients for an interaction. We have two independent variables (weather and plans) and a dependent variable (rating of the weather).

If plans called for camping, "no snow" is rated good, but if plans called for skiing, "no snow" is rated bad. To complete the possibilities, campers rated snow bad and skiers rated it good. Table 11.4 summarizes this paragraph. Study it before going on.

Table 11.4

Illustration of a Significant Interaction—Good and Bad Refer to Ratings of the Weather

		Kind of Weather	
		Snow	No Snow
Plans for Weekend	Camping	Bad	Good
	Skiing	Good	Bad

Here is a similar example in which there is *no* interaction. Suppose you wanted to know how people would rate the weather, which again could be snow or no snow. This time, however, the people are divided into camping enthusiasts and rock-climbing enthusiasts. For both groups, snow would rate as bad weather. You might make up a version of Table 11.4 that describes this second example. It will help you follow our summary explanation below.

An interaction between two independent variables exists when the results found for one independent variable *depend on* which level of the other independent variable you are looking at. Thus, for Table 11.4, the rating of the variable weather (snow or no snow) depends on whether you plan to camp or ski. For campers, a change from snow to no snow brings joy; for skiers, the same change brings unhappiness.

In our second example there is no interaction. The rating of the weather does not depend on a person's plans for the weekend. A change from snow to no snow brings joy to the hearts of both groups. You can see this in the table you constructed.

MAIN EFFECTS AND INTERACTION

In a factorial ANOVA, the comparison of the levels of Factor A is called a **main effect**. Likewise, the comparison of the levels of factor B is a **main effect**. The extent to which scores on Factor A depend on Factor B is the **interaction**. Thus, comparisons for main effects are like one-way ANOVAs, and information about the interaction is a bonus that comes with the factorial design.

Table 11.5 shows a 2×3 factorial ANOVA. The numbers in each of the six cells are cell means. Examine those six means and see for yourself what effect changing from A_1 to A_2 to A_3 has. Make the same examination for B_1 to B_2. (Incidentally, this kind of preliminary examination is very valuable for any set of data.)

Table
11.5

A 2 × 3 Factorial Design (The number in each cell represents the mean for all the subjects in the cell.)

		Factor A			
		A_1	A_2	A_3	Factor B Means
Factor B	B_1	10	20	60	30
	B_2	50	60	100	70
	Factor A Means	30	40	80	Grand Mean 50

We will use Table 11.5 to show you what goes on in a factorial ANOVA's analysis of the main effects and the interaction. Look at the comparison between the mean of B_1 (30) and the mean of B_2 (70). A factorial ANOVA will give the probability that the two means came from populations with identical means. Thus, with this probability you can decide whether the null hypothesis,

$$H_0: \mu_{B_1} = \mu_{B_2}$$

should be rejected or not. A factorial ANOVA will also give you the probability of getting means of 30 (A_1), 40 (A_2), and 80 (A_3) from populations with identical means, if chance only is at work. Thus, with the same factorial ANOVA you can decide whether the null hypothesis,

$$H_0: \mu_{A_1} = \mu_{A_2} = \mu_{A_3}$$

should be rejected or not.

Notice that a comparison of B_1 and B_2 satisfies the requirements of an experiment (p. 154) in that, except for getting either B_1 or B_2, the groups were treated alike. That is, Group B_1 and Group B_2 are alike in that both groups have equal numbers of subjects who received A_1. Whatever the effect is of receiving A_1, it will occur as much in the B_1 group as it does in the B_2 group; the only way B_1 and B_2 differ is in their levels of Factor B. It would be worthwhile to you to apply this same line of reasoning to the question of whether comparisons of A_1, A_2, and A_3 satisfy the requirement for an experiment.

In Table 11.5, there is *no* interaction. The effect of changing from level A_1 to A_2 is to increase the mean score by 10 points. This is true at *both* level B_1 and level B_2. The effect of changing from A_2 to A_3 is to increase the mean score by 40 points at both B_1 and B_2. The same constancy is found in the columns; the effect of changing from B_1 to B_2 is to increase the mean score 40 points, and this is true at *all three levels* of A. There is no interaction; the effect of changing from B_1 to B_2 is to increase the score 40 points *regardless* of the level of A.

It is common to display an interaction (or lack of one) with a graph. There is a good reason for this. A graph is the best way of arriving at a clear interpretation of an interaction. We urge you to always draw a graph of the interaction on the factorial problems you work. Figure 11.1 graphs the data in Table 11.5. The result

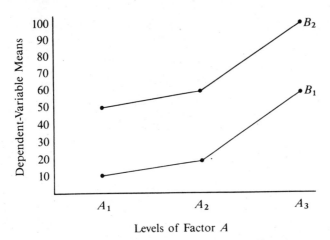

Figure 11.1   Graphic representation of data in Table 11.5; no interaction exists.

is two parallel curves. Parallel curves mean that there is no interaction between two factors.

   We can also graph the data in Table 11.5 with each curve representing a level of A. Figure 11.2 is the result. Again, the parallel lines indicate that there is no interaction.

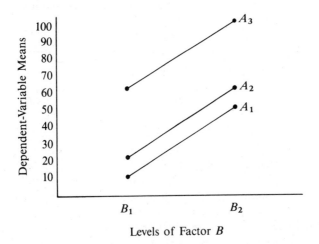

Figure 11.2   Graphic representation of data in Table 11.5; no interaction exists.

   Table 11.6 shows a 2×3 factorial design in which there *is* an interaction between the two independent variables. The main effect of Factor A is indicated by the overall means along the bottom. The average effect of a change from A_1 to A_2 to A_3 is to *reduce* the mean score by 10 (main effect). But look at the cells. For B_1, the effect of changing from A_1 to A_2 to A_3 is to *increase* the mean score by 10 points. For B_2, the effect is to *decrease* the score by 30 points. These data illustrate an interaction because the effect of one factor *depends* upon which level of the other factor you administer.

We will describe this interaction in Table 11.6 another way. Look at the difference between Cell A_1B_1 and Cell A_2B_1—a difference of -10. If there were no interaction, we would predict this same difference (-10) for the difference between Cells A_1B_2 and A_2B_2. But this latter difference is $+30$; it is in the opposite direction. Something about B_2 reverses the effect of changing from A_1 to A_2 that was found under the condition B_1.[3]

Table 11.6   A 2 × 3 Factorial Design with an Interaction between Factors (The numbers represent the means of all scores within each cell.)

		Factor A			
		A_1	A_2	A_3	Factor B Means
Factor B	B_1	10	20	30	20
	B_2	100	70	40	70
	Factor A Means	55	45	35	Grand Mean 45

This interaction is illustrated graphically in Figure 11.3. You can see that B_1 increases across the levels of Factor A but B_2 decreases. The lines for the two levels of B are not parallel.

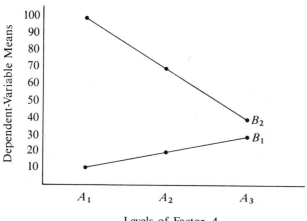

Figure 11.3   Graphic representation of the interaction of Factors A and B from Table 11.6.

[3]Students often have trouble with the concept of interaction. Usually, having the same idea presented in different words facilitates understanding. Two good references are Roger E. Kirk, *Introductory Statistics*, Monterey: Brooks/Cole, 1978, pp. 325–328; and G. A. Ferguson, *Statistical Analysis in Psychology* (5th ed.), New York: McGraw-Hill, 1981, p. 256.

Finally, we will graph in Figure 11.4 the example of rating the weather by skiers and campers. Again, the lines are not parallel.

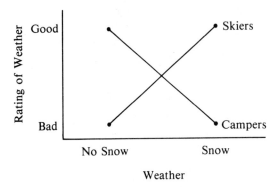

Figure 11.4   Graphic representation of the "rating of the weather" example indicating an interaction. Whether snow or no snow is rated highest depends on whether the ratings came from skiers or campers.

RESTRICTIONS AND LIMITATIONS

We have tried to emphasize throughout this book the limitations that go with each statistical test you learn. For the factorial analysis of variance presented in this chapter, the restrictions are the same as those given on page 228 for a one-way ANOVA plus the following:

1. The number of scores in each cell must be equal. For techniques dealing with unequal N's, see Kirk (1982), Ferguson (1981), or Winer (1971).

2. The cells must be independent. This is usually accomplished by randomly assigning a subject to only one of the cells. This restriction means that these techniques should not be used with any type of correlated-samples design. For factorial designs that use correlated samples, see Keppel (1982) or Winer (1971).

3. The levels of both factors are chosen by the experimenter. The alternative is that the levels of one or both factors be chosen at random from several possible levels of the factor. The techniques of this chapter are used when the levels are *fixed* by the experimenter and not chosen randomly. For a discussion of fixed and random models of ANOVA, see Winer (1971) or Ferguson (1981).

Problems

1. Identify the following factorial designs, using $R \times C$ notation.
   a. Three methods of teaching Spanish were evaluated using either a grade report or a credit/no credit report.
   b. Four strains of mice were injected with three types of virus.
   c. Males and females worked on a complex problem at either 7 A.M. or 7 P.M.
   d. This one is a little different. Four different species of pine were grown with three different kinds of fertilizer in three different soil types.

2. What is a main effect?

3. What is an interaction?

4. In a *t* test, are you testing for a main effect or for an interaction?
5. In a one-way ANOVA, are you testing for a main effect or for an interaction?
6. The following five tables present cell means for some ANOVAs. For each of them, do the following.
   a. Graph the means.
   b. Decide whether an interaction appears to be present. In these problems, you may state that there appears to be *no* interaction or that there appears to be one. Interactions, like main effects, are subject to sampling variations.
   c. Write a statement of explanation of the interaction or lack of one.
   d. Decide whether any main effects appear to be present.
   e. Write a statement of explanation of the main effects or lack of them.

I. The dependent variable for this ANOVA is attitude scores toward playing games of house or baseball.

Factor A (*Sex*)

		A_1 (Boys)	A_2 (Girls)
Factor B (*Games*)	B_1 (House)	50	75
	B_2 (Baseball)	75	50

II. This is a study of the effect of dosage of a new drug on three types of schizophrenics. The dependent variable is scores on a "happiness" test.

A (*Diagnosis*)

		A_1 (Depressive)	A_2 (Paranoid)	A_3 (Hebephrenic)
B (*Dose*)	B_1 (small)	5	10	70
	B_2 (medium)	20	25	85
	B_3 (large)	35	40	100

III. The dependent variable in this one is the number of cars sold by General Motors in each of the three cities.

A (*Cities*)

		Atlanta	Omaha	Tulsa
B (*Air Conditioning*)	B_1 (without)	15	60	30
	B_2 (with)	75	120	90

IV. In this problem, the dependent variable is the number of months a piece of "indoor-outdoor" carpeting lasted.

A (*Brands*)

		A_1 (Brand X)	A_2 (Brand Y)	A_3 (Brand Z)
B (*Location*)	B_1 (outdoors)	10	20	30
	B_2 (indoors)	5	35	65

V. Here, the dependent variable is attitude toward lowering the taxes on profits from investments.

A (*Socioeconomic Status*)

		A_1 (low)	A_2 (middle)	A_3 (high)
B	B_1 (males)	10	35	60
(*Sex*)	B_2 (females)	15	35	55

7. Name the six requirements that must be met if a factorial ANOVA is to be analyzed using techniques outlined in this chapter.

A SIMPLE EXAMPLE OF A FACTORIAL DESIGN

As you read the following story, try to pick out the two factors and to identify the levels of each factor.

Two groups of hunters, six squirrel hunters and six quail hunters, met in a bar. An argument soon began over the marksmanship required for the two kinds of hunting.

"Squirrel hunters are just better shots," barked a biased squirrel hunter.

"Poot, poot, and balderdash!" heartily swore a logic-oriented quail hunter. "It takes a lot better eye to hit a moving bird than it does to hit a still squirrel."

"Hold it a minute, you guys," demanded an empirically minded squirrel hunter. "We can settle this easily enough on our hunting-club target range. We'll just see if you six quail hunters can hit the target as often as we can."

"O.K.," agreed a quail hunter. "What kind to trap throwers do you have out there?"

"What kind of what? Oh, you mean those gadgets that throw clay pigeons into the air? Gee, yeah, there are some out there, but we never use them."

"Well, if you want to shoot against us, you will have to use them this time," the quail hunter insisted. "It's one thing to hit a still target, but hitting a target flying through the air above you is something else. We'll target shoot against you guys, but let's do it the fair way. Three of us and three of you will shoot at still targets and the other six will shoot at clay pigeons."

"Fair, enough," the squirrel hunters agreed; and all 12 men took up their shotguns and headed for the target range.

This yarn establishes conditions for a 2×2 factorial ANOVA with three scores per cell. The cell means and computation of the sums of squares are illustrated in Table 11.7. The dependent variable is the number of hits made by each man out of 15 shots. Factor A is the type of target and it has two levels: A_1 for still targets and A_2 for moving targets. Factor B is the type of hunter and it also has two levels: B_1 for squirrel hunters and B_2 for quail hunters. Before going to the next section of the table, which shows calculations, simply *look* at the data in Table 11.7. Compare the mean number of hits for moving ($\bar{X}_{A_2}$) and

Table 11.7

Cell Means and Computation of Sums of Squares for the Hunters' Contest

		Targets (Factor A)			
		Still (A_1)		Moving (A_2)	
		X	X^2	X	X^2
Hunters (Factor B)	Squirrel B_1	12	144	8	64
		11	121	7	49
		10	100	7	49

$$\Sigma X_g = 33. \qquad \Sigma X_g = 22. \qquad \Sigma X_{B_1} = 55.$$
$$\Sigma X_g^2 = 365. \qquad \Sigma X_g^2 = 162. \qquad \Sigma X_{B_1}^2 = 527.$$
$$\bar{X}_g = 11.00. \qquad \bar{X}_g = 7.3333. \qquad \bar{X}_{B_1} = 9.1667.$$

		Still (A_1)		Moving (A_2)	
	Quail B_2	12	144	10	100
		10	100	9	81
		9	81	8	64

$$\Sigma X_g = 31. \qquad \Sigma X_g = 27. \qquad \Sigma X_{B_2} = 58.$$
$$\Sigma X_g^2 = 325. \qquad \Sigma X_g^2 = 245. \qquad \Sigma X_{B_2}^2 = 570.$$
$$\bar{X}_g = 10.3333. \qquad \bar{X}_g = 9.00. \qquad \cdot\bar{X}_{B_2} = 9.6667.$$

$$\Sigma X_{A_1} = 64. \qquad \Sigma X_{A_2} = 49. \qquad \Sigma X_{tot} = 113.$$
$$\Sigma X_{A_1}^2 = 690. \qquad \Sigma X_{A_2}^2 = 407. \qquad \Sigma X_{tot}^2 = 1097.$$
$$\bar{X}_{A_1} = 10.6667. \qquad \bar{X}_{A_2} = 8.1667. \qquad \bar{X}_{tot} = 9.4167.$$

$$SS_{tot} = \Sigma X_{tot}^2 - \frac{(\Sigma X_{tot})^2}{N_{tot}} = 1097 - \frac{(113)^2}{12} = 1097 - 1064.0833 = 32.9167.$$

$$SS_{bg} = \Sigma \left[\frac{(\Sigma X_g)^2}{N_g} \right] - \frac{(\Sigma X_{tot})^2}{N_{tot}} = \frac{(33)^2}{3} + \frac{(22)^2}{3} + \frac{(31)^2}{3} + \frac{(27)^2}{3} - \frac{(113)^2}{12}$$
$$= 363 + 161.3333 + 320.3333 + 243 - 1064.0833 = 23.5833.$$

$$SS_{targets} = \frac{(\Sigma X_{A_1})^2}{N_{A_1}} + \frac{(\Sigma X_{A_2})^2}{N_{A_2}} - \frac{(\Sigma X_{tot})^2}{N_{tot}} = \frac{(64)^2}{6} + \frac{(49)^2}{6} - \frac{(113)^2}{12}$$
$$= 682.6667 + 400.1667 - 1064.0833 = 1082.8334 - 1064.0833 = 18.7501.$$

$$SS_{hunters} = \frac{(\Sigma X_{B_1})^2}{N_{B_1}} + \frac{(\Sigma X_{B_2})^2}{N_{B_2}} - \frac{(\Sigma X_{tot})^2}{N_{tot}} = \frac{(55)^2}{6} + \frac{(58)^2}{6} - \frac{(113)^2}{12}$$
$$= 504.1667 + 560.6667 - 1064.0833 = 1064.8334 - 1064.0833 = .7501$$

$$SS_{AB} = N_g[(\bar{X}_{A_1B_1} - \bar{X}_{A_1} - \bar{X}_{B_1} + \bar{X}_{tot})^2 + (\bar{X}_{A_2B_1} - \bar{X}_{A_2} - \bar{X}_{B_1} + \bar{X}_{tot})^2$$
$$+ (\bar{X}_{A_1B_2} - \bar{X}_{A_1} - \bar{X}_{B_2} + \bar{X}_{tot})^2 + (\bar{X}_{A_2B_2} - \bar{X}_{A_2} - \bar{X}_{B_2} + \bar{X}_{tot})^2]$$
$$= 3[(11.00 - 10.6667 - 9.1667 + 9.4167)^2$$
$$+ (7.3333 - 8.1667 - 9.1667 + 9.4167)^2$$
$$+ (10.3333 - 10.6667 - 9.6667 + 9.4167)^2$$
$$+ (9.00 - 8.1667 - 9.6667 + 9.4167)^2]$$
$$= 4.0836.$$

Check: $SS_{AB} = SS_{bg} - SS_A - SS_B = 23.5833 - 18.7501 - .7501 = 4.0831.$

Table 11.7 cont.

$$SS_{wg} = \Sigma \left[\Sigma X_g{}^2 - \frac{(\Sigma X_g)^2}{N_g} \right]$$

$$= \left[365 - \frac{(33)^2}{3} \right] + \left[162 - \frac{(22)^2}{3} \right] + \left[325 - \frac{(31)^2}{3} \right] + \left[245 - \frac{(27)^2}{3} \right]$$

$$= (365 - 363) + (162 - 161.3333) + (325 - 320.3333) + (245 - 243)$$

$$= 2 + .6667 + 4.6667 + 2 = 9.3334.$$

Check: $SS_{tot} = SS_{bg} + SS_{wg}.$

$$32.9167 = 23.5833 + 9.3334.$$

---

still $(\bar{X}_{A_1})$ targets. Compare the mean hits for the quail hunters $(\bar{X}_{B_2})$ with that for the squirrel hunters $(\bar{X}_{B_1})$. Look at the cell means. Does there appear to be an interaction? This kind of preliminary inspection of the data will improve your understanding of a study and will help you catch any gross computational errors. For example, in Table 11.7, the difference between targets is larger than the difference between hunters. This will be reflected in a larger F value for Factor A than for Factor B.

Sources of Variance and Sums of Squares

Remember that in Chapter 10 you identified three sources of variance in the one-way analysis of variance. They were: (1) the total variance, (2) the between-groups variance, and (3) the within-groups variance. In a factorial design with two factors, the same sources of variance can be identified. However, the between-groups variance may now be partitioned into three components. These are the two *main effects* and the *interaction*. Thus, of the variability among the means of the four groups in Table 11.7, some can be attributed to the A main effect, some to the B main effect, and the rest to the interaction.

Total sum of squares.   This calculation will be easy for you, since it is the same as SS_{tot} in the one-way analysis. It is defined as $\Sigma (X - \bar{X}_{tot})^2$ or the sum of the squared deviations of all the scores in the experiment from the grand mean of the experiment. To actually compute SS_{tot}, use the formula

$$SS_{tot} = \Sigma X_{tot}{}^2 - \frac{(\Sigma X_{tot})^2}{N_{tot}}.$$

For the hunters' contest,

$$SS_{tot} = 1097 - \frac{(113)^2}{12} = 32.9167.$$

Between-groups sum of squares.[4]   In order to find the main effects and interaction, you must first find the between-groups variability, and then partition it into

---

[4]Some texts call this the *between-cells sum of squares*.

its components parts. As in the one-way design, SS_{bg} is defined $\sum [N_g(\bar{X}_g - \bar{X}_{tot})^2]$. A "group" in this context is a group of participants treated alike; therefore, for example, squirrel hunters shooting at still targets constitute a group. In other words, a group is composed of those scores in the same cell.

The computational formula for SS_{bg} is the same as that for one-way analysis.

$$SS_{bg} = \sum \left[\frac{(\sum X_g)^2}{N_g} \right] - \frac{(\sum X_{tot})^2}{N_{tot}}.$$

For the hunters' contest,

$$SS_{bg} = \frac{(33)^2}{3} + \frac{(22)^2}{3} + \frac{(31)^2}{3} + \frac{(27)^2}{3} - \frac{(113)^2}{12} = 23.5833.$$

After SS_{bg} is obtained, it is partitioned into its three components: the A main effect, the B main effect, and the interaction.

The sum of squares for each main effect is somewhat like a one-way ANOVA. The sum of squares for Factor A ignores the existence of Factor B and considers the deviations of the Factor A means from the grand mean. Thus,

$$SS_A = N_{A_1}(\bar{X}_{A_1} - \bar{X}_{tot})^2 + N_{A_2}(\bar{X}_{A_2} - \bar{X}_{tot})^2,$$

where N_{A_1} = the total number of scores in the A_1 cells.

For Factor B,

$$SS_B = N_{B_1}(\bar{X}_{B_1} - \bar{X}_{tot})^2 + N_{B_2}(\bar{X}_{B_2} - \bar{X}_{tot})^2.$$

Computational formulas for the main effects also look like formulas for SS_{bg} in a one-way design.

$$SS_A = \frac{(\sum X_{A_1})^2}{N_{A_1}} + \frac{(\sum X_{A_2})^2}{N_{A_2}} - \frac{(\sum X_{tot})^2}{N_{tot}}.$$

And for the hunters' contest,

$$SS_{targets} = \frac{(64)^2}{6} + \frac{(49)^2}{6} - \frac{(113)^2}{12} = 18.7501.$$

The computational formula for the B main effect simply substitutes B for A in the previous formula. Thus,

$$SS_B = \frac{(\sum X_{B_1})^2}{N_{B_1}} + \frac{(\sum X_{B_2})^2}{N_{B_2}} - \frac{(\sum X_{tot})^2}{N_{tot}}.$$

For the hunters' contest,

$$SS_{hunters} = \frac{(55)^2}{6} + \frac{(58)^2}{6} - \frac{(113)^2}{12} = .7501.$$

To find the sum of squares for the interaction, use this formula.

$$SS_{AB} = N_g[(\bar{X}_{A_1B_1} - \bar{X}_{A_1} - \bar{X}_{B_1} + \bar{X}_{tot})^2$$
$$+ (\bar{X}_{A_2B_1} - \bar{X}_{A_2} - \bar{X}_{B_1} + \bar{X}_{tot})^2$$
$$+ (\bar{X}_{A_1B_2} - \bar{X}_{A_1} - \bar{X}_{B_2} + \bar{X}_{tot})^2$$
$$+ (\bar{X}_{A_2B_2} - \bar{X}_{A_2} - \bar{X}_{B_2} + \bar{X}_{tot})^2].$$

For the hunters' contest,

$$SS_{AB} = 3[(11.00 - 10.6667 - 9.16667 + 9.4167)^2$$
$$+ (7.3333 - 8.1667 - 9.1667 + 9.4167)^2$$
$$+ (10.3333 - 10.6667 - 9.6667 + 9.4167)^2$$
$$+ (9.00 - 8.1667 - 9.6667 + 9.4167)^2]$$
$$= 4.0836.$$

Since SS_{bg} contains only the components SS_A, SS_B, and the interaction SS_{AB}, we can also obtain SS_{AB} by subtraction. This serves as your check.

$$SS_{AB} = SS_{bg} - SS_A - SS_B.$$

For the hunters' contest,

$$SS_{AB} = 23.5833 - 18.7501 - .7501 = 4.0831.$$

Within-groups sum of squares.   As in the one-way analysis, the within-groups variability is due to the fact that subjects treated alike differ from one another on the dependent variable. Since all were treated the same, this difference must be due to uncontrolled variables and is sometimes called **error variance** or the error term. SS_{wg} for a 2×2 design is defined as

$$SS_{wg} = \sum (X_{A_1B_1} - \bar{X}_{A_1B_1})^2 + \sum (X_{A_1B_2} - \bar{X}_{A_1B_2})^2$$
$$+ \sum (X_{A_2B_1} - \bar{X}_{A_2B_1})^2 + \sum (X_{A_2B_2} - \bar{X}_{A_2B_2})^2.$$

In words, SS_{wg} is the sum of the sums of squares for each *cell* in the experiment. The computational formula is

$$SS_{wg} = \sum \left[\sum X_g^2 - \frac{(\sum X_g)^2}{N_g} \right].$$

For the hunters' contest,

$$SS_{wg} = \left[365 - \frac{(33)^2}{3} \right] + \left[162 - \frac{(22)^2}{3} \right]$$
$$+ \left[325 - \frac{(31)^2}{3} \right] + \left[245 - \frac{(27)^2}{3} \right] = 9.3333.$$

The computational check for the hunters' contest is

$$32.9166 = 23.5833 + 9.3333.$$

We will interrupt the analysis of the hunters' contest so that you may practice what you have learned about the sums of squares in a factorial ANOVA.

Error Detection

The computational check for a factorial ANOVA is the same as that for the one-way classification: $SS_{tot} = SS_{bg} + SS_{wg}$. As before, this check will not catch errors in $\sum X$ or $\sum X^2$.

Problems

8. The dependent variable in this hypothetical study was the number of pounds gained during the 20 weeks following weaning. The subjects were male and female puppies fed three different diets.

	Diets (Factor A)		
Sex (Factor B)	A_1	A_2	A_3
Males (B_1)	7	4	9
	6	4	8
	4	3	7
Females (B_2)	6	7	5
	4	5	4
	4	5	3

a. Identify the design, using $R \times C$ notation.
b. Name the independent variables.
c. Calculate SS_{tot}, SS_{bg}, SS_{diets}, SS_{sex}, SS_{AB}, and SS_{wg}.

9. An educational psychologist was interested in the effect that three kinds of teacher response had on children's final achievement in arithmetic. This psychologist was also interested in whether girls or boys were better in arithmetic. A third interest was whether the kind of response by the teacher affected girls and boys differently. Three classrooms of children, each classroom containing ten randomly selected boys and ten randomly selected girls, were used in the experiment. In one classroom, the teacher's response was one of "neglect." The children were not even observed during the time they were working. In a second classroom, the teacher's response was "reproof." The children were observed and errors were corrected and criticized. In the third classroom, children were observed and "praised" for correct answers. Incorrect answers were ignored. The numbers that follow are the numbers of errors made on a comprehensive examination of arithmetic achievement.

a. Identify the design, using $R \times C$ notation.
b. Name the independent variables.
c. Name the dependent variable.
d. Calculate SS_{tot}, SS_{bg}, $SS_{response}$, SS_{sex}, SS_{AB}, and SS_{wg}.

Teacher Response

Sex	Neglect (A_1)	Reproof (A_2)	Praise (A_3)
Boys (B_1)	25	20	18
	22	17	17
	20	17	16
	19	16	14
	19	15	14
	18	15	13
	16	13	10
	13	10	7
	10	8	5
	7	5	4
Girls (B_2)	20	23	16
	18	22	14
	17	20	13
	16	18	13
	16	17	12
	15	16	11
	13	16	9
	12	14	9
	10	11	7
	9	10	5

10. Given the following summary statistics, calculate SS_{tot}, SS_{bg}, SS_A, SS_B, SS_{AB}, and SS_{wg}.

	A_1	A_2
B_1	$\sum X = 25.$ $\sum X^2 = 155.$ $N = 5.$	$\sum X = 57.$ $\sum X^2 = 750.$ $N = 5.$
B_2	$\sum X = 43.$ $\sum X^2 = 400.$ $N = 5.$	$\sum X = 20.$ $\sum X^2 = 100.$ $N = 5.$

11. Examine the following summary statistics and explain why the techniques of this chapter cannot be used for this factorial design.

	A_1	A_2
B_1	$\sum X = 405.$ $\sum X^2 = 8503.$ $N = 20.$	$\sum X = 614.$ $\sum X^2 = 15540.$ $N = 25.$
B_2	$\sum X = 585.$ $\sum X^2 = 12060.$ $N = 30.$	$\sum X = 246.$ $\sum X^2 = 3325.$ $N = 20.$

Degrees of Freedom, Mean Squares, and F tests

Now that you are skilled at calculating sums of squares, we can proceed with the rest of the analysis of the hunters' contest. Mean squares, as before, are found by dividing the sums of squares by their appropriate degrees of freedom. Degrees of freedom for the sources of variance are

In general:

$df_{tot} = N_{tot} - 1.$

$df_A = A - 1.$

$df_B = B - 1.$

$df_{AB} = (A - 1)(B - 1).$

$df_{wg} = N_{tot} - (A)(B).$

For the hunters' contest:

$df_{tot} = 12 - 1 = 11.$

$df_{targets} = 2 - 1 = 1.$

$df_{hunters} = 2 - 1 = 1.$

$df_{AB} = (1)(1) = 1.$

$df_{wg} = 12 - (2)(2) = 8.$

In the equations above, A and B stand for the number of levels of Factor A and Factor B, respectively.

Error Detection

$$df_{tot} = df_A + df_B + df_{AB} + df_{wg}.$$

$$\text{Also, } df_{bg} = df_A + df_B + df_{AB}.$$

Mean squares for the hunters' contest are

$$MS_{targets} = \frac{SS_{targets}}{df_{targets}} = \frac{18.7501}{1} = 18.7501.$$

$$MS_{hunters} = \frac{SS_{hunters}}{df_{hunters}} = \frac{.7501}{1} = .7501.$$

$$MS_{AB} = \frac{SS_{AB}}{df_{AB}} = \frac{4.0831}{1} = 4.0831.$$

$$MS_{wg} = \frac{SS_{wg}}{df_{wg}} = \frac{9.3333}{8} = 1.1667.$$

F is computed, as usual, by dividing each mean square by MS_{wg}.

$$F_{targets} = \frac{MS_{targets}}{MS_{wg}} = \frac{18.7501}{1.1667} = 16.07.$$

$$F_{hunters} = \frac{MS_{hunters}}{MS_{wg}} = \frac{.7501}{1.1667} < 1.0.$$

$$F_{AB} = \frac{MS_{AB}}{MS_{wg}} = \frac{4.0831}{1.1667} = 3.50.$$

Again, you should refer to Table F to determine the significance of these *F* values. You have 1 degree of freedom in the numerator and 8 degrees of freedom in the denominator for $F_{targets}$. An *F* value of 11.26 is required to reject the null hypothesis at the .01 level and an *F* value of 5.32 to reject at the .05 level. Since 16.07 is larger than 11.26, it is significant beyond the .01 level, and the null hypothesis that $\mu_{still} = \mu_{moving}$ is rejected. Thus, you may conclude that the hunters hit significantly fewer moving targets than still targets. $F_{hunters}$ was not computed because its value is less than 1, and values of *F* that are less than 1 are never significant. Thus, there was no significant difference in the mean number of targets hit by the two kinds of hunters. $F_{AB}(3.50)$ is less than 5.32 and is, therefore, not significant. There was no significant interaction between kind of hunter and kind of target. Although squirrel hunters were the best on still targets and the worst on moving targets, with the quail hunters intermediate, this departure from parallel performance was not great enough to reach significance.

Results of a factorial ANOVA are usually presented in a summary table. Table 11.8 is the example of the hunters' contest.

Table 11.8

ANOVA Summary Table for the Hunters' Contest

Source	df	SS	MS	F	p
A (Targets)	1	18.7501	18.7501	16.07	<.01
B (Hunters)	1	.7501	.7501	<1.0	>.05
AB	1	4.0831	4.0831	3.50	>.05
Within groups	8	9.3333	1.1667		
Total	11	32.9166			

Problems

12. For the data in Problem 8, compute *df*, *MS*, and *F* values. Arrange these in a summary table. Plot the cell means. Tell what the results mean.
13. For the data in Problem 9, compute *df*, *MS*, and *F* values. Arrange these in a summary table. Plot the cell means. Tell what the results mean.
14. For the data in Problem 10, compose a summary table and plot the cell means.

ANALYSIS OF A 3 × 3 DESIGN

This section describes the analysis of a 3 × 3 design. The procedures are exactly like those for the other designs you have analyzed. This section will emphasize the interpretation of results.

Two experimenters were interested in the Gestalt principle of closure—the drive to have things finished, or closed. An illustration of this drive is the fact that people often see a circle with a gap as closed, even if they are looking for the gap. These experimenters thought that the strength of the closure drive in an anxiety-arousing situation would depend on the subjects' general anxiety level. Thus, the experimenters hypothesized an interaction between the anxiety of a person and the kind of situation he or she is in.

The independent variables for this experiment were (1) anxiety level of the subject[5] and (2) kind of situation the person was in—that is, whether it was anxiety arousing or not. As you probably realized from the title of this section, there were three levels for each of these independent variables. The dependent variable was a measure of closure drive.

To get subjects, the experimenters administered the Taylor Manifest Anxiety Scale (Taylor, 1953) to a large group of randomly selected college students. From this large group, they selected the 15 lowest scorers, 15 of the middle scorers, and the 15 highest scorers as participants in the study. The first factor in the experiment, then, was anxiety, with three levels: low (A_1), middle (A_2), and high (A_3).

The second factor was the kind of situation. The three kinds were dim illumination (B_1), normal illumination (B_2), and very bright illumination (B_3). The assumption was that dim and bright illumination would create more anxiety than would normal illumination.

Participants viewed 50 circles projected on a screen. Ten of the circles were closed, ten contained a gap at the top, ten a gap at the bottom, ten a gap on the right, and ten a gap on the left. Participants simply stated whether the gap was at the top, bottom, right, or left, or whether the circle was closed. The experimenters recorded as the dependent variable the number of circles reported as closed by each participant.

The hypothetical data and its analysis are reported in Table 11.9. Read the experiment over again and work through the analysis of the data in Table 11.9.

Table 11.10 is the ANOVA summary table. The probabilities in Table 11.10 are from Table F in the Appendix. F_A and F_B have two degrees of freedom in the numerator and 36 df in the denominator. The critical value of F for 2,36 df with $\alpha = .01$ is 5.25. Thus, for the factor anxiety, reject the null hypothesis. It would seem that a person's closure drive *is* related to his or her anxiety level. Also, for the illumination variable, reject the null hypothesis. Again it would seem that the level of illumination has an effect on the number of circles that are seen as closed. For the interaction, the critical value of F for 4,36 df with $\alpha = .01$ is 3.89. (The change in critical value results from the increase in df.) Thus, reject the hypothesis that the illumination conditions affected high-, medium-, and low-anxious participants in the same way. Conclude that there was an interaction between anxiety level and illumination. As you will soon see, this significant interaction affects the interpretation of the main effects.

The interaction can be seen in Figure 11.5. For participants who had high anxiety scores, the dim illumination *and* the bright illumination caused more circles to be seen as closed. As you recall, the experimenter expected both the dim and bright illuminations to be anxiety arousing. Thus, the significant interaction in this case is statistical confirmation of the hypothesis that closure drive is very great in high-anxious persons placed in an anxiety-arousing situation.

Look again at Figure 11.5, the graph of the significant interaction effect from the closure study. The significant F for the anxiety factor indicates that the

---

[5]For an experiment that manipulated only this variable, with a two-group design, see J. Calhoun and J. O. Johnston, "Manifest Anxiety and Visual Acuity," *Perceptual and Motor Skills*, 1968, *27*, 1177–1178.

Table 11.9   **ANOVA of Hypothetical Study of Effects of Anxiety and Illumination on Closure Drive**

Illumination (Factor B)	Anxiety (Factor A)						Summary Values
	LOW (A_1)		MIDDLE (A_2)		HIGH (A_3)		
	X_{A_1}	$X_{A_1}^2$	X_{A_2}	$X_{A_2}^2$	X_{A_3}	$X_{A_3}^2$	
DIM (B_1)	17	289	14	196	21	441	$\sum X_{B_1} = 206.$
	15	225	14	196	19	361	$\sum X_{B_1}^2 = 3048.$
	13	169	12	144	17	289	$\bar{X}_{B_1} = 13.7333.$
	10	100	10	100	16	256	
	8	64	7	49	13	169	
Totals	63	847	57	685	86	1516	
Means	12.60		11.40		17.20		
NORMAL (B_2)	14	196	12	144	12	144	$\sum X_{B_2} = 158.$
	13	169	11	121	10	100	$\sum X_{B_2}^2 = 1708.$
	11	121	11	121	10	100	$\bar{X}_{B_2} = 10.5333.$
	11	121	9	81	9	81	
	9	81	8	64	8	64	
Totals	58	688	51	531	49	489	
Means	11.60		10.20		9.80		
BRIGHT (B_3)	12	144	11	121	18	324	$\sum X_{B_3} = 178.$
	11	121	11	121	17	289	$\sum X_{B_3}^2 = 2228.$
	10	100	10	100	15	225	$\bar{X}_{B_3} = 11.8667.$
	10	100	9	81	14	196	
	9	81	9	81	12	144	
Totals	52	546	50	504	76	1178	
Means	10.40		10.00		15.20		

Summary Values

$\sum X_{A_1} = 173.$ $\sum X_{A_2} = 158.$ $\sum X_{A_3} = 211.$ $\sum X_{tot} = 542.$

$\sum X_{A_1}^2 = 2081.$ $\sum X_{A_2}^2 = 1720.$ $\sum X_{A_3}^2 = 3183.$ $\sum X_{tot}^2 = 6984.$

$\bar{X}_{A_1} = 11.5333.$ $\bar{X}_{A_2} = 10.5333.$ $\bar{X}_{A_3} = 14.0667.$ $\bar{X}_{tot} = 12.0444.$

Sums of Squares

$$SS_{tot} = \sum X_{tot}^2 - \frac{(\sum X_{tot})^2}{N_{tot}} = 6984 - \frac{(542)^2}{45} = 455.9111.$$

$$SS_{bg} = \sum \left[\frac{(\sum X_g)^2}{N_g} \right] - \frac{(\sum X_{tot})^2}{N_{tot}} = \frac{(63)^2}{5} + \frac{(57)^2}{5} + \frac{(86)^2}{5} + \frac{(58)^2}{5} + \frac{(51)^2}{5} + \frac{(49)^2}{5}$$

$$+ \frac{(52)^2}{5} + \frac{(50)^2}{5} + \frac{(76)^2}{5} - \frac{(542)^2}{45} = 263.9111.$$

$$SS_A = \frac{(\sum X_{A_1})^2}{N_{A_1}} + \frac{(\sum X_{A_2})^2}{N_{A_2}} + \frac{(\sum X_{A_3})^2}{N_{A_3}} - \frac{(\sum X_{tot})^2}{N_{tot}} = \frac{(173)^2}{15} + \frac{(158)^2}{15} + \frac{(211)^2}{15} - \frac{(542)^2}{45} = 99.5111.$$

$$SS_B = \frac{(\sum X_{B_1})^2}{N_{B_1}} + \frac{(\sum X_{B_2})^2}{N_{B_2}} + \frac{(\sum X_{B_3})^2}{N_{B_3}} - \frac{(\sum X_{tot})^2}{N_{tot}} = \frac{(206)^2}{15} + \frac{(158)^2}{15} + \frac{(178)^2}{15} - \frac{(542)^2}{45} = 77.5111.$$

$$SS_{AB} = N_g[(\bar{X}_{A_1 B_1} - \bar{X}_{A_1} - \bar{X}_{B_1} + \bar{X}_{tot})^2 + (\bar{X}_{A_1 B_2} - \bar{X}_{A_1} - \bar{X}_{B_2} + \bar{X}_{tot})^2$$

$$+ (\bar{X}_{A_1 B_3} - \bar{X}_{A_1} - \bar{X}_{B_3} + \bar{X}_{tot})^2 + \cdots + (\bar{X}_{A_3 B_3} - \bar{X}_{A_3} - \bar{X}_{B_3} + \bar{X}_{tot})^2]$$

$$= 5[(12.60 - 11.5333 - 13.7333 + 12.0444)^2 + (11.60 - 11.5333 - 10.5333 + 12.0444)^2$$

$$+ (10.40 - 11.5333 - 11.8667 + 12.0444)^2 + (11.40 - 10.5333 - 13.7333 + 12.0444)^2$$

$$+ (10.20 - 10.5333 - 10.5333 + 12.0444)^2 + (10.00 - 10.5333 - 11.8667 + 12.0444)^2$$

Table 11.9 cont.

$$+ (17.20 - 14.0667 - 13.7333 + 12.0444)^2 + (9.80 - 14.0667 - 10.5333 + 12.0444)^2$$
$$+ (15.20 - 14.0667 - 11.8667 + 12.0444)^2] = (5)(17.3778) = 86.8890.$$

Check: $SS_{AB} = SS_{bg} - SS_A - SS_B = 263.9111 - 99.5111 - 77.5111 = 86.8889.$

$$SS_{wg} = \Sigma \left[\Sigma X_g^2 - \frac{(\Sigma X_g)^2}{N_g} \right]$$

$$= \left[847 - \frac{(63)^2}{5} \right] + \left[688 - \frac{(58)^2}{5} \right] + \left[546 - \frac{(52)^2}{5} \right] + \left[685 - \frac{(57)^2}{5} \right]$$

$$+ \left[531 - \frac{(51)^2}{5} \right] + \left[504 - \frac{(50)^2}{5} \right] + \left[1516 - \frac{(86)^2}{5} \right]$$

$$+ \left[489 - \frac{(49)^2}{5} \right] + \left[1178 - \frac{(76)^2}{5} \right] = 192.$$

Check: $455.9111 = 263.9111 + 192.$

Degrees of Freedom

$df_A = A - 1 = 3 - 1 = 2.$

$df_B = B - 1 = 3 - 1 = 2.$

$df_{AB} = (A - 1)(B - 1) = (2)(2) = 4.$

$df_{wg} = N_{tot} - (A)(B) = 45 - (3)(3) = 36.$

$df_{tot} = N - 1 = 45 - 1 = 44.$

Mean Squares   *F Values*

$$MS_A = \frac{SS_A}{df_A} = \frac{99.51}{2} = 49.76. \qquad F_A = \frac{MS_A}{MS_{wg}} = \frac{49.76}{5.33} = 9.33.$$

$$MS_B = \frac{SS_B}{df_B} = \frac{77.51}{2} = 38.76 \qquad F_B = \frac{MS_B}{MS_{wg}} = \frac{38.76}{5.33} = 7.27.$$

$$MS_{AB} = \frac{SS_{AB}}{df_{AB}} = \frac{86.89}{4} = 21.72. \qquad F_{AB} = \frac{MS_{AB}}{MS_{wg}} = \frac{21.72}{5.33} = 4.07.$$

$$MS_{wg} = \frac{SS_{wg}}{df_{wg}} = \frac{192}{36} = 5.33.$$

Table 11.10

ANOVA Summary Table for the Closure Study

Source	df	SS	MS	F	p
A (Anxiety)	2	99.51	49.76	9.33	<.01
B (Illumination)	2	77.51	38.76	7.27	<.01
AB	4	86.89	21.72	4.07	<.01
Within groups	36	192.00	5.33		
Total	44	455.91			

$F_{.01}(2,36\ df) = 5.25.$ $F_{.01}(4,36\ df) = 3.89.$

three anxiety groups differed in closure. However, it would appear that this difference may be due entirely to the performance of the high-anxious group under conditions of dim and bright illumination. Similarly, the significant illumination main effect indicates that the three different amounts of light produced three sets of scores that do not appear to have a common population mean.

Figure 11.5 shows that this significant main effect may be due primarily to the high-anxious subjects and not to subjects in general.

When an interaction is significant, the interpretation of the main effects is usually not simple and straightforward. A main effect is a comparison of the average of each level with the grand mean. A significant interaction indicates that the averages may be misleading.

In summary, when an interaction is significant, main effects must be interpreted in the light of the interaction. For a problem like the closure study, summarized in Table 11.10 and Figure 11.5, the experienced researcher would probably make statistical comparisons among the *cell* means. Such comparisons are called **simple effects** and are beyond the scope of this book.[6] However, when an interaction is significant, you can often correctly interpret the results simply by drawing a graph of cell means and examining it.

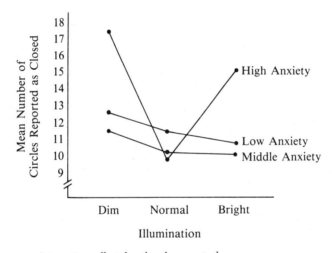

Figure 11.5   Interaction effect for the closure study.

Problems

15. The following three problems are designed to help you learn to *interpret* the results of a factorial experiment. Consider as statistically significant any effect that has a probability less than .05.

   a. A clinical psychologist has been looking at the relationship between humor and aggression. The subject's task in these experiments is to think up as many captions as possible for four cartoons. The test lasts eight minutes, and the psychologist simply counts the number of captions produced. This clinical psychologist's latest experiment is one in which a subject is either insulted or treated in a neutral way by an experimenter. The subject then responds to the cartoons at the request of either the same experimenter or a different experimenter. The cell means and the summary table are included below. Identify the independent and dependent variables, fill in the probability values in the summary table, and interpret the results.

[6]For information on simple effects, see Kirk (1982) or Winer (1971).

Subject was

Experimenter was		insulted	treated neutrally
	same	31	18
	different	19	17

Source	df	MS	F	p
A (between treatments)	1	675.00	9.08	
B (between experimenters)	1	507.00	6.82	
AB	1	363.00	4.88	
Within groups	44	74.31		

b. An educational psychologist was interested in response biases. You exhibit a response bias when you make a decision on the basis of an irrelevant stimulus. For example, a response bias is shown if the grade an English teacher puts on a theme written by a tenth grader is influenced by the student's name or by the occupation of his father. For this particular experiment, the psychologist picked a "high-prestige" name (David) and a "low-prestige" name (Elmer). In addition, she made up two biographical sketches that differed only in the occupation of the student's father (research chemist or unemployed). The subjects in this experiment were given the task of grading a theme on a scale of 50 to 100. They had been told the name of the writer and had read his biographical sketch. The same theme was given to each subject in the experiment. Identify the independent and dependent variables, fill in the probability values in the summary table, and interpret the results.

Name

Occupation of Father		David	Elmer
	Research Chemist	86	81
	Unemployed	80	88

Source	df	MS	F	p
A (between names)	1	22.50	<1.00	
B (between occupations)	1	2.50	<1.00	
AB	1	422.50	8.20	
Within groups	36	51.52		

c. A social psychologist was in charge of a large class in community psychology. During the course, 20 outside speakers holding community-service jobs each gave a short talk. Ten of these speakers were male, ten female. Five of the males had been at their jobs two years or more, and the other five had been at their jobs less than six months. This same job-experience variable also divided the females into two groups of five. The social psychologist wanted to see whether these two variables (sex and job experience) were related to the amount of attention the speakers received. She arranged a wide-angle camera to take a picture of the class when the speaker had completed the third minute of his or her talk. The dependent variable was the number of persons who were looking directly at the speaker when the picture was taken. Fortunately, the same number of people were in class on every occasion. The cell means and the summary table are included below. Identify the independent and dependent variables, fill in the probability values, and interpret the results.

		Sex of Speaker	
		Female	Male
Job Experience	Less Than Six Months	23	28
	More Than Two Years	31	39

Source	df	MS	F	p
A (between sex)	1	211.25	4.59	
B (between experience)	1	451.25	9.80	
AB	1	11.25	.24	
Within groups	16	46.03		

COMPARING LEVELS WITHIN A FACTOR

Procedures for deciding which levels within a factor are significantly different from the others are similar to those used in one-way ANOVA. The general formulas are the same. You may want to go back to Chapter 10 at this point and review pages 220–228. These comparisons are like main effects; they detect differences between the averages of two or more levels. Thus, such comparisons are appropriate only when the interaction effect is not significant. If the interaction effect is significant, comparisons among the levels of a factor are usually not made. We will illustrate the method of comparing means within a factor by making three comparisons taken from the neglect-reproof-praise study—which you analyzed while working Problems 9 and 13. In that study the interaction was not significant.

In that study, the experimenter planned two *a priori* comparisons. Of particular interest was the comparison between the praised and reproved groups. The null hypothesis was $H_0: \mu_2 - \mu_3 = 0$. The neglect group was included as a control group to determine whether either praise or reproof had any effect. The null hypothesis for this test is

$$H_0: \mu_1 - \frac{\mu_2 + \mu_3}{2} = 0.$$

Notice that these two tests are orthogonal.

		Coefficients		
Hypothesis	Neglect	Reproof	Praise	Σ
$H_0: \mu_2 - \mu_3 = 0$	0	1	-1	0
$H_0: \mu_1 - \dfrac{\mu_2 + \mu_3}{2} = 0$	2	-1	-1	0
Products	0	-1	1	0

First, we will compare the mean of the reproved class with the mean of the praised class. The formula is the same as that given in Chapter 10, p. 225.

$$ t = \frac{c_2 \bar{X}_2 + c_3 \bar{X}_3}{\sqrt{MS_{wg}\left[\dfrac{(c_2)^2}{N_2} + \dfrac{(c_3)^2}{N_3}\right]}} $$

$$ = \frac{(1)(15.15) + (-1)(11.35)}{\sqrt{19.89\left[\dfrac{(1)^2}{20} + \dfrac{(-1)^2}{20}\right]}} = 2.69. $$

Degrees of freedom for this test are $N - A$ (where A is the total number of levels of Factor A). For this test, $60 - 3 = 57$. The critical t values from Table D for $\alpha = .05$ are 2.000 for 60 df and 2.021 for 40 df. For $\alpha = .01$, the critical values are 2.660 for 60 df and 2.704 for 40 df. The experimenter could simply report a significant difference with $p < .05$ or interpolate to find the critical value of F at $\alpha = .01$ for 57 df. If interpolation is used, the critical value will be $(3/20)(2.704 - 2.660) + 2.660 = 2.667$. Thus, reject the null hypothesis that $\mu_2 - \mu_3 = 0$ at the .01 level (since $2.69 > 2.667$). The praised group made significantly fewer errors than the reproved group.

Next, we will compare the control group (neglect) with the average of the other two groups.

$$ t = \frac{c_1 \bar{X}_1 + c_2 \bar{X}_2 + c_3 \bar{X}_3}{\sqrt{MS_{wg}\left[\dfrac{(c_1)^2}{N_1} + \dfrac{(c_2)^2}{N_2} + \dfrac{(c_3)^2}{N_3}\right]}} $$

$$ = \frac{(2)(15.75) + (-1)(15.15) + (-1)(11.35)}{\sqrt{19.89\left[\dfrac{(2)^2}{20} + \dfrac{(-1)^2}{20} + \dfrac{(-1)^2}{20}\right]}} $$

$$ = \frac{5}{2.4427} = 2.047. $$

Critical values for this comparison are the same as for the comparison between praise and reproof. Therefore, the difference is significant beyond the .05 level.

At this point in the analysis, the typical experimenter would go data snooping in an attempt to find more information of interest. Be that experimenter and examine the means of the groups.

$$\bar{X}_{neglect} = 15.75 \text{ errors.}$$

$$\bar{X}_{reproof} = 15.15 \text{ errors.}$$

$$\bar{X}_{praise} = 11.35 \text{ errors.}$$

Two questions came to our minds. Is praise significantly better than neglect? Is reproof significantly better than neglect? These are questions suggested by the data so *a posteriori* Scheffé tests are in order. The method and formulas are the same as those you used in Chapter 10.

To find the critical value of F' for Scheffé tests on factorial designs, use the formula

$$F'_\alpha = (A - 1)(F_\alpha)$$

where: A = the total number of levels of the independent variable
          being examined,
and F_α = the critical value of F used to test for the main effect of A.

For neglect vs. praise,

$$F'_{ob} = \frac{(c_1\bar{X}_1 + c_3\bar{X}_3)^2}{MS_{wg}\left[\dfrac{(c_1)^2}{N_1} + \dfrac{(c_3)^2}{N_3}\right]}$$

$$= \frac{[(1)(15.75) + (-1)(11.35)]^2}{19.89\left[\dfrac{(1)^2}{20} + \dfrac{(-1)^2}{20}\right]} = 8.45.$$

$$F'_{.05} = (A - 1)(F_{.05}) = (3 - 1)(3.18) = 6.36.$$

Thus, children who were praised made significantly fewer arithmetic errors than those who were neglected.

For reproof vs. neglect,

$$F'_{ob} = \frac{(c_1\bar{X}_1 + c_2\bar{X}_2)^2}{MS_{wg}\left[\dfrac{(c_1)^2}{N_1} + \dfrac{(c_2)^2}{N_2}\right]} = \frac{[(1)(15.75) + (-1)(15.15)]^2}{19.89\left[\dfrac{(1)^2}{20} + \dfrac{(-1)^2}{20}\right]}$$

$$= \frac{.36}{1.989} = <1.00.$$

Since F values less than 1.00 are never significant, the null hypothesis is retained. There is no evidence that reproof results in fewer errors than neglect does.

Problems

16. Suppose you were asked to determine whether there is a significant difference between the low anxiety scores and the high anxiety scores in the study of the effects of anxiety and illumination on closure drive. Could you use an orthogonal comparisons test for this? A Scheffé test?

17. Two social psychologists asked freshmen and senior college students to write an essay supporting recent police action on campus. (The students were known to be against the police action.) The students were given 50 cents, one dollar, five dollars, or ten dollars for their essay. Later, each student's attitude toward the police was measured on a scale of 1–20. Perform an ANOVA, fill out a summary table, and write an interpretation of the analysis at this point. Decide whether an orthogonal comparison of the 50-cent reward and the one-dollar reward is appropriate. If so, do it. Now, determine with an *a posteriori* test whether the ten-dollar reward differs significantly from the combination of the 50-cent and the one-dollar rewards. (See Brehm and Cohen, 1962, for a similar study of Yale students.) Write an overall interpretation of the results of the study.

Reward

		$10	$5	$1	50¢
Freshmen	$\sum X$	73	83	99	110
	$\sum X^2$	750	940	1290	1520
	N	8	8	8	8
Seniors	$\sum X$	70	81	95	107
	$\sum X^2$	820	910	1370	1610
	N	8	8	8	8

18. Here is the final problem for this chapter—one that gives you raw data so that you start from scratch and work through the whole procedure. Gumbo U.'s

Teachers (*Factor A*)

		A_1 Elementary School	A_2 Middle School	A_3 High School
Colleges (*Factor B*)	B_1 (Gumbo U.)	6.8	6.2	6.6
		6.6	6.1	6.5
		6.6	6.0	6.3
		6.5	5.7	6.0
		6.3	5.4	5.7
		6.2	4.8	5.6
		5.9	4.5	5.4
		5.7	4.2	5.1
		5.4	3.6	4.9
		4.3	3.2	4.3
	B_2 (Others)	6.6	6.5	6.7
		6.5	6.4	6.5
		6.5	6.4	6.4
		6.1	6.3	6.1
		5.9	6.2	5.8
		5.7	5.9	5.5
		5.2	5.7	5.4
		4.8	5.4	5.0
		4.6	5.1	4.8
		3.7	4.4	4.4

Teacher Education Department is preparing for an accreditation visit by the National Council for Accreditation of Teacher Education (NCATE). In order to demonstrate that its graduates are performing well as teachers, the department randomly selected 30 of these graduates to be evaluated by their principals. Ten were teaching in elementary schools, ten in middle or junior high schools, and ten in high schools. Principals evaluated their performance on 34 teaching competencies by rating them on each competency using a 7-point scale, with 1 the poorest performance and 7 the highest.

Thirty teachers who had graduated from schools other than Gumbo U. were also randomly selected to be evaluated by their principals on the same scale. Mean ratings for each teacher are given in the table. Analyze the data and determine what the department can truthfully say in its report to NCATE.

TRANSITION PAGE

In your study of inferential statistics, you have used two families of curves. The normal curve is appropriate when sampling is random and you know σ or can estimate σ reliably from a large sample (Chapters 6–8). The t and F distributions are appropriate when sampling is random and population scores are normally distributed and have equal variances (Chapters 9–11). In the next two chapters, you will learn about some statistical tests that do not require knowledge or estimates of σ, assumptions about the form of the population distribution, or homogeneity of variance. Random sampling, however, will still be required.

In Chapter 12, "The Chi Square Distribution," you will learn to analyze frequency count data. Such data exist when observations are classified into categories and the frequencies in each category are counted. In Chapter 13, "Nonparametric Statistics," you will learn four techniques for analyzing scores that are ranks or are reduced to ranks.

The techniques in these next two chapters are often described as "less powerful." This means that *if* the population scores satisfy the assumptions of normality and homogeneity of variance, a t or F test is more likely than a chi square test or a nonparametric test to reject H_0 if it should be rejected. To put this same idea another way, t and F tests have a smaller probability of a Type II error if the population scores are normally distributed and have equal variances.

12 THE CHI SQUARE DISTRIBUTION

Objectives for Chapter 12: After studying the text and working the problems in this chapter, you should be able to:

1. tell when the chi square distribution may be used in hypothesis testing,
2. use the chi square distribution to test how well empirical frequency data fit a theoretical model (goodness of fit),
3. use the chi square distribution to test the independence of two variables, and
4. calculate chi square from a 2×2 table using the shortcut method.

A sociology class was deep into a discussion of methods of population control. As the class ended, the topic was abortion; it was clear that there was strong disagreement about using abortion as a method of population control. Both sides in the argument had declared that "educated people" supported their side. Two empirically minded students left the class determined to get actual data on attitudes toward abortion. They solicited the help of a friendly psychology professor, who encouraged them to distribute a questionnaire to his General Psychology class. The questionnaire asked, "Do you consider abortion to be an acceptable or unacceptable method of population control?" The questionnaire also asked the respondent's sex. The results were presented to the sociology class in the following tabular form.

| | *Abortion is* | |
	Acceptable	*Unacceptable*
Females	59	29
Males	15	37
Σ	74	66
Percent	53%	47%

    The general conclusion of the class was that college students were pretty evenly divided on this issue, 53 percent to 47 percent.

    The attitude data, then, did not seem to indicate that "educated people" were clearly on one side or the other on this issue. The class discussion resumed, with personal opinions dominating, until an observant student said "Look! Look at that table! The girls are for abortion, and the guys are against it. I wonder whether that is a real difference between males and females, or whether it is just a chance event that occurred in this sample."

    Does that question sound familiar? Indeed it sounds like the kind of question that could be answered by comparing the observed results with a sampling distribution based on chance results. In this case, a proportion of .67 of the females (59 out of 88) considered abortion to be acceptable but only .29 of the males did so (15 out of 52). Could these two statistics have come from the same population, or should they be considered different and independent?[1]

    This is yet another chapter on hypothesis testing; to retain or reject the null hypothesis—that is the question. In this chapter, however, the data are different. In previous chapters, the dependent variables were phenomena that could be measured on a fairly continuous scale, such as IQ, racial attitudes, or time. These are all things that exist in degrees. In this chapter, the phenomena either exist or do not exist. There are no intermediate degrees, and "measurement" results in a "score" of "Yes, it's there," or "No, it's not there." What the experimenter does in this case is count the frequency with which the phenomenon occurs. Chi square (pronounced "ki" as in kite, "square," and symbolized χ^2) is a sampling distribution that can be used in this situation. Frequencies like those in the story just told are distributed approximately as chi square, so the chi square distribution can be used to answer the question at the end of the story.

---

[1] Casting the chi square problems in this chapter into proportions will help you in interpreting the results. However, the test of whether two or more proportions, obtained from separate samples, have a common population proportion, is not covered in this book. (See Kirk, 1978, pp. 342–344.)

Chi square is a very common statistic in psychology, education, biology, agriculture, sociology, economics, and political science. It was developed by Karl Pearson (of Pearson product-moment correlation fame) and published in 1900. According to his son, Professor E. S. Pearson (an eminent statistician himself), the symbol χ^2 was used sometime earlier for a more complex problem and, when the test you are considering was developed, the symbol χ^2 was used again (personal communication, 1974).

Basically, a χ^2 test is appropriate when you want to test the null hypothesis that there is no difference between the observed and the theoretical frequency of events. The observed events are those that the experimenter notes. The theoretical events are those predicted by chance or a theory.

THE CHI SQUARE DISTRIBUTION

The **chi square distribution** is a theoretical distribution, just as t and F are. Like them, as the number of degrees of freedom increases, there is a change in the shape of the distribution. Figure 12.1 shows four different chi square distributions, one each for 1, 3, 5, and 10 df. As you can see, χ^2, like the F distribution, is a positively skewed curve. Table E in the Appendix gives five critical values for χ^2: .10, .05, .02, .01, and .001. We will symbolize these critical values of χ^2 by giving the α level, df, and the χ^2 value. Thus, $\chi^2_{.05}$ (1 df) = 3.84. Notice that, with chi square, as degrees of freedom increase (looking down columns), larger and larger values of χ^2 are required to reject the null hypothesis. *This is just the opposite of what occurs in the t and F distributions.*

When sets of observed and expected frequencies are being compared, a value for chi square may be computed using the formula

$$\chi^2 = \Sigma \left[\frac{(O - E)^2}{E} \right],$$

where: O = observed frequency
and E = expected frequency.

Obtaining the observed frequency is simple enough—count the events in each category. Finding the expected frequency is a bit more complex. There are two methods that can be used to determine the expected frequency. One method is used if the problem is one of "goodness of fit" and the other method is used if the problem is a "test of independence." We'll discuss these two tests in some detail.

CHI SQUARE AS A TEST FOR GOODNESS OF FIT

If you have a hypothesis or theory that predicts the expected frequency, χ^2 can be used to test how well the observed frequency conforms to that expected frequency—that is, how good the "fit" is between the actual (observed) frequency and the expected frequency. In testing for goodness of fit, *the null hypothesis is*

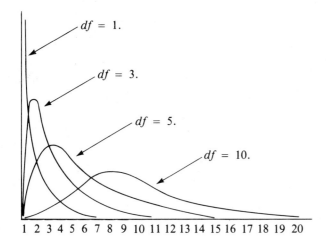

df = 1.

df = 3.

df = 5.

df = 10.

1 2 3 4 5 6 7 8 9 10 11 12 13 14 15 16 17 18 19 20

Values of χ^2

Figure 12.1   Chi square distribution for four different degrees of freedom. Adapted with permission from Cornell (1956, p. 198).

that the actual data do fit the expected data. Thus, rejection of H_0 means that the data *do not* fit the theory; the theory is inadequate. If H_0 is retained then you must, once again, word your interpretation carefully. A retained H_0 means that the data are not at odds with the theory. Such results do not prove the theory (other theories may also predict such a result) but they do support the theory.

The goodness-of-fit application of χ^2 is frequent in population genetics, where ratios such as 3:1 or 1:2:1 or 9:3:3:1 are predicted on the basis of Mendelian laws. The theory may predict that a certain hybrid will produce three times as many smooth pea seeds as wrinkled seeds (3:1 ratio). If you perform the experiment, you might get 316 smooth seeds and 84 wrinkled seeds, for a total of 400. How well do these actual frequencies "fit" the expected frequencies of 300 and 100? Chi square will give you a probability figure that will help you decide if the theory accounts for the observed data.

Although the mechanics of χ^2 require you to manipulate raw frequency counts (like 316 and 84), you can best understand this test by thinking of proportions. Thus, 316 and 84 out of 400 represent proportions of .79 and .21. How likely are such proportions if the population proportions are .75 and .25? Chi square provides you with the probability of the observed proportions and you can decide "good fit" and "poor fit."

In the genetics example, the expected frequencies were predicted by a theory. Sometimes, in a goodness-of-fit test, the expected frequencies are predicted by chance. In such a case, a rejected H_0 means that something besides chance is at work. When you interpret the analysis you should identify what that "something" is.

Here is an example. Suppose you were so interested in sex stereotypes and job discrimination that you conducted the following experiment.[2] You made up

[2]For a more sophisticated experiment dealing with a similar problem, read Harriet N. Mischel, "Sex Bias in the Evaluation of Professional Achievements," *Journal of Educational Psychology*, 1974, *66*, 157–166.

four one-page résumés and four fictitious names—two female, two male. The résumés and names were then paired in all possible combinations. The design was balanced so that each name appeared equally often on each résumé. Subjects in the experiment were shown four résumés and asked to "hire" one of the "applicants" for a management trainee position. The theory (or model) to be tested is a "no discrimination" theory. This theory (a null hypothesis) says that sex is *not* being used as a basis for hiring and, therefore, equal numbers of males and females will be hired.

Suppose you had 120 subjects, and they "hired" 75 males and 45 females. Is there statistical evidence that sex stereotypes are leading to discrimination? That is, given the hypothesized result of 60 male, 60 female, how good a fit is the observed data of 75 male, 45 female?

Applying the χ^2 formula

$$\chi^2 = \Sigma \left[\frac{(O - E)^2}{E} \right] = \frac{(75 - 60)^2}{60} + \frac{(45 - 60)^2}{60} = 3.75 + 3.75$$

$$= 7.50.$$

The number of degrees of freedom for this problem is the number of categories minus 1. There are two categories here—hired males and hired females. Thus, $2 - 1 = 1$ *df*. Looking in Table E, you'll find in Row 1 that, if $\chi^2 \geqslant 6.64$, you may reject the null hypothesis at the .01 level. Thus, these data *do not* fit the "no discrimination" model, and so the model may be rejected. The conclusion from the experiment is that discrimination in favor of males was demonstrated.

The term *degrees of freedom* implies that there are some restrictions. In the case of χ^2, one restriction is always that the sum of the expected events must be equal to the sum of the observed events; that is, $\Sigma E = \Sigma O$. If you manufacture a set of expected frequencies from a model, their sum must be equal to the sum of the observed frequencies. In our sex-discrimination example, $\Sigma O = 75 + 45 = 120$, and $\Sigma E = 60 + 60 = 120$.

Error Detection

The sum of the expected frequencies must be equal to the sum of the observed frequencies: $\Sigma E = \Sigma O$.

Problems

1. Using the data on wrinkled and smooth pea seeds on page 263, test how well the data fit a 3:1 hypothesis. Set $\alpha = .05$.
2. Here is a problem that illustrates some work John B. Watson, the behaviorist (1879–1958), did on the question of whether there are any basic, inherited emotions. Watson (1924) thought there were three: fear, rage, and love. He suspected that the wide variety of emotions experienced by adults was learned over the years. Proving his suspicion would be difficult because "unfortunately there are no facilities in maternity wards for keeping mother and child under

close observation for years." So Watson attacked the apparently simpler problem of showing that the emotions of fear, rage, and love could be distinguished in infants. (He recognized the difficulty of proving that these were the *only* basic emotions, and he made no such claim.) Fear could be elicited by dropping the child onto a soft feather pillow (but not by the dark, dogs, white rats, or a pigeon fluttering its wings in the baby's face). Rage could be elicited by holding the child's arms tightly at its sides, and love by "tickling, shaking, gentle rocking, and patting" among other things.

Here is an experiment reconstructed from the conclusions Watson described in his book.

A child was stimulated in a way designed to elicit fear, rage, or love. An observer then looked at the child and judged the emotion the child was experiencing. Each judgment was scored as correct or incorrect—correct meaning that the judgment (say, fear) corresponded to the stimulus (say, dropping). Sixty observers made one judgment each with the following results.

Correct	Incorrect
32	28

To find the expected frequencies, think about the chance of being correct or incorrect when there are three possible outcomes. Analyze the data with a χ^2 and write your conclusion in a sentence.

CHI SQUARE AS A TEST OF INDEPENDENCE

Another use of χ^2 is to test the *independence* of two variables on which frequency data are available. Table 12.1, the work of empirically minded students, shows the abortion-questionnaire data. Such a table is called a *contingency table*; here, the question is whether a person's attitude toward abortion is contingent upon sex. *The null hypothesis is that attitudes and sex in the population are independent*—that knowing a person's sex gives you *no* clue to his or her attitude and vice versa. Rejection of the null hypothesis will support the alternate hypothesis, which is that sex and attitude are not independent but related—that knowing a person's sex does help you to predict his or her attitude and that attitude is contingent upon sex.

Table 12.1 gives the observed frequencies, with the expected frequencies in parentheses. Pay careful attention to the logic behind the calculation of the expected frequencies. We will start with an explanation of the expected frequency in the upper left corner, 46.51, the expected number of females who think abortion is acceptable. Of all the 140 subjects, 88 were female, so if we chose a subject at random, the probability that the person would be female is $88/140 = .6286$. In a similar way, of the 140 subjects, 74 thought that abortion was acceptable. Thus, the probability that a randomly chosen person would think that abortion is acceptable is $74/140 = .5286$. If we now ask what the probability is that a person chosen at random is a female who thinks that abortion is acceptable, we can find

Table 12.1   Hypothetical Data on Attitudes toward Abortion (Expected frequencies are in parentheses.)

| | Abortion is | | Σ |
	Acceptable	Unacceptable	
Females	59 (46.51)	29 (41.49)	88
Males	15 (27.49)	37 (24.51)	52
Σ	74	66	140

the answer by multiplying together the probability of the two separate events.[3] Thus, $(.6286)(.5286) = .3323$.

Finally, we can find the expected frequency of such people by multiplying the probability of such a person by the total number of people. Thus, $(.3323)(140) = 46.52$. Notice what happens to the arithmetic if these steps are combined.

$$\left(\frac{88}{140}\right)\left(\frac{74}{140}\right)(140) = \frac{(88)(74)}{140} = 46.51.$$

The slight difference in the two methods is due to rounding (46.52 vs. 46.51). The second answer is the correct one.

In a similar way, the expected frequency of males with an attitude that abortion is unacceptable is the probability of a male times the probability of an attitude of unacceptable, times the number of subjects. For all the data, here are the calculations of the expected values.

$$\frac{(88)(74)}{140} = 46.51. \qquad \frac{(88)(66)}{140} = 41.49.$$

$$\frac{(52)(74)}{140} = 27.49. \qquad \frac{(52)(66)}{140} = 24.51.$$

Once the expected values are calculated, a table like Table 12.2 can be constructed that will lead to a χ^2 value.

To determine the *df* for any $R \times C$ (rows by columns) table like Table 12.1, use the formula $(R - 1)(C - 1)$. In this case, $(2 - 1)(2 - 1) = 1$. To explain this surprising state of affairs ("Hold it! Before, with two categories, there was a 1 *df* and now, with four categories, I still only have 1 *df*???"), we will resort to our "freedom to vary" explanation. For any $R \times C$ table, the *marginal* totals of the expected frequencies are fixed so that they equal the observed frequency.[4] In a 2×2 table, this leaves only one cell free to vary. Once a value is fixed for one cell, the other cell values are determined if you are to maintain the marginal

---

[3]This is like determining the chances of obtaining two heads in two tosses of a coin. For each toss, the probability of a head is $\frac{1}{2}$. The probability of two heads in two tosses is obtained by multiplying the two probabilities: $(\frac{1}{2})(\frac{1}{2}) = \frac{1}{4} = .25$. See page 114 for a review of this topic.

[4]In Table 12.1, the marginal total of 88 must be the sum of the two expected frequencies, 46.51 and 41.49, as well as the sum of the obtained frequencies, 59 and 29.

Table 12.2

Calculation of χ^2 for the Data in Table 12.1

O	E	$O - E$	$(O - E)^2$	$\dfrac{(O - E)^2}{E}$
59	46.51	12.49	156.00	3.35
15	27.49	−12.49	156.00	5.67
29	41.49	−12.49	156.00	3.76
37	24.51	12.49	156.00	6.36
				$\chi^2 = \overline{19.14}.$

totals. The general rule again: $df = (R - 1)(C - 1)$, where R and C refer to the number of rows and columns.

To determine the significance of a $\chi^2 = 19.14$ with 1 df, look at Table E. In the first row, you will find that, if $\chi^2 = 10.83$, the null hypothesis may be rejected at the .001 level of significance. Since our χ^2 exceeds this, you can conclude that the attitudes toward abortion are definitely influenced by sex; that is, sex and attitudes toward abortion in the population are not independent but related. By examining the proportions, you can conclude that females are more likely to consider abortions to be acceptable.

This test of independence is essentially a test of whether there is an interaction between the two variables—in this case, sex and attitude. The interpretation of a χ^2 test of independence in an $R \times C$ table is the same as the interpretation of an interaction in a two-way ANOVA: the result for one independent variable *depends* upon the level of the other independent variable.

SHORTCUT FOR ANY 2 × 2 TABLE

When you have a 2 × 2 table to analyze and a calculator, the following shortcut will save you time. With this shortcut, it is unnecessary to calculate the expected frequencies. For the general case of a 2 × 2 table,

A	B	$A + B$
C	D	$C + D$
$A + C$	$B + D$	N

$$\chi^2 = \frac{N(AD - BC)^2}{(A + B)(C + D)(A + C)(B + D)}.$$

The term in the numerator, $AD - BC$, is the difference between the cross-products. The denominator is the product of the four marginal totals.

To illustrate the equivalence of the shortcut method and the method explained in the previous section, we will calculate χ^2 on the data in Table 12.1.

Translating cell letters into cell totals: $A = 59$, $B = 29$, $C = 15$, $D = 37$, and $N = 140$.

$$\chi^2 = \frac{(140)[(59)(37) - (29)(15)]^2}{(88)(52)(74)(66)} = 19.14.$$

Both methods yield $\chi^2 = 19.14$.

Problems

3. Two botanists were investigating the resistance of two species of pine trees to the dread "pine borer." In a walk through a mixed forest, they found the following frequencies of infestation.

		Evidence of the Dread Pine Borer	
		Yes	No
Species	A	13	38
	B	19	41

   Test the hypothesis that the two species are equally subject to infestation. Write a conclusion. Calculate expected frequencies or use the shortcut method.

4. A student of sociology was interested in the effect of group size on likelihood of joining an informal group. On one part of the campus, he placed a group of *two* people looking intently up into a tree. On a distant part of the campus, a group of *five* stood looking up into a tree. Single passersby were classified as joiners or nonjoiners. If a passerby stopped and looked up for five or more seconds or made some comment to the group, he or she was classified as a joiner. The following data were obtained. Use a χ^2 test to determine whether group size had an effect on joining. Use the method you did not choose for Problem 3. Write a conclusion.

	Group Size		
	2	5	Σ
Joiners	9	26	35
Nonjoiners	31	34	65
Σ	40	60	100

You have probably heard that salmon return to the stream in which they hatched in order to spawn. According to the story, after years of maturing in the ocean, the salmon arrive at the mouth of a river and swim upstream, choosing at each fork the stream that leads to the pool in which they hatched. Arthur D. Hasler did a number of interesting experiments on this homing instinct. He reported these in *Underwater Guideposts* (1966). Here are two sets of data that he gathered.

5. These data help answer the question of whether salmon really do make consistent choices at the fork of a stream in their return upstream. Forty-six salmon were captured from the Issaquah Creek (in Washington State) and 27 from its East Fork. All salmon were released below the confluence of these two streams. All

of the 46 captured in the Issaquah were recaptured there. Of the 27 captured in the East Fork, 19 were recaptured from the East Fork and eight from the Issaquah. Use a χ^2 to determine if salmon make consistent choices at the confluence of two streams. Write a sentence explanation of the results.

6. Hasler believed that the salmon were making the sequence of choices at each fork on the basis of olfactory (smell) cues. He thought that young salmon become imprinted on the particular mix of dissolved mineral and vegetable molecules in their home stream. As adults, they simply make choices at forks on the basis of where the smell of home is coming from. To test this, he captured 70 salmon from the two streams, plugged their nasal openings, and released them downstream. The fish were recaptured above the fork in one stream or the other. Compute a χ^2 and write a sentence about Hasler's hypothesis.

		Recapture Site	
		Issaquah	East Fork
Capture Site	Issaquah	39	12
	East Fork	16	3

CHI SQUARE WITH MORE THAN ONE DEGREE OF FREEDOM

So far the problems we have considered have been analyzed with a χ^2 distribution with 1 df. In this section, you will analyze more complex designs that require χ^2 distributions with more than 1 df.

Suppose you asked the following question of 128 members of a freshman psychology course in the United States, all of whom were born and raised in the United States. "Imagine that you have volunteered to have a foreign roommate next year, and you must tell whether you prefer an African, an Asian, or a European. Which would you choose?" The hypothetical responses could be compiled into the following form.

Preference			
African	Asian	European	Total
40	28	60	128

We may use χ^2 to test the hypothesis that such frequencies could have arisen by chance. That is, we will test these data for goodness of fit to a model in which each of the three geographical areas is equally preferred. In this case, the chance expectation for each cell is one-third of 128, or 42.67. The application of χ^2 is a simple extension of what you have already learned.

Options	O	E	$O - E$	$(O - E)^2$	$\dfrac{(O - E)^2}{E}$
African	40	42.67	−2.67	7.1289	.1671
Asian	28	42.67	−14.67	215.2089	5.0436
European	60	42.67	17.33	300.3289	7.0384
					$\chi^2 = \overline{12.2491}.$

What is the df for this χ^2 value? The marginal total of 128 is fixed, and, since $\sum E$ must equal $\sum O$, only two of the expected cell frequencies are free to vary. Once two are determined, the third is restricted to whatever value will make $\sum E = \sum O$. Thus, there are $3 - 1 = 2\ df$. For this design, the number of df is the number of categories minus 1. From Table E, for $\alpha = .01$, χ^2 with 2 $df = 9.21$. Therefore, reject the hypothesis of independence and conclude that the students were responding in a non-chance manner to the questionnaire.

One of the advantages of χ^2 is its **additive** nature. Each $(O - E)^2/E$ value is a measure of deviation of the data from the model, and the final χ^2 is simply the sum of these measures. Because of this, you can examine the $(O - E)^2/E$ values and see which deviations are contributing the most to χ^2. For the roommate-preference data, it is a greater number of European choices and a lesser number of Asian choices that make the final χ^2 significant. The African choices are about what would be predicted by the "equal likelihood" model.

Suppose you showed your roommate-preference data to a few of your friends, and one of them (a senior psychology major) said "That's typical of data in psychology—it's based only on college students! If you do any more studies, why don't you find out what the noncollege population thinks? Try some high school kids; try some business types downtown." Since this attack seems to you to have some validity, you decide to include all three of the suggested groups in your study, one on family planning. This time the questionnaire reads "A couple, just starting a family this year should plan to have 0 or 1 (   ), 2 or 3(   ), 4 or 5 (   ), child(ren). Check one."

The compiled data look like this;

Subjects	Number of Children			
	0–1	2–3	4–5	$\sum$
High School	26	57	9	92
College	61	38	18	117
Business	17	35	14	66
$\sum$	104	130	41	275

For this problem, there is no theoretical expectancy. The question is whether the three groups (high school, college, and business) differ in their responses. What is required is a test of independence; it will tell us whether attitudes toward family size are related to groups or are independent of group affiliation. Set $\alpha = .05$.

To get the expected frequency for each cell, assume independence (H_0) and apply the reasoning we explained earlier about multiplying probabilities.

$$\frac{(92)(104)}{275} = 34.793. \qquad \frac{(92)(130)}{275} = 43.491. \qquad \frac{(92)(41)}{275} = 13.716.$$

$$\frac{(117)(104)}{275} = 44.247. \qquad \frac{(117)(130)}{275} = 55.309. \qquad \frac{(117)(41)}{275} = 17.444.$$

$$\frac{(66)(104)}{275} = 24.960. \qquad \frac{(66)(130)}{275} = 31.200. \qquad \frac{(66)(41)}{275} = 9.840.$$

O	E	$O - E$	$(O - E)^2$	$\dfrac{(O - E)^2}{E}$
26	34.793	−8.793	77.317	2.222
57	43.491	13.509	182.493	4.196
9	13.716	−4.716	22.241	1.622
61	44.247	16.753	280.663	6.343
38	55.309	−17.309	299.602	5.417
18	17.444	.556	.309	.018
17	24.960	−7.960	63.362	2.539
35	31.200	3.800	14.440	.463
14	9.840	4.160	17.306	1.759
				$\chi^2 = 24.579.$

The *df* for a contingency table such as this is obtained from the formula

$$df = (R - 1)(C - 1).$$

Thus, $df = (R - 1)(C - 1) = (3 - 1)(3 - 1) = 4.$

For $df = 4$, a $\chi^2 = 9.488$ is required to reject H_0. The obtained χ^2 exceeds this, so reject H_0 and conclude that attitudes toward family size are related to the groups sampled.[6]

By examining the right-hand column, you can see that 6.343 and 5.417 constitute a large portion of the final χ^2 value. By working backward on those rows, you will discover that college students chose the 0–1 category more often than expected and the 2–3 category less often than expected. The interpretation then is that college students think that families should be smaller than high school students and business people do.

Problems

7. Professor Stickler always grades "on the curve." "Yes, I see to it that my grades are distributed as 7% *A*'s, 24% *B*'s, 38% *C*'s, 24% *D*'s, and 7% flunks," he explained to a young colleague. The colleague, convinced that the professor was really a softer touch than he sounded, looked at Professor Stickler's grade

---

[6]The statistical analysis for the problem is correct, but the experimental design makes interpretation of the significant χ^2 difficult. Since the three groups differ in both age and education, we are unsure whether the differences in attitude are related only to age, only to education, or to both of these variables. A better design would have groups that differed only on the age or only on the education variable.

distribution for the past year. He found frequencies of 20, 74, 120, 88, and 38, respectively, for the five categories, *A* through *F*. Run a χ^2 test for the colleague at the .05 level. What would *you* conclude about Professor Stickler?

8. Is Problem 7 a problem of goodness-of-fit or of independence?

9. "Snake-eyes", a local gambler, let a group of sophomores know they would be welcome in a dice game. After three hours the sophomores went home broke. However, one sharp-minded, sober lad had recorded the results on each throw. He decided to see if the results of the evening would fit an "unbiased dice" model. Conduct the test and write a sentence summary of your conclusions.

Number of Spots	Frequency
6	195
5	200
4	220
3	215
2	190
1	180

10. Identify Problem 9 as a test of independence or goodness of fit.

11. A political science student was struck by the strong differences in American opinion that resulted from Soviet invasions of other countries (for example, Afghanistan). She decided to see if she could isolate some variables that would help account for these differences. She categorized possible U.S. responses as seen in the table and then asked a total of 170 people in her university what they thought the U.S.'s response should be. The young students were less than 22 years old and the older students were all over 40. The faculty, of course, varied in age. Analyze the data in her table and write a conclusion.

	Young Students	Older Students	Faculty
Do Nothing	29	6	11
Publicly Denounce the Invasion	31	10	10
Impose Economic Sanctions	20	13	12
Intervene with Military Force	7	14	7

12. Is Problem 11 one of independence or goodness of fit?

SMALL EXPECTED FREQUENCIES

The theoretical chi square distribution that comes from a mathematical formula is a continuous function that can have any positive numerical value. Chi square values that are calculated from frequencies, however, do not work this way. They change in discrete steps and, when the expected frequencies are very small, the discrete steps are quite large. It is clear to mathematical statisticians that as the expected frequencies approach zero, the theoretical chi square distribution becomes a less and less reliable way to estimate probabilities. In particular, the fear is that such chi square analyses will reject the null hypothesis more often than warranted.

The question for researchers who hope to analyze their data with chi square has always been, "Those expected frequencies—how small is too small?" For years the usual answer was "less than 5" or, according to some, "less than 10." Recent studies, however, have led to a revision of the answer.

When $df = 1$

Several studies have used a computer to draw thousands of samples from a known population for which the null hypothesis is true. Each sample is analyzed with a chi square and the proportion of rejections of the null hypothesis (Type I errors) can be calculated. If this proportion is approximately equal to α, the chi square analysis is clearly appropriate. The general trend of these studies has been to show that the theoretical chi square distribution gives accurate conclusions even when the expected frequencies are considerably below 5.[7] (See Camilli and Hopkins, 1978, for an analysis of a 2×2 test of independence.)

When $df > 1$

When $df > 1$, the same uncertainty exists if one or more expected frequencies is small. Is the chi square test still advisable? Bradley, Bradley, McGrath, and Cutcomb (1979) addressed some of the cases in which $df > 1$. Again, they used a computer to draw thousands of samples and analyzed each sample with a chi square test. The proportion of cases in which the null hypothesis was mistakenly rejected was calculated. Nearly all were in the .03 to .06 range but a few fell outside this range, especially when the sample sizes were small ($N = 20$).

An Important Consideration

Now, let's back away from these trees we've been examining and visualize the forest that we are trying to understand. The big question, as always, is, "What is the nature of the population?" The question is answered by studying a sample, which may lead you to reject or to retain the null hypothesis. Either decision could be in error. The concern of this section has been whether the chi square test is keeping the percentage of mistaken rejections (Type I errors) near an acceptable level (.05).

Since the big question is, "What is the nature of the population?" you should be reminded of the other error you make: retaining the null hypothesis when it should be rejected (a Type II error). How likely is a Type II error when expected frequencies are small? Overall (1980) has addressed this question and his answer is, "Very."

---

[7]Yates' correction (reducing each expected frequency by $\frac{1}{2}$) used to be recommended for 2×2 tables that had one expected frequency less than 5. The effect of this was to reduce the size of the obtained chi square. According to current thinking, Yates' correction overcorrected, resulting in α levels that were much smaller than the investigator claimed to be using. Therefore, we do not recommend that Yates' correction be used.

You may recall that this issue was discussed earlier in Chapter 8 under the topic of power. Overall's warning to us is that if the effect of the variable being studied is not just huge, then small expected frequencies have a high probability of dooming us to make the mistake of failing to discover a real difference.

How can this mistake be avoided? Use large Ns. The larger the N, the more power you have to detect any real differences. And in addition, the larger the N, the more confident you are that the actual probability of a Type I error is your adopted α level.

Combining Categories

Combining categories is a technique that allows you to ensure that your chi square analysis will be more accurate. (It reduces the probability of a Type II error and keeps the probability of a Type I error close to α.) We will illustrate this technique with some data from Chapter 6 on the normal distribution.

In Chapter 6, we attempted to convince you that IQ scores are distributed normally by showing you a frequency polygon, Figure 6.7. We asked you simply to look at it and note that this empirical distribution appeared to fit the normal curve. Now you are in a position to use a statistical goodness-of-fit test rather than a casual glance to determine whether a normal curve fits the distribution. For those IQ scores, the mean was 101 and the standard deviation was 13.4. The question this goodness-of-fit test will answer is "Will a theoretical normal curve with a mean of 101 and a standard deviation of 13.4 fit the data graphed in Figure 6.7?"

To find the expected frequency of each class interval, we constructed Table 12.3. The first three columns show the observed scores arranged into a grouped frequency distribution and the lower limit of each class interval. The z score column shows the z score for the lower limit of the class interval, based on a

Table 12.3   **IQ Scores and Expected Frequencies for 261 Fifth-Grade Students**

Class Interval	Observed Frequency	Lower Limit of Class Interval	z Score	Proportion of Normal Curve Within Class Interval	Expected Frequency (261 × proportion)
130 up	4	129.5	2.13	.0166	4.3
125–129	5	124.5	1.75	.0235	6.1
120–124	18	119.5	1.38	.0437	11.4
115–119	19	114.5	1.01	.0724	18.9
110–114	25	109.5	.63	.1081	28.2
105–109	28	104.5	.26	.1331	34.7
100–104	36	99.5	−.11	.1464	38.2
95–99	41	94.5	−.49	.1441	37.6
90–94	31	89.5	−.86	.1172	30.6
85–89	28	84.5	−1.23	.0856	22.3
80–84	14	79.5	−1.60	.0545	14.2
75–79	7	74.5	−1.98	.0309	8.1
70–74	4	69.5	−2.35	.0145	3.8
65–69	1	64.5	−2.72	.0061	1.6
−64	0		below −2.72	.0033	0.9
Sum	261			1.0000	260.9

mean of 101 and a standard deviation of 13.4. The next column was obtained by entering the normal curve table with the z score and finding the proportion of the curve above the z score. From this proportion we subtracted the proportion of the curve associated with class intervals with higher scores (for example, from the proportion above $z = 1.75$, .0401, we subtracted .0166 to get .0235). Finally, the proportion in the interval was multiplied by 261, the number of students, to get the expected frequency.

Now we have the ingredients necessary for a χ^2 test. The observed frequencies are in the second column. The frequencies expected if the data were distributed as a normal curve with a mean of 101 and a standard deviation of 13.4 are seen in the last column. We could combine these observed and expected frequencies into the now familiar $(O - E)^2/E$ formula, except that several categories have suspiciously small expected frequencies.

The solution to this problem is to combine some adjacent categories. If we make our top category 125 and up, our expected frequency becomes 10.4(4.3 + 6.1). In a similar fashion, we can make our bottom category 74 and below and have an expected frequency of 6.3(0.9 + 1.6 + 3.8).[8] We have made these combinations in Table 12.4, a familiar χ^2 table with 12 categories. The χ^2 value is 7.98.

Table 12.4   **A Goodness-of-Fit Test of IQ Data to a Normal Curve**

Class Interval	O	E	$O - E$	$(O - E)^2$	$\dfrac{(O-E)^2}{E}$
125 up	9	10.4	−1.4	1.96	.189
120–124	18	11.4	6.6	43.56	3.821
115–119	19	18.9	.1	.01	.001
110–114	25	28.2	−3.2	10.24	.363
105–109	28	34.7	−6.7	44.89	1.294
100–104	36	38.2	−2.2	4.84	.127
95–99	41	37.6	3.4	11.56	.307
90–94	31	30.6	.4	.16	.005
85–89	28	22.3	5.7	32.49	1.457
80–84	14	14.2	−.2	.04	.003
75–79	7	8.1	−1.1	1.21	.149
−74	5	6.3	−1.3	1.69	.268
Σ	261	260.9			$\chi^2 = 7.98$

What is the df for this problem? This is a case in which df is not equal to the number of categories minus 1. The reason is that we placed more than our usual one restriction on the expected frequencies. Usually, we simply require $\Sigma E = \Sigma O$, which costs 1 df. This time, however, we also required that the mean and standard deviation of the expected frequency be equal to the observed mean and observed standard deviation. Thus, df in this case is the number of categories minus 3, and $12 - 3 = 9$ df.

---

[8]We have to confess that the guidelines we used for how much combining to do was the old notion about not having expected frequencies of less than five.

By consulting Table E, you can see that a χ^2 of 16.92 would be required to reject the null hypothesis at the .05 level. The null hypothesis for a goodness-of-fit test is that the data *do* fit the model. Thus, to reject H_0 in this case is to show that the data are not normally distributed. However, since the obtained $\chi^2 = 7.98$ is less than the critical value of 16.92, we can state that these data do not differ significantly from a normal-curve model in which $\mu = 101$ and $\sigma = 13.4$.

There are other situations in which small expected frequencies may be combined. For example, in an opinion survey, the response categories might be "strongly agree," "agree," "no opinion," "disagree," and "strongly disagree." If there are few respondents who check the "strongly" categories, those who do might simply be combined with their "agree" or "disagree" neighbors—reducing the table to three cells instead of five. The basic rule on combinations is that they must "make sense." It wouldn't make sense to combine the "strongly agree" and the "no opinion" categories in this example.

WHEN YOU MAY USE CHI SQUARE

1. Chi square, like all the methods of statistical inference that you have studied, requires you to have random (or at least representative) samples from the population to which you wish to generalize.
2. Chi square is appropriate when the data are frequency counts rather than some form of quantitative measurement. Chi square is not appropriate to test a difference between means or a difference between medians.
3. The sampling distribution of χ^2 can be used to test for (a) goodness-of-fit and (b) independence of two variables. In addition, χ^2 has other applications, which you are likely to encounter if you take a second course in statistics.
4. The procedures presented in this book require that each event be independent of the other events. *Independent* here means uncorrelated; *independent* is used in the same sense as it was in the assumption of independent samples for t and F. As one example, if each subject contributes two or more observations to the study, the observations will *not* be independent. In addition, if the outcome of one event depends upon the outcome of an already recorded event, the events are not independent. For example, if people are asked for their preference in a foreign roommate in the presence of others who have already responded, their stated preference is not independent of the other preferences.

Problems

13. A researcher believed that schizophrenia (a form of severe mental illness) has a simple genetic cause. In accordance with the theory, one-fourth (a $1:3$ ratio) of the offspring of a certain selected group of parents would be expected to be diagnosed as schizophrenic. After several months of gathering data, the results were as shown.

Schizophrenic	Not Schizophrenic	Total
23	107	130

Test these data for goodness-of-fit to a $1:3$ model. Set α at .01.

14. What is the difference between the goodness-of-fit test and the test of independence?

15. Suppose a friend of yours decides to gather his own data on sex stereotypes and job discrimination. He makes up ten résumés and ten fictitious names, five male and five female. He asks his 24 subjects each to "hire" five of the candidates rather than just one. The data show that 65 males and 55 females were hired. He asks you for advice on χ^2. What would you say to your friend?

16. As an assignment for a political science class, students were asked to gather data to establish a "voter profile" for backers of each of the two candidates for Senator (Hill and Dale). One student stationed herself at a busy intersection and categorized the cars that turned left according to whether they signaled the turn or not and whether they had a bumpers sticker for either of the candidates. After two hours, she left with the following frequencies.

Signaled Turn	Sticker	Frequency
No	Hill	11
No	Dale	2
Yes	Hill	57
Yes	Dale	31

Test these data with a χ^2 to see whether there is a significant difference at the .01 level. Use the expected-frequency procedure rather than the shortcut. Carefully phrase the student's contribution to the voter profile.

17. Another friend, who knows that you are almost finished with this book, comes to you for advice on the statistical analysis of the following data. He wants to know whether there is a significant difference between males and females. What do you tell your friend?

	Males	Females
Mean Score	123	206
N	18	23

18. Our political science student from Problem 16 has launched a new experiment to see whether affluence is related to voter choice. This time she looks for yard signs for the two candidates and then classifies the houses as brick or one-story frame. Her reasoning is that the one-story frame homes represent less affluent voters in her city. Houses that do not fit the categories are ignored. After three hours of driving and three-fourths of a tank of gas, the frequency counts are as shown.

Type of House	Yard Signs	Frequency
Brick	Hill	17
Brick	Dale	88
Frame	Hill	59
Frame	Dale	51

(cont.)

Test the hypothesis that candidate preference and affluence (as measured by style of home) are independent. You select the α level. Carefully write a statement of the results that can be used by the political science class in the voter profile.

19. When using χ^2 what is the proper *df* for the following tables?

a. 1×4

b. 4×5

c. 2×4

d. 6×3

13 NONPARAMETRIC STATISTICS

Objectives for Chapter 13: After studying the text and working the problems in this chapter, you should be able to:

1. describe the rationale behind nonparametric statistical tests,
2. determine when a Mann-Whitney U test is appropriate, perform the test, and interpret the results,
3. determine when a Wilcoxon matched-pairs signed-ranks test is appropriate, perform the test, and interpret the results,
4. determine when a Wilcoxon-Wilcox multiple-comparisons test is appropriate, perform the test, and interpret the results, and
5. calculate a Spearman's r_s correlation coefficient and determine the probability that the coefficient came from a population with a correlation of zero.

Two child psychologists were talking shop over coffee one morning. (Much research begins with just such "bull sessions.") The topic was the effect of intensive early training in athletics. Both psychologists were convinced that such training made the child less sociable as an adult, but one psychologist went even further. "I think that really intensive training of young kids is ultimately detrimental to their performance in the sport. Why, I'll bet that, among the top ten men's singles tennis players, those with intensive early training are not in the highest ranks."

"Well, I certainly wouldn't go that far," said the second psychologist. "I think all that early intensive training would be quite helpful."

"Good. In fact, great. We disagree and we may be able to decide who is right. Let's get the ground rules straight. For tennis players, how early is early, and what is intensive?"

"Oh, I'd say early is starting by age 7 and intensive is playing every day for two or more hours."[1]

"That seems reasonable. Now, let's see, our population is 'excellent tennis players' and these top ten will serve as our representative sample."

"Yes, indeed. What we would have among the top ten players would be two groups to compare. One had intensive early training, and the other didn't. The dependent variable is the player's rank. What we need is some statistical test that will tell us whether the difference in average ranks of the two groups is statistically significant."[2]

"Right. Now, a *t* test won't give us an accurate probability figure because *t* tests assume that the population of dependent variable scores is normally distributed. A distribution of ranks is rectangular with each score having a frequency of one."

"I think there is a nonparametric test that would be proper to use on such data."

So, here is a new category of tests, a category often called nonparametric tests, that can be used to analyze experiments in which the dependent variable is ranks. Here is the rationale of these tests.

THE RATIONALE OF NONPARAMETRIC TESTS

Suppose you drew two samples of equal size (for example, $N_1 = N_2 = 10$) from the same population.[3] You then arranged all the scores from both samples into one overall ranking, from 1 to 20. Since the samples are from the same population, the sum of the ranks of one group should be equal to the sum of the ranks of the second group. In this case, the expected sum for each group is 105. (With a little figuring, you can prove this for yourself now, or you can wait for our explanation later in the chapter.) Any difference between the actual sum and

[1]Since the phrase *intensive early training* can mean different things to different people, the first psychologist has provided the second with an operational definition. An **operational definition** is a definition that specifies a concrete meaning for a term. A concrete meaning is one everyone understands. "Seven years old" and "Two or more hours of practice every day" are concepts that everyone understands.

[2]This is how the experts convert vague questionings into comprehensible ideas that can be communicated to others—they identify the independent and the dependent variables.

[3]As always, drawing two samples from one population is statistically the same as starting with two identical populations and drawing a random sample from each.

105 would be the result of sampling fluctuation. Clearly, a sampling distribution of such differences could be constructed.

Now, you are ready to experiment. Tentatively, assume the null hypothesis is true. Conduct an experiment on two groups. Find the probability of the obtained result from the sampling distribution. If the obtained result has a probability of less than .05, reject chance as an explanation of the obtained results. Sound familiar?

Even if the two sample sizes are unequal, the same logic will still work. A sampling distribution can be constructed that will show the expected variation in sums of ranks for one of the two groups.

In this chapter, you will learn four new techniques. The first three are examples of hypothesis testing, which determine whether samples came from the same population. These three employ the rationale described before, although only one of the tests (the Wilcoxon-Wilcox test) uses the arithmetic in exactly the way the rationale suggests. Also, for each sampling distribution, only a few points are given to you in the tables. Like the tables for t and F, only values that experimenters use as α levels (critical values) are given. The fourth technique in this chapter is a descriptive statistic, a correlation coefficient for ranked data symbolized r_s.

The four nonparametric techniques in this chapter and their functions are listed in Table 13.1. In earlier chapters, you studied parametric tests that have similar functions. They are listed on the right side of the table. Study this table carefully now.

Table 13.1   **The Function of Some Nonparametric and Parametric Tests**

Nonparametric Test	Function	Parametric Test
Mann-Whitney U	Tests for a significant difference between two independent samples	Independent samples t test
Wilcoxon matched-pairs signed-ranks	Tests for a significant difference between two correlated samples	Correlated samples t test
Wilcoxon-Wilcox multiple-comparisons	Tests for significant differences among all possible pairs of independent samples	One-way ANOVA and Scheffé tests
r_s	Describes the degree of correlation between two variables	Pearson product-moment correlation coefficient, r

COMPARISON OF NONPARAMETRIC AND PARAMETRIC TESTS

In what ways are nonparametric tests similar to parametric tests (t tests, and ANOVA), and in what ways are they different? They are similar in that both kinds of tests have the same goal: to determine whether samples came from the same population. Both kinds of tests require you to have random samples from the population (or at least to assign subjects randomly to subgroups). Both kinds of tests are based on the logic of testing the null hypothesis. (If you can show that H_0 is very unlikely, you are left with the alternate hypothesis.) As you will see, though, the null hypotheses are different for the two kinds of tests.

As for differences, the t test and ANOVA assume that the scores in the populations that are sampled are normally distributed and have equal variances, but no such assumptions are necessary if you run a nonparametric test. Also, with parametric tests, the null hypothesis is that the population *means* are the same (H_0: $\mu_1 = \mu_2$). In nonparametric tests, the null hypothesis is that the population *distributions* are the same. Since distributions can differ in form, variability, central value, or all three, the interpretation of a rejection of the null hypothesis may not be quite so clear-cut with a nonparametric test.

Recommendations on how to choose between a parametric and a nonparametric test have varied over the years. Two of the issues have been the scale of measurement and the power of the tests.

In the 1950s and after, some texts recommended that the nonparametric tests be used if the scale of measurement were nominal or ordinal. After a period of controversy (see Chapter 2 of Kirk, 1972, or Gardner, 1975), this consideration was dropped.

The other issue, power, remains. Power, you may recall, is how likely a test is to reject the null hypothesis if it *should* be rejected. Clearly, if the populations that are being sampled are normally distributed and have equal variances, then parametric tests are more powerful than nonparametric ones. If the populations don't have these characteristics, recommendations are not so clear. In the past, parametric tests were considered robust and able to handle large departures from such population characteristics (but see p. 201). Now, this robustness is being called into question. For example, Blair, Higgins, and Smitley (1980) show that a nonparametric test (Mann-Whitney U) is generally more powerful than its parametric counterpart (t test) for the non-normal distribution they tested.

We are sorry to leave this issue without giving you more specific advice, but the field of statistics is unsettled on this issue. One thing, however, is agreed upon. If the data are ranks, use a nonparametric test.

Finally, there is not even a satisfactory name for these tests. Besides nonparametric, they are also referred to as distribution-free statistics. Although nonparametric and distribution-free mean different things to statisticians, the two words are used almost interchangeably by research workers. Ury (1967) suggested a third term, *assumption-freer* tests, which conveys the fact that these tests have fewer restrictive assumptions than parametric tests. Some texts have adopted Ury's term (for example, Kirk, 1978). We will use the term nonparametric tests. We will examine them in the order given in Table 13.1.

THE MANN-WHITNEY U TEST

The Mann-Whitney U test is used to determine whether two sets of data based on two *independent* samples came from the same population. Thus, it is the appropriate test for the child psychologists to use to test the difference in ranks of tennis players. The Mann-Whitney U test is identical to the **Wilcoxon rank-sum test**, which is not covered in this book. Wilcoxon published his test first

(1945). However, when Mann and Whitney (1947) independently published a test based on the same logic, they provided tables *and a name* for the statistic (U). Currently, the Mann-Whitney U test appears to be referred to more often than the Wilcoxon rank-sum test.

The Mann-Whitney U test produces a statistic, U, which is evaluated by consulting the sampling distribution of U. Like all the distributions you have encountered that can be used in the analysis of small samples, the sampling distribution of U depends on sample size. Table G, which you will learn to use later, gives you critical values of U. Use Table G if neither of your two samples is as large as 20.

When the number of scores in one of the samples is over 20, the statistic U is distributed approximately as a normal curve. In this case, a z score is calculated and familiar values like 1.96 and 2.58 are used as critical values for $\alpha = .05$ and $\alpha = .01$.

Mann-Whitney U Test for Small Samples

To give us some data to illustrate the Mann-Whitney U test, we invented some information about the intensive early training of the top ten male singles tennis players (Table 13.2).

Table 13.2 **Early Training of Top Ten Male Singles Tennis Players**

Initials	Rank	Intensive Early Training
Y.O.	1	No
U.E.	2	No
X.P.	3	No
E.C.	4	Yes
T.E.	5	No
D.H.	6	Yes
U.M.	7	No
O.R.	8	Yes
H.E.	9	Yes
R.E.	10	No

$\sum R_{yes} = 4 + 6 + 8 + 9 = 27$
$\sum R_{no} = 1 + 2 + 3 + 5 + 7 + 10 = 28$

There are two groups: one had intensive early training ($N_{yes} = N_1 = 4$), and a second did not ($N_{no} = N_2 = 6$).

The sum of the ranks for each group is shown at the bottom of the table. A U value can be calculated for each group and then the smaller of the two is used to enter Table G. For the "yes" group, the U value is

$$U = (N_1)(N_2) + \frac{N_1(N_1 + 1)}{2} - \sum R_1$$

$$= (4)(6) + \frac{(4)(5)}{2} - 27 = 7.$$

For the "no" group, the U value is

$$U = (N_1)(N_2) + \frac{N_2(N_2 + 1)}{2} - \sum R_2$$

$$= (4)(6) + \frac{(6)(7)}{2} - 28 = 17.$$

A convenient way to check your calculation of your U values is to know that the sum of the two U values is equal to $(N_1)(N_2)$. For our example, $7 + 17 = 24 = (4)(6)$.

Now, please examine Table G. It appears on two pages. To enter the table use the values for N_1 (to find the correct column) and N_2 (to find the correct row). Each intersection of N_1 and N_2 has two critical values. On the first page of the table the Roman type gives the critical values for α levels of .01 (one-tailed test) and .02 (two-tailed test). The numbers in boldface type are critical values for $\alpha = .005$ (one-tailed test) and $\alpha = .01$ (two-tailed test). In a similar way the second page handles larger α values for both one- and two-tailed tests. Now we are almost ready to enter Table G to find a probability figure for the smaller U value of 7.

From the conversation of the two child psychologists, it is clear that a two-tailed test is appropriate; they would be interested in knowing if intensive, early training *helps* or *hinders* players. Since an α level wasn't discussed, we will do what they would do—see if the difference is significant at the .05 level, and if it is, see if it is also significant at some smaller α level. Thus, in Table G we will begin by looking for the critical value of U for a two-tailed test with $\alpha = .05$. This number is on the second page in boldface type at the intersection of $N_1 = 4$, $N_2 = 6$. The critical value is 2. Since our obtained value of U is 7, we must *retain* the null hypothesis and conclude that there is no evidence from our sample that the distribution of players trained early and intensively is significantly different from the distribution of those without such training.

Note that in Table G you reject H_0 when the obtained U value is *smaller than* the tabled critical value.

Although you can easily find a U value using the method above and quickly go to Table G and reject or retain the null hypothesis, it would help your understanding of this test to think about small values of U. Under what conditions would you get a small U value? What kind of samples would give you a U value of zero? By examining the formula for U, you can see that $U = 0$ when the members of one sample all rank lower than every member of the other sample. Under such conditions, rejecting the null hypothesis seems reasonable. By playing with numbers in this manner, you can move from the rote memory level to the understanding level.

Assigning Ranks and Tied Scores

When your dependent variable comes to you as a set of ranks, a nonparametric test is the proper one to use. Many times, however, the dependent variable is scores from a test, time measures, or some readings from a dial. If you decide that a nonparametric test is in order, you will have to rank the scores. Two questions often arise. Is the largest or the smallest score ranked 1, and what should I do about the ranks for scores that are tied?

You will find the answer to the first question to be very satisfactory. It doesn't make any difference whether you call the largest or the smallest score 1.

Ties are handled by giving all tied scores the same rank. This rank is the mean of the ranks the tied scores would have if no ties had occurred. For example, if a distribution of scores was 12, 13, 13, 15, and 18, the corresponding ranks would be 1, 2.5, 2.5, 4, 5. The two scores of 13 would have been 2 and 3 if they had not been tied and 2.5 is the mean of 2 and 3. As another example, the scores 23, 25, 26, 26, 26, 29 would have ranks of 1, 2, 4, 4, 4, 6. Ranks of 3, 4, and 5 average out to be 4.

Ties do not affect the value of U if they are in the same group. If there are several ties that involve both groups, a correction factor may be advisable.[4]

Mann-Whitney U Test for Larger Samples

When one sample size is 21 or more, the normal curve should be used to assess probability. The z value is obtained by the formula

$$z = \frac{(U + c) - \mu_U}{\sigma_U}$$

where

$c = .5$, a correction factor explained below.

$$\mu_U = \frac{(N_1)(N_2)}{2}$$

and

$$\sigma_U = \sqrt{\frac{(N_1)(N_2)(N_1 + N_2 + 1)}{12}}.$$

c is a correction for continuity. It is used because the normal curve is a continuous function but the values of z that we may obtain in this test are discrete.

U, as before, is the smaller of the two possible U values.

Once a z score is obtained, the decision rules are the ones you have used in the past. For a two-tailed test, reject H_0 if $z \geq 1.96$ ($\alpha = .05$). For a one-tailed test, reject H_0 if $z \geq 1.65$ ($\alpha = .05$). The corresponding values for $\alpha = .01$ are $z \geq 2.58$ and $z \geq 2.33$.

Here is a problem for which the normal curve is necessary. An undergraduate psychology major was devoting a year to the study of memory. The principal independent variable was sex. Among her several experiments was one in which

---

[4]See Kirk (1978, p. 355) for the correction factor.

she asked the students in a General Psychology class to write down everything they remembered that was unique to the previous day's class, during which a guest had lectured. Students were encouraged to write down every detail they remembered. This class was routinely videotaped so it was easy to check each recollection for accuracy and uniqueness.

The scores, their ranks, and the statistical analysis are presented in Table 13.3.

Table 13.3

Number of Items Recalled by Males and Females, Ranks, and a Mann-Whitney Analysis

Males (N = 24)		Females (N = 17)	
Items Recalled	Rank	Items Recalled	Rank
70	3	85	1
51	6	72	2
40	9	65	4
29	13	52	5
24	15	50	7
21	16.5	43	8
20	18.5	37	10
20	18.5	31	11
17	21	30	12
16	22	27	14
15	23	21	16.5
14	24.5	19	20
13	26.5	14	24.5
13	26.5	12	28.5
11	30.5	12	28.5
11	30.5	10	33
10	33	10	33
9	35.5		
9	35.5		$\sum R_2 = 258$
8	37.5		
8	37.5		
7	39		
6	40		
3	41		
$\sum R_1 = 603$			

$$U = (N_1)(N_2) + \frac{N_1(N_1 + 1)}{2} - \sum R_1 = (24)(17) + \frac{(24)(25)}{2} - 603 = 105.$$

$$\mu_U = \frac{(N_1)(N_2)}{2} = \frac{(24)(17)}{2} = 204.$$

$$\sigma_U = \sqrt{\frac{(N_1)(N_2)(N_1 + N_2 + 1)}{12}} = \sqrt{\frac{(24)(17)(42)}{12}} = 37.79.$$

$$z = \frac{(U + c) - \mu_U}{\sigma_U} = \frac{105 + .5 - 204}{37.79} = -2.61.$$

Since the distribution of the number of recollections was very positively skewed, this student decided to run a Mann-Whitney test. (A plot of the scores in Table 13.3 will show this skew.) The z score of -2.61 led to rejection of the null hypothesis so she returned to the original data in order to interpret the results. Since the mean rank of the females, 15 (258 ÷ 17), is higher than that of the males, 25 (603 ÷ 24), and since higher ranks (those closer to 1) mean more recollections, she concluded that females recalled significantly more items than the males did.

Her conclusion is one that singles out central value for emphasis. On the average, females did better than males. The Mann-Whitney test, however, is one that compares distributions. What our undergraduate has done is what most researchers who use the Mann-Whitney do: she has assumed that the two populations have the same form but differ in central value. Thus, when significant U value is found, it is common to attribute it to a difference in central value.

Error Detection

Here are two checks you can easily make. First, the lowest rank will be the sum of the two N's. In Table 13.3, $N_1 + N_2 = 41$, which is the lowest rank. Second, when $\sum R_1$ and $\sum R_2$ are added together, they will equal $N(N + 1)/2$, where N is the total number of scores. In Table 13.3, $603 + 258 = (41)(42)/2$.

Now you can see how we figured the expected sum of 105 in the section on the rationale of nonparametric tests. There were 20 scores, so the overall sum of the ranks is

$$\frac{(20)(21)}{2} = 210.$$

Half of this total should be found in each group, so the *expected* sum of ranks of each group, both of which came from the same population, is 105.

Problems

1. In an experiment to determine the effects of estrogen on dominance, seven rats were given injections of that female hormone. The control group ($N = 8$) was injected with sesame oil, which is inert. Each rat was paired once with every other in a narrow runway. The rat that pushed the other one out was considered the more dominant of the two (see Work and Rogers, 1972). The number of bouts each rat won is listed next. Those with asterisks were injected with estrogen. Analyze the results with a Mann-Whitney test and write a conclusion about the effects of estrogen.

   14, 13, 12, 12, *11, 9, 8, *7, *6, 4, 3, *2, *2, *1, *0

2. Grackles, commonly referred to as blackbirds, are hosts for a number of parasites. One variety of parasite is a thin worm that lives in the tissue around the brain. To see if the incidence of this parasite was changing, four and twenty blackbirds were captured one winter from a pine thicket. The number of brain parasites was recorded for each bird. These data were compared with the infestation of

16 birds that had been captured from the same pine thicket ten years earlier. Analyze the data with a Mann-Whitney test and write a conclusion. Be careful and systematic in assigning ranks. Errors here are quite frustrating. Use an α level of .05.

Present Day

20	16	12	11	51	8	23	68	23	44	0	78
0	28	53	20	44	20	36	32	64	16	101	0

Ten Years Earlier

16	19	43	90	16	72	29	62
103	39	70	29	110	32	87	57

3. A friend of yours is trying to convince a mutual friend that the ride in an automobile built by [AMC, Chrysler (you choose)] is quieter than the ride in an automobile built by [Chrysler, AMC (no choice this time; you had just one degree of freedom—once the first choice was made, the second was determined)]. This friend has arranged to borrow six fairly new cars—three made by each company—and to drive your mutual friend around in the six cars (labeled *A–F*) blindfolded, until a stable ranking for quietness is achieved. So convinced is your friend that he insists on adopting $\alpha = .01$, "so that only the recalcitrant will not be convinced."

   Since you are the statistician in the group, you decide to do a Mann-Whitney U test on the results. However, being the type who thinks through the statistics *before* gathering any experimental data, you look at the appropriate subtable in Table G and find that the experiment, as designed, is doomed to retain the null hypothesis. Write an explanation for your friend, explaining why the experiment must be redesigned.

4. Suppose your friend talked three more persons out of their cars for an afternoon, labeled the cars *A–I*, changed α to .05, and conducted his "quiet" test. The results are shown in the following table.

Car	Company Y Ranks	Car	Company Z Ranks
B	1	H	3
I	2	A	5
F	4	G	7
C	6	E	8
		D	9

Perform a Mann-Whitney U test.

THE WILCOXON MATCHED-PAIRS SIGNED-RANKS TEST

The Wilcoxon matched-pairs signed-ranks test (1945) is appropriate for testing the difference between two correlated samples. In Chapter 9, we discussed three kinds of correlated-samples designs: natural pairs, matched pairs, and repeated measures (before and after). In each of these designs, a score in one group is logically paired with a score in the other group. If you are not sure of your understanding of the difference between a correlated-samples and an independent-samples design, you should review that section in Chapter 9. Being sure of the difference is necessary in order to make a proper choice between a Mann-Whitney U test and a Wilcoxon matched-pairs signed-ranks test.

The result of a Wilcoxon matched-pairs signed-ranks test is a T value,[5] which is interpreted using Table H. We will illustrate the rationale and calculation of T, using the four pairs of scores in Table 13.4.

First, the difference (D) between each pair of scores is found. The *absolute values* of these differences are then ranked, with the smallest difference given the rank of 1, the next smallest a rank of 2, and so on. The original sign of the difference is then given to the rank, and the positive ranks and the negative ranks are each summed separately. T is the *smaller of the absolute values* of the two sums.[6] For Table 13.4, $T = 4$. Note for this test it is the *differences* that are ranked, and not the scores themselves.

Table 13.4

Illustration of How to Calculate T

Pair	Variable 1	Variable 2	D	Rank	Signed Rank
A	16	24	−8	3	−3
B	14	17	−3	1	−1
C	23	18	5	2	2
D	23	9	14	4	4

Σ (positive ranks) = 6.

Σ (negative ranks) = −4.

$T = 4$.

The rationale is that, *if* there is no true difference between the two groups, the absolute value of the negative sum should be equal to the positive sum, with any deviations being due to sampling fluctuations.

Table H shows the critical values by sample size for both one- and two-tailed tests. Reject H_0 when T is equal to or *smaller* than the critical value in the table.[7]

We will illustrate the calculation and interpretation of a Wilcoxon matched-pairs signed-ranks test with an experiment based on some early work of Muzafer Sherif (1935). Sherif was interested in whether a person's basic perception could be influenced by others. The basic perception he used was a judgment of the size

---

[5]Be alert when you see a capital T in your outside readings; it has uses other than to symbolize the Wilcoxon matched-pairs signed-ranks test. Also note that this T is capitalized whereas the t in the t test and t distribution is not capitalized except by some computer printers that do not have lowercase letters.

[6]The Wilcoxon test is like the Mann-Whitney test in that you have a choice of two values for your test statistic. For both tests, choose the smaller value.

[7]Like the Mann-Whitney test your obtained statistic (T) must be smaller than the tabled critical value if you are to reject H_0.

of the *autokinetic effect.* The autokinetic effect is obtained when a person views a stationary point of light in an otherwise dark room. After a few moments, the light appears to move erratically. Sherif asked his subjects to judge how many inches the light moved. Under such conditions, judgments differ widely between individuals but they are fairly consistent for each individual. After establishing a stable mean for each subject, other observers were brought into the room. These new observers were confederates of the experimenter who always judged the movement of the light to be somewhat less than the subject did. Finally, the confederates left and the subject again made judgments until a stable mean was achieved. The before and after scores and the Wilcoxon matched-pairs signed-ranks test are shown in Table 13.5. The D column is simply the pretest minus the post-test. These D scores are then ranked by absolute size and the sign of the difference attached in the Signed-Ranks column. Notice that when $D = 0$, that pair of scores is dropped from further analysis and N is reduced by 1. The negative ranks have the smaller sum, so $T = 4$.

Table 13.5   **Data and a Wilcoxon Matched-Pairs Signed-Ranks Analysis of the Effects on Perception of Others' Judgments**

Subject	Mean Movement (inches) Before	After	D	Signed Ranks
1	3.7	2.1	1.6	4.5
2	12.0	7.3	4.7	10
3	6.9	5.0	1.9	6
4	2.0	2.6	−.6	−3
5	17.6	16.0	1.6	4.5
6	9.4	6.3	3.1	8
7	1.1	1.1	.0	eliminated
8	15.5	11.4	4.1	9
9	9.7	9.3	.4	2
10	20.3	11.2	9.1	11
11	7.1	5.0	2.1	7
12	2.2	2.3	−.1	−1

Check: $62 + 4 = 66$ $\sum$ positive $= 62.$

and $\dfrac{11(12)}{2} = 66$ $\sum$ negative $= -4.$

$T = 4.$

$N = 11.$

Since this T is smaller than the T value of 5 shown in Table H under $\alpha = .01$ (two-tailed test) for $N = 11$, the null hypothesis is rejected. The after scores represent a distribution that is different from the before scores. Now let's interpret this in terms of the experiment.

By examining the D column, you can see that all scores but two are positive. This means that, after hearing others give judgments smaller than one's own, the amount of movement seen was less. Thus, you may conclude (as did Sherif) that even basic perceptions tend to conform to perceptions expressed by others.

Tied Scores and $D = 0$

Ties among the D scores are handled in the usual fashion of assigning to each tied score the mean of the ranks that would have been assigned if there had been no ties. Ties do not affect the probability of the rank sum unless they are numerous (10 percent or more of the ranks are tied). In the case of numerous ties, the probabilities in Table H associated with a given critical T value may be too large. In a situation with numerous ties, the test is described as too conservative because it may fail to ascribe significance to differences that are in fact significant (Wilcoxon and Wilcox, 1964).

As you already know, when *one* of the D scores is zero, it is not assigned a rank and N is reduced by 1. When *two* of the D scores are tied at zero, each is given the average rank of 1.5. Each is kept in the computation with one being assigned a plus sign and the other a minus sign. If *three* D scores are zero, one is dropped, N is reduced by 1, and the remaining two are given signed ranks of +1.5 and −1.5.

Wilcoxon Matched-Pairs Signed-Ranks Test for Large Samples

When the number of pairs exceeds 50, the T statistic may be evaluated using the normal curve. The test statistic is

$$z = \frac{(T + c) - \mu_T}{\sigma_T}$$

where: T = smaller sum of the signed ranks,

   $c = .5,$

   $\mu_T = \dfrac{N(N + 1)}{4},$

   $\sigma_T = \sqrt{\dfrac{N(N + 1)(2N + 1)}{24}}.$

and N = number of pairs.

Problems

5. A private consultant was asked to evaluate a government job-retraining program. As part of the evaluation, she gathered information on 112 individuals' income before retraining and after retraining. She found a T value of 4077. Complete the analysis and draw a conclusion. Be especially careful in wording your conclusion.

6. Six industrial workers were chosen for a study of the effects of rest periods on production. Output was measured for one week before the new rest periods were instituted and again during the first week of the new schedule. Perform an appropriate statistical test on the following results.

Worker	Without Rests	With Rests
1	2240	2421
2	2069	2260
3	2132	2333
4	2095	2314
5	2162	2297
6	2203	2389

7. A political science student was interested in differences between two sets of Americans in attitudes toward government regulation of business. One set of Americans was made up of Canadians and the other of people from the United States. At a model UN session, the student managed to get 30 participants to fill out his questionnaire. High scores on this questionnaire indicated a favorable attitude toward government regulation of business. Analyze the following data with the appropriate nonparametric test and write a conclusion.

Canadians	United States
12	16
39	19
34	6
29	14
7	20
10	13
17	28
5	9
27	26
33	15
34	14
18	25
31	21
17	30
8	3

8. On the first day of class, one professor always gave his General Psychology class a Beginning-Point Test. The purpose was to find out the students' beliefs about punishment, reward, mental breakdowns, and so forth—topics that would be

Student	Before	After
1	18	4
2	14	14
3	20	10
4	6	9
5	15	10
6	17	5
7	29	16
8	5	4
9	8	8
10	10	4
11	26	15
12	17	9
13	14	10
14	12	12

covered during the course. Many of the items were phrased so that they represented a common misconception. For example, "mental breakdowns run in families and are usually caused by defective genes" was one item on the test. At the end of the course, the same test was given again. High scores mean lots of misconceptions. Analyze the before and after data and write a conclusion about the effect the General Psychology course had on misconceptions.

9. A health specialist conducted an eight-week workshop on weight control during which all 17 of the people who completed the course lost weight. In order to assess the long-term effects of the workshop, she weighed the participants again ten months later. The weight lost or gained is shown with a positive sign for those who continued to lose weight and a negative sign for those who gained some back. What can you conclude about the long-term effects of the workshop?

$$-5, \quad 24, \quad 0, \quad 13, \quad 9, \quad 6, \quad -7, \quad 31, \quad 2$$
$$-10, \quad -16, \quad 7, \quad 12, \quad -19, \quad -4, \quad 8, \quad 15$$

THE WILCOXON AND WILCOX MULTIPLE-COMPARISONS TEST

So far in this chapter on the analysis of ranked data, we have covered both designs for the two-group case (independent and correlated samples). The next step is to analyze results from three or more groups. The method presented here is one that allows you to compare all possible *pairs* of groups, regardless of the number of groups in the experiment. This is the nonparametric equivalent of a one-way ANOVA followed by Scheffé tests. A direct analogue of the overall F test is the Kruskal-Wallis one-way ANOVA on ranks, which is explained in many elementary statistics texts.

The **Wilcoxon and Wilcox multiple-comparisons test** (1964) is a method that allows you to compare all possible pairs of treatments. This is like running several Mann-Whitney tests, with a test for each pair of treatments. However, the Wilcoxon-Wilcox multiple-comparisons test keeps your α level at .05 or .01 regardless of how many pairs you have. The test is an extension of the procedures in the Mann-Whitney U test, and like it, requires independent samples. (Remember that Wilcoxon devised a test identical to the Mann-Whitney U test.)

The Wilcoxon and Wilcox method requires you to order the scores from the K samples into one overall ranking. Then the sum of the ranks in each group is computed. The rationale is that these sums should all be equal and that large differences in sums must reflect samples from different populations. Of course, the larger K is, the greater the likelihood of large differences by chance alone, and this is taken into account in the table of critical values, Table J.

The Wilcoxon and Wilcox test can be used only when N's for all groups are equal. A common solution to the problem of unequal N's is to reduce the too-large group(s) by throwing out one or more randomly chosen scores. A better solution is to design the experiment so that you have equal N's.

The data in Table 13.6 represent the results of an experiment conducted on a solar collector by two designer/entrepreneurs. These two had designed and built a 4-foot by 8-foot solar collector they planned to market, and they wanted to know the optimal rate at which to pump water through the collector. Since the rule of thumb for this is one-half gallon per hour per square foot of collector, they chose values of 14, 15, 16, and 17 gallons per hour for their experiment. Starting with the reservoir full of ice water, the water was pumped for one hour through the collector and back to the reservoir. At the end of the hour, the temperature of the water in the reservoir was measured in degrees centigrade. Then the water was replaced with ice water, the flow rate was changed and the process was repeated. The numbers in the body of Table 13.6 are the temperature measurements (to the nearest tenth of a degree).

Table 13.6

Experiment on Flow Rates in a Solar Collector

				Flow Rate (gal/hr)			
14	Rank	15	Rank	16	Rank	17	Rank
28.7	7	28.9	3	25.1	14	24.7	15
28.8	4	27.7	8	25.3	13	23.5	18
29.4	1	26.2	9	23.7	17	22.6	19
29.0	2	28.6	5	25.9	11	21.7	20
28.3	6	26.0	10	24.2	16	25.8	12
Σ (ranks)	20		35		71		84

Check: $20 + 35 + 71 + 84 = 210$,

$$\frac{N(N+1)}{2} = \frac{20(21)}{2} = 210.$$

There are six ways to make pairs of the four groups. The rate of 14 gallons per hour can be paired with 15, 16, and 17; the rate of 15 can be paired with 16 and 17; and the rate of 16 can be paired with 17. For each pair, a difference in the sum of ranks is found and the absolute value of that difference is compared with the critical value in Table J to see if it is significant.

Table J appears on two pages—one for the .05 level and one for the .01 level. In both cases, critical values are given for a two-tailed test. In the case of the data in Table 13.6, where $K = 4$, $N = 5$, you will find in Table J that rank-sum differences of 48.1 and 58.2 are required to reject H_0 at the .05 and .01 levels respectively.

A convenient summary table for the Wilcoxon-Wilcox multiple-comparisons test is shown in Table 13.7. The table displays, for each pair of independent variable values, the difference in the sum of the ranks. The next task is simply to see if any of these obtained differences is *larger* than the critical values of 48.1 or 58.2. At the .05 level, rates of 14 and 16 are significantly different from each other, as are 15 and 17. In addition, a rate of 14 is significantly different from a rate of 17 at the .01 level. What does all this mean for our two designer/entrepreneurs? Let's listen to their explanation to their old statistics professor.

"How did the flow-rate experiment come out, fellows?" inquired the kindly old gentleman.

"O.K., but we are going to have to do a followup experiment using different flow rates. We know that 16 and 17 gallons per hour are not as good as 14, but we don't know if 14 is optimal for our design. Fourteen was the best of the rates we tested, though. On our next experiment, we are going to test rates of 12, 13, 14, and 15."

The professor stroked his beard and nodded thoughtfully. "Typical experiment. You know more after it than you did before, ... but not quite enough."

Table 13.7 **Summary Table for Differences in the Sums of Ranks in the Flow-Rate Experiment**

Flow Rates		*Flow Rates*		
		14	15	16
	$\sum R$	(20)	(35)	(71)
	15 (35)	15		
	16 (71)	51*	36	
	17 (84)	64**	49*	13

$^{*}p < .05$
$^{**}p < .01$

Problems

10. Given the following summary data, test all possible comparisons. $N = 8$ for each group.

	Sum of Ranks for Six Groups					
	1	2	3	4	5	6
$\sum R$	196	281	227	214	93	165

11. The effect of three types of leadership on group productivity and satisfaction was investigated.[8] Groups of five children were randomly constituted and assigned an authoritarian, a democratic, or a laissez-faire leader. Nine groups were formed—three with each type of leader. The groups worked for a week on various projects. During this time, four different measures of each child's personal satisfaction with his/her group were taken. These were pooled to give one score for each child. The data are presented in the table. Test all possible comparisons with a Wilcoxon-Wilcox test.

[8]For a summary of a similar investigation, see Lewin (1958).

Leaders		
Authoritarian	*Democratic*	*Laissez-Faire*
120	108	100
102	156	69
141	92	103
90	132	76
130	161	99
153	90	126
77	105	79
97	125	114
135	146	141
121	131	82
100	107	84
147	118	120
137	110	101
128	132	62
86	100	50

12. List the tests presented so far in this chapter and the design for which each is appropriate. Be sure you can do this from memory.

CORRELATION OF RANKED DATA

We will begin this section with a short review of what you learned about correlation in Chapter 5.

1. Correlation requires a logical pairing of scores (a bivariate distribution).
2. Correlation is a method of describing the degree of relationship between two variables—that is, the degree to which high scores on one variable are associated with low or high scores on the other variable.
3. Correlation coefficients range in value from +1.00 (perfect positive) to −1.00 (perfect negative). A value of .00 indicates that there is no relationship between the two variables.
4. Statements about causal relations may not be made on the basis of a correlation coefficient alone.

In 1901, Charles Spearman (1863–1945) was "inspired by a book by Galton" and began experimenting at a little village school nearby. He wanted to see if there was a relationship between intellect (school grades) and sensory ability (detecting musical discord). He thought there was a relationship and he wanted to determine its *degree.* So, he developed a coefficient to express this degree. Later, he found that others were already ahead of him in developing a coefficient (Spearman, 1930).

Spearman's name is attached to the coefficient that is used to show the degree of correlation between two sets of *ranked* data. He used the Greek letter, ρ (rho), as the symbol for his coefficient. Later statisticians began to reserve Greek letters to indicate parameters, so the modern symbol for Spearman's statistic has become r_s, the s honoring Spearman.

r_s is a special case of the Spearman product-moment correlation coefficient and is most often used when the number of pairs of scores is small (less than 20).

Actually, r_s is a *descriptive statistic* that could have been introduced in the first part of this book. We waited until just before the end of the part of the book on inferential statistics to introduce it because r_s is a rank-order statistic, and this is a chapter about ranks. We will show you how to calculate this descriptive statistic and then show you how the statistical significance of r_s can be determined.

Calculation of r_s

The formula for r_s is:

$$r_s = 1 - \frac{6 \sum D^2}{N(N^2 - 1)},$$

where: D = the difference in ranks of a pair of scores
and N = the number of pairs of scores.

We started this chapter with speculation about men tennis players; we will end it with data about women tennis players. Suppose you were interested in the relationship between age and rank among professional women tennis players. r_s will give you a numerical index of the degree of the relationship. A high positive r_s would mean that, the older the player, the higher her rank. A high negative r_s would mean that, the older the player, the lower her rank. A zero or near zero r_s would indicate that there is no relationship between age and rank.

Table 13.8 shows the ten top-ranked women tennis players for 1982, their age as a rank score among the ten, and an $r_s = .22$. Now you can ask how much it means. Can you say with confidence to a friend, "There is a distinct tendency for older women to rank higher in tennis." Or, perhaps an $r_s = .22$ is not trustworthy and reliable. Phrasing the question in statistical language, "Is it likely that such an r_s would come from a population in which the true correlation is zero?" You will recall that you answered this kind of question for a Pearson r at the end of Chapter 9.

Testing the Significance of r_s

Table K in the Appendix gives values of r_s that are significant at the .05 and .01 levels when the number of pairs is 10 or less. The tennis data in Table 13.8 produced an $r_s = .22$ based on 10 pairs. Table K shows that a correlation of .648 (either positive or negative) is required for significance at the .05 level. Thus, a correlation of .22 is not statistically significant.

Notice in Table K that rather large correlations are required for significance. As with r, not much confidence can be placed in low or moderate correlation coefficients that are based on only a few pairs of scores.

For samples larger than 10, you may test the significance of r_s by comparing it to the tabled values in Table A, just as you did for a Pearson r.

Table 13.8   **The Top Ten Women Tennis Players in 1982. Their Rank in Age, and the Calculation of Spearman's r_s**

Player	Rank in Tennis	Rank in Age	D	D^2
Navratilova	1	3	−2	4
Lloyd	2	2	0	0*
Jaeger	3	10	−7	49
Mandlikova	4	6	−2	4
Shriver	5	7	−2	4
Turnbull	6	1	5	25
Potter	7	5	2	4
Bunge	8	9	−1	1
Austin	9	8	1	1
Hanika	10	4	6	36
				$\Sigma = 128$

$$r_s = 1 - \frac{6 \sum D^2}{N(N^2 - 1)} = 1 - \frac{6(128)}{10(99)} = .22$$

*Note that N is *not* reduced for r_s when a difference is zero.

Tied Ranks

The formula for r_s that you are working with is not designed to handle ties. With r_s, ties are troublesome. A tedious procedure has been devised to overcome ties, but your best solution is to arrange your data collection so that no ties occur. Sometimes, however, you are stuck with tied ranks, perhaps as a result of working with someone else's data. Kirk (1978) recommends assigning average ranks to ties, as you did for the three other procedures in this chapter, and then computing a Pearson r on the data.

Problems   13. Calculate r_s for the following hypothetical data. Determine whether it is significant at the .001 level of significance.

Years	Number of Marriages (in 10,000s)	Total Grain Crop (in Billions of Dollars)
1918	131	1.7
1919	142	2.5
1920	125	1.4
1921	129	1.7
1922	145	2.8
1923	151	3.3
1924	142	2.6
1925	160	3.7
1926	157	3.4
1927	163	3.5
1928	141	2.6
1929	138	2.6
1930	173	3.6
1931	166	3.9

14. With $N = 16$ and $\sum D^2 = 308$, calculate r_s and test its significance at the .05 level.

15. Two members of the Department of Philosophy (Locke and Kant) had the principal responsibility for hiring a new professor. Each privately ranked from 1 to 10 the ten candidates who met the objective requirements (degree, specialty, and so forth). The rankings are shown. Calculate an r_s. If the correlation is low, the philosophers probably have different sets of criteria of what's important. If so, they should discuss the two different sets. On the other hand, a high correlation probably means they each have about the same set of criteria.

| | Professors | |
Candidates	Locke	Kant
A	7	8
B	10	10
C	3	5
D	9	9
E	1	1
F	8	7
G	5	3
H	2	4
I	6	6
J	4	2

16. You are once again asked to give advice to a friend who comes to you for criticism of an experimental design. This friend has four pairs of scores obtained randomly from a population. Her intention is to calculate a correlation coefficient r_s and decide whether there is any significant correlation in the population. Give advice.

17. For each situation described, tell whether you would use a Mann-Whitney test, a Wilcoxon matched-pairs signed-ranks test, a Wilcoxon-Wilcox multiple-comparisons test, or an r_s.

a. A limnologist (a scientist who studies freshwater streams and lakes) measured algae growth in a lake before and after the construction of a nuclear reactor to see what effect the reactor had.

b. An educational psychologist compared the sociability scores of first-born children with scores of their next-born brother or sister to see if the two groups differed in sociability.

c. A child psychologist wanted to determine the degree of relationship between IQ scores obtained at age 3 and scores obtained from the same individuals at age 12.

d. A nutritionist randomly and evenly divided boxes of cornflakes cereal into three groups. One group was stored at 40° F, one group at 80° F, and one group alternated from day to day between the two temperatures. After 30 days, the vitamin content of each box was assayed.

e. The effect of STP gas treatment on gasoline mileage was assessed by driving six cars over a ten-mile course, adding STP, and then again driving over the same ten-mile course.

18. Fill in the descriptions for the following table.

	Symbol of Statistic (if any)	Appropriate for What Design?	Calculated Statistic Must Be (Larger, Smaller) Than The Tabled Statistic To Reject H_0
Mann-Whitney U Test			
Wilcoxon Matched-Pairs Signed-Ranks Test			
Wilcoxon-Wilcox Multiple-Comparisons Test			

OUR FINAL WORD

The basic idea of statistics is to let numbers stand for things of interest, manipulate the numbers according to the rules of statistics, translate the numbers back into the things, and finally, describe the relationship between the things of interest. The question of "What things are better understood when translated into numbers?" was not raised in this book, but it is important nonetheless. Here is an anecdote by E. F. Schumacher (1979) that helps emphasize that importance.

I will tell you a moment in my life when I almost missed learning something. It was during the war and I was a farm laborer and my task was before breakfast to go to yonder hill and to a field there and count the cattle. I went and I counted the cattle—there were always thirty-two—and then I went back to the bailiff, touched my cap, and said, "Thirty-two, sir," and went and had my breakfast. One day when I arrived at the field an old farmer was standing at the gate, and he said, "Young man, what do you do here every morning?" I said, "Nothing much. I just count the cattle." He shook his head and said, "If you count them every day they won't flourish." I went back, I reported thirty-two, and on the way back I thought, Well, after all, I am a professional statistician, this is only a country yokel, how stupid can he get. One day I went back, I counted and counted again, there were only thirty-one. Well, I didn't want to spend all day there so I went back and reported thirty-one. The bailiff was very angry. He said, "Have your breakfast and then we'll go up there together." And we went together and we searched the place and indeed, under a bush, was a dead beast. I thought to myself, Why have I been counting them all the time? I haven't prevented this beast dying. Perhaps that's what the farmer meant. They won't flourish if you don't look and watch the quality of each individual beast. Look him in the eye. Study the sheen on his coat. Then I might have gone back and said, "Well, I don't know how many I saw but one looks mimsey."

14

TALES OF DISTRIBUTIONS— PAST, PRESENT, AND FUTURE

In the preface, we said that we thought you would like this book. We hope we were right. We also hope that, through the use of this book, you have acquired a good basic knowledge of statistical methods and concepts and that you will now use this knowledge as a basis for designing and conducting experiments. We hope you are more attuned to defects in research designs and less likely to be influenced by conclusions coming from defective studies (or inappropriate conclusions from good studies). We hope that your study of statistics will not stop here, but that you will go on to more advanced courses for which this book has prepared you.

TALES OF DISTRIBUTIONS—PAST

If you have been reading for integration, the kind of reading in which you are continually asking the question, "How does this stuff relate to what I learned before?," you may have clearly in your head a principal thread that ties this book together—from Chapter 3 onward. That thread is named distributions. It is the importance of distributions that justifies the subtitle of this book, *Tales of Distributions*. We want to summarize explicitly some of what you have learned about distributions to help your integration of this material and to provide the groundwork that will allow us to describe some other statistical techniques that were not covered in this course but that you might encounter in future courses or in future encounters with statistics.

In Chapter 3, you were confronted with a disorganized array of scores on some variable. You learned to organize those scores into a frequency distribution, graph the distribution, and find the central values. In Chapter 4, you learned about the variability of distributions—especially the standard deviation—and how to express scores as z scores. In Chapter 5, you were introduced to bivariate distributions, where you learned to express the relationship between variables with a correlation coefficient and to predict scores on one variable from scores on another. In Chapter 6, you learned about the normal distribution and how to use means and standard deviations to find probabilities associated with distributions. In Chapter 7, you were introduced to the important concept of the sampling distribution. You learned there that distributions of statistics (like the mean) could be used to estimate parameters. Chapter 8 presented you with the idea that a difference between means (the heart of an experiment) had a sampling distribution and that conclusions about experiments were made on the basis of such sampling distributions. In Chapter 9, you learned that some experiments require a new kind of sampling distribution (t distribution) in order to get accurate probability figures. In Chapter 10, you learned about still another sampling distribution (F distribution) that is used when experiments with two or more groups are analyzed. The next step (Chapter 11) was an even more complex experiment, one with *two* independent variables for which probability figures could be obtained with the F distribution. Chapter 12 taught you how to use the chi square distribution when your data consisted of frequencies. In Chapter 13, you learned how to use distributions that do not require assumptions of normally distributed variables and equality of variances.

TALES OF DISTRIBUTIONS—FUTURE

There are several paths that might be taken from here. One more or less obvious next step is the analysis of an experiment with three independent variables. The three main effects and the several interactions are tested with an F distribution. A second possibility would be the analysis of experiments with correlated groups. Such experiments are often referred to as repeated-measures designs and there are repeated-measures analogues for simple ANOVA designs and factorial ANOVA designs. Again, an F distribution is the appropriate sampling distribution. A third possibility is to study techniques for the analysis of experiments that have more than one *dependent* variable. Such techniques are called multivariate statistics. Many of these techniques are analogous to those you have studied in this book, except there are two or more dependent variables instead of just one. For example, Hotelling's T^2 is analogous to the t test; one-way multiple analysis of variance (MANOVA) is analogous to one-way ANOVA; and higher-order MANOVA is analogous to factorial ANOVA. [For a good non-technical introduction to many of the techniques of multivariate statistics, see Chapter 1 in Harris (1975) or Chapter 1 in Johnson and Wichern (1982).]

TALES OF DISTRIBUTIONS—PRESENT

An almost universal condition at the end of a first course in statistics is a feeling of "almost understanding." Most students feel that they can analyze data and interpret results of experiments using any of the techniques they have studied, but they lack a "feel" for the whole course. When faced with the description of an experiment, they often have difficulty deciding upon the appropriate method of analysis.

Our experience in teaching statistics has taught us that the best way for you to bring all the bits and pieces you have learned together is to reread the whole book, giving particular emphasis to the parts that you recognize as especially important or that you have forgotten, and less emphasis to the parts that are less important or are well remembered. It will take a considerable amount of time for you to do this—probably 8 to 16 hours. However, the benefits are great, according to students who have invested the time. A typical quote is, "Well, during the course I would read a chapter and work the problems without too much trouble; but I didn't relate that to the earlier stuff. When I went back over the whole book, though, I got the big picture. All the pieces do fit together."

To assist you in getting the "big picture," we have posed some problems for you. If you have reread this book, you will find these problems much easier to solve.

The problems are divided into two sets. The first set requires you to decide what statistical test or descriptive statistic will answer the question the investigator is asking. The second set requires you to interpret the results of an analyzed experiment. You will probably be glad to find out that no number crunching is required for these problems. You can put away your calculator.

Here is a word of warning. For a few of the problems, the statistical techniques necessary for analysis are not covered in this book. Thus, part of your task is to recognize what problems you are prepared to solve and what problems would require some digging into more advanced statistics textbooks.

Problems   **Set A.** Determine what descriptive statistic or inferential test is appropriate for each problem in this set.

1. A company had three separate divisions. Based on the capital invested, one

year Division A made 10 cents per dollar, Division B made 20 cents per dollar, and Division C made 30 cents per dollar. How can the overall profit for the company be found?

2. Reaction time scores tend to be severely skewed. The majority of the scores are quick responses and there are diminishing numbers of scores in the slower categories. A student wanted to find out the effects of alcohol on reaction time, so he found the reaction time of each subject under both conditions—alcohol and no alcohol.

3. As part of a two-week treatment for phobias, a therapist measured the general anxiety level of each client four times: before, after one week of treatment, at the end of treatment, and four months later.

4. "I want a number that will describe the typical score on this test. Most did very well, some scored in the middle, and a very few did quite poorly."

5. The designer for a city park had some benches built that were 16.7 inches high. The mean distance from the bottom of the shoe to the back of the knee is 18.1 inches for females. The standard deviation is .70 inch. What proportion of the women who use the bench will have their feet dangle (unless they sit forward on the bench)?

6. A promoter of low-cost, do-it-yourself buildings found the insulation scores (R values) for 12 samples of concrete mixed with clay and for 12 samples of concrete mixed with hay. Each sample was 6 inches thick. He wanted to know if clay or straw was superior as an insulating additive.

7. A large school district wanted to know the range of scores within which they could expect the mean reading achievement of their sixth-grade students to be. A random sample of 50 of these students and their scores on a standardized reading achievement test was available.

8. Based on several large studies, a researcher knew that 40 percent of the public in the U.S. favored capital punishment, 35 percent were against it, and 25 percent expressed no strong feelings. The researcher polled 150 Canadians on this question with the intention of comparing the two countries in attitudes toward capital punishment.

9. What descriptive statistic could be used to identify the most typical choice of major at your college?

10. An experimentalist wanted to express with a correlation coefficient the strength of the relationship between stimulus intensity and pleasantness. This experimentalist worked with music and he knew that both very low and very high intensities are not pleasant and that the middle range of intensity produces the highest pleasantness ratings. Would you recommend a Pearson r or an r_s?

11. A consumer-testing group compared Boraxo and Tide to determine which got laundry whiter. White towels that had been subjected to a variety of filthy treatments were identified on each end and were cut in half. Each half was then washed in either Boraxo or Tide. After washing, each half was tested with a photometer for the amount of light reflected.

12. An investigator wanted to predict a male adult's height from his length at birth. He obtained records of both measures from a sample of male military personnel.

13. An experimenter was interested in the effect of expectations and drugs on alertness. Each subject was given an amphetamine (stimulant) or a placebo (an inert substance). In addition, half the subjects in each group were told that the drug taken was a stimulant, and half that the drug was a depressant. An alertness score for each subject was obtained from a composite of measures which included a questionnaire and observation of the subjects.

14. A teacher told a class that she found a correlation coefficient of .20 between

the alphabetical order of a person's last name and that person's cumulative point total for a General Psychology course. One student in the class wondered whether the correlation coefficient was reliable; that is, would a nonzero correlation be likely to be found in other classes. How can a probability figure be found without calculating an r on another class?

15. As part of a test for some advertising copy, a company assembled 120 people who rated four toothpastes. They then read several ads (including the ad being tested). Then they rated all four toothpastes again. The data to be analyzed consisted of the number of people who rated Crest highest before reading the ad and the number who rated it highest after reading the ad.

16. Most light-bulb manufacturing companies claim that their 60-watt bulbs have an average life of 1000 hours. A skeptic with some skill as an electrician wired up some sockets and timers and burned 40 bulbs until they failed. The life of each bulb was recorded automatically. The skeptic wanted to know if the companies' claim was justified.

17. An investigator wanted to know the degree to which a person's education was related to his or her satisfaction in life. The investigator had a way to measure both of these variables.

18. In an effort to find out the effect of a "terrible drug" on reaction time, an investigator administered 0, 25, 50, or 75 mg to four groups of volunteer rats. The quickness of their response to a tipping platform was measured in seconds.

19. The experimenters (a male and a female) staged 140 "shoplifting" incidents in a grocery store. In each case, the "shoplifter" (one experimenter) picked up a carton of cigarettes in full view of a customer and then walked out. Half the time the experimenter was well dressed, and half the time sloppily dressed. Obviously, half the incidents involved a male shoplifter and half involved a female. For each incident, the experimenters (with the cooperation of the checkout person) simply noted whether the shoplifter was reported or not.

20. A psychologist wanted to illustrate for a group of seventh graders the fact that items in the middle of a series are more difficult to learn than those at either end. The students learned an eight-item series. The psychologist found the mean number of repetitions necessary to learn each of the items.

21. A nutritionist, with the help of an anthropologist, gathered data on the incidence of cancer among 21 different cultures. The incidence data were quite skewed, with many cultures having low incidences and the rest scattered among the higher incidences. In eight of the cultures, red meat was a significant portion of the diet. For the other 13 groups, the diet consisted primarily of cereals.

22. A military science teacher wanted to find if there was any relationship between the rank in class of graduates of West Point and their military rank in the service at age 45.

23. The boat dock operators at a lake sponsored a fishing contest. Prizes were to be awarded to the persons who caught the largest bass, the largest crappie, and the largest bream, and to the overall winner. The problem in deciding the overall winner is that the bass are by nature larger than the other two species of fish. Describe an objective way to find the overall winner that will be fair to all three kinds of contestants.

Set B. Read each problem, look at the result of the statistical analysis, and write an appropriate conclusion. For each conclusion, use the terms used in the problem. That is, rather than saying "the null hypothesis was rejected at the .05 level," say "those with medium anxiety solved problems more quickly than those with high anxiety."

For some of the problems, the design of the study has flaws. That is, there are uncontrolled extraneous variables that would make it impossible to draw precise conclusions about the experiment. It is a very good thing for you to be able to recognize such design flaws, but what we want you to do in these problems is to interpret the statistics. Thus, don't let design flaws keep you from drawing a conclusion based on the statistic.

For all problems, use a two-tailed test with $\alpha = .05$. If the results are significant at the .01 or .001 level, report that, but treat any difference that has a probability greater than .05 as being due to chance.

24. Four kinds of herbicides were compared for their weed-killing characteristics. Eighty plots were randomly but evenly divided and one herbicide was applied to each. Since the effects of two of the herbicides were quite variable, the dependent variable was reduced to ranks and a Wilcoxon-Wilcox multiple-comparisons test was run. The plot with the fewest weeds remaining was given a rank of 1. The summary table is shown next.

	A (936.5)	B (316.5)	C (1186)
B (316.5)	620		
C (1186)	249.5	869.5	
D (801)	135.5	482.5	385

25. In a particular recycling process, the break even point for each batch was when 54 kg of the raw material was unusable foreign matter. An engineer developed a screening process that left a mean of 38 kg of foreign material. The upper and lower limits of a 95 percent interval were 30 and 46 kg.

26. The inference theory of the serial position effect predicts that subjects who are given extended practice will perform more poorly on the initial items of a series than will subjects who are not given extended practice. An experiment was done and the mean number of errors on initial items is shown in the table. A t test with 54 df produced a value of 3.88.

	Extended Practice	No Extended Practice
Mean Number of Errors	13.7	18.4

27. A developmental psychologist developed a theory that predicted the proportion of children who would, during a period of stress, cling to their mother, attack the mother, or attack a younger sibling. The stress situation was set up and the responses of 50 children recorded. A χ^2 was calculated and found to be 5.30.

28. An experimental psychologist at a VA hospital obtained clearance from the ethics committee to conduct a study of the efficacy of Elavil for treatment of depression. Patients whose principal problem was depression were chosen for the study. The experiment lasted 30 days, during which one group was given placebos, one was given a low dose of Elavil, and one was given a high dose of Elavil. At the end of the experiment, the patients were observed by two outside psychologists who rated several behaviors (eye contact, posture, verbal output, activity, and such) for degree of depression. A composite score was obtained on each patient, with low scores indicating depression.

The mean scores were: placebo, 9.86; low dose, 23.00; and high dose, 9.21. An ANOVA summary table and two Scheffé comparisons are shown in the table.

Source	df	F
Between groups	2	21.60
Within groups	17	

F'_{ob} (placebo vs. low) = 28.69
F'_{ob} (placebo vs. high) = .19

29. One sample of 16 third graders had been taught to read by the "look-say" method. A second sample of 18 had been taught by the phonics method. Both were given a reading achievement test and a Mann-Whitney U test was performed on the results. A U value of 51 was found.

30. Sociologists sometimes make a profile of the attitudes of a group by asking each individual to rate his or her tolerance of people with particular characteristics. For example, members of a group might be asked to assess, on a scale of 1–7, their tolerance of prostitutes, atheists, former mental patients, intellectuals, college graduates, and so on. The mean score for each category is then calculated and the categories are ranked from low to high. When 18 categories of people were ranked from low to high by a group of people who owned businesses and by a group of college students, an r_s of .753 was found.

31. In a large high school, one group of juniors took an English course that included a nine-week unit on poetry. Another group of juniors studied plays during that nine-week period. Afterward, both groups completed a questionnaire on their attitudes toward poetry. High scores mean favorable attitudes. The means and variances are presented in the table.

	Attitudes toward Poetry	
	Studied Poetry	Studied Plays
Mean	23.7	21.7
Variance	51.3	16.1

To compare the means, a t test was run: $t(78\ df) = 0.48$. To compare the variances, an F test was performed: $F(37,41) = 3.62$. (See footnote 6, Chapter 10.)

32. Here is an example of the results of experiments that asked the question: Can you change the attitudes of an audience by presenting just one side of an argument claiming that it is correct, or should you present both sides and then claim that one side is correct? In the following experiment, a second independent variable is also examined: level of education. The first table contains the mean change in attitude and the second is an ANOVA summary table.

	Presentation	
Level of Education	One Side	Both Sides
Less than a high school diploma	4.3	2.7
At least one year college	2.1	4.5

Source	df	F
Between presentations	1	2.01
Between education	1	1.83
Presentations × education	1	7.93
Within groups	44	

33. A college student was interested in the relationship between handedness and verbal ability. She gave three classes of third-grade students a test of verbal ability that she had devised and then classified each child as right-handed, left-handed, or mixed. The obtained F value was 2.63.

	Handedness		
	Left	Right	Mixed
Mean verbal ability score	21.6	25.9	31.3
N	16	36	12

34. As part of a large-scale study on alcoholism, the alcoholics and the control group were classified according to when they were toilet-trained as children. The three categories were early, normal, or late. A χ^2 of 7.90 was found.

APPENDIXES

1. References *310*
2. Tables *314*
3. Glossary of Words *333*
4. Glossary of Formulas *338*
5. Answers to Problems *344*

REFERENCES

Austin, A. W. *American freshmen: National norms for fall, 1982.* Los Angeles, Calif., American Council of Education, 1982.

Barker, R. B., Dembo, T., & Lewin, K. Frustration and regression: An experiment with young children, *University of Iowa Studies in Child Welfare*, 1941, *18*, No. 1.

Barker, R. B., Dembo, T., Lewin, K., & Wright, M. E. Experimental studies of frustration in young children. In T. M. Newcomb & E. L. Hartley (Eds.), *Readings in social psychology*, New York: Holt, Rinehart and Winston, 1947.

Birch, H. G.; & Rabinowitz, H. S. The negative effect of previous experience on productive thinking. *Journal of Experimental Psychology*, 1951, *41*, 121–125.

Blair, R. C., Higgins, J. J., & Smitley, W. D. S. *On the relative power of the U and t tests. British Journal of Mathematical and Statistical Psychology*, 1980, *33*, 114–120.

Bradley, D. R., Bradley, T. D., McGrath, S. G., & Cutcomb, S. D. Type I error rate of the chi-square test of independence in R & C tables that have small expected frequencies. *Psychological Bulletin*, 1979, *86*, 1290–1297.

Bradley, J. V. Robustness? *British Journal of Mathematical Statistical Psychology*, 1978, *31*, 144–152.

Brady, J. V., Porter, R. W., Conrad, D. G., & Mason, J. W. Avoidance behavior and the development of gastroduodenal ulcers. *Journal of the Experimental Analysis of Behavior*, 1958, *1*, 69–73.

Brehm, J. W., & Cohen, A. R. *Explorations in cognitive dissonance.* New York: Wiley, 1962.

Bureau of Labor Statistics. Union wages and hours of motor truck drivers and helpers. July 1, 1944, *Monthly Labor Review*, December 1944.

Calhoun, J. P., & Johnston, J. O. Manifest anxiety and visual acuity. *Perceptual and Motor Skills*, 1968, *27*, 1177–1178.

Camilli, G., & Hopkins, K. D. Applicability of chi-square to 2×2 contingency tables with small expected cell frequencies. *Psychological Bulletin*, 1978, *85*, 163–167.

Campbell, S. K. *Flaws and fallacies in statistical thinking.* Englewood Cliffs, N.J.: Prentice-Hall, 1974.

Cornell, Francis G. *The essentials of educational statistics.* New York: John Wiley, 1956.

Downie, N. M., & Heath, R. W. *Basic statistical methods* (5th ed.), New York: Harper & Row, 1983.

Edwards, A. L. *An introduction to linear regression and correlation.* San Francisco: Freeman, 1976.

Edwards, A. L. *Experimental design in psychological research* (4th ed.). New York: Holt, Rinehart and Winston, 1972.

Ellis, Willis D. *A source book of Gestalt psychology.* London: Routledge & Kegan Paul, 1938.

Ferguson, G. A. *Statistical analysis in psychology and education* (5th ed.). New York: McGraw-Hill, 1981.

Fisher, R. A., & Yates, F. *Statistical tables for biological, agricultural, and medical research* (6th ed.). Edinburgh: Oliver and Boyd, 1963.

Forbs, R., & Meyer, A. B. *Forestry handbook.* New York: Ronald Press, 1955.

Games, P. A., & Klare, G. R. *Elementary statistics.* New York: McGraw-Hill, 1967.

Gardner, P. L. Scales and statistics. *Review of Educational Research*, 1975, *45*, 43–57.

Garfield, S. Research problems in clinical diagnosis. *Journal of Consulting and Clinical Psychology*, 1978, *46*, 596–607.

Greene, J. E. (Ed.-in-chief) *McGraw-Hill modern men of science.* New York: McGraw-Hill, 1966.

Guilford, J. P. *The nature of human intelligence.* New York: McGraw-Hill, 1967.

Guilford, J. P., & Fruchter, B. *Fundamental statistics in psychology and education* (6th ed.). New York: McGraw-Hill, 1978.

Harris, R. J. *A primer of multivariate statistics.* New York: Academic Press, 1975.

Hasler, A. D. *Underwater guideposts.* Madison: University of Wisconsin Press, 1966.

Howell, D. C. *Statistical methods for psychology.* Boston: Duxbury, 1982.

Huff, D. *How to lie with statistics.* New York: Norton, 1954.

Jenkins, J. G., & Dallenbach, K. M. Obliviscence during sleep and waking. *American Journal of Psychology*, 1924, *35*, 605–612.

Johnson, R. A., & Wichern, D. W. *Applied multivariate statistical analysis.* New York: Prentice-Hall, 1982.

Johnston, J. O., & Maertens, N. W. Effects of parental involvement in arithmetic homework. *School Science and Mathematics*, 1972, *72*, 117–126.

Johnston, J. O., & Spatz, K. C. Racial attitudes and self-esteem levels among Southeast Arkansas public school students. *Educational Research in Arkansas*, 1971–72. Little Rock: Arkansas Department of Higher Education, 1973.

Kendall, M. G. *Rank correlation methods* (4th ed.). London: Charles Griffin, 1970.

Keppel, Geoffrey. *Design and analysis: A researcher's handbook.* Englewood Cliffs: Prentice-Hall, 1982.

Kirk, Roger E. *Introductory statistics.* Monterey, Calif.: Brooks/Cole, 1978.

Kirk, Roger E. *Experimental design: Procedures for the behavioral sciences* (2nd ed.). Monterey, Calif.: Brooks/Cole, 1982.

Kirk, Roger E. (Ed.). *Statistical issues: A reader for the behavioral sciences.* Monterey, Calif.: Brooks/Cole, 1972.

Kish, L. *Survey sampling.* New York: John Wiley, 1965.

Lewin, K. Group decision and social change. E. E. Maccoby, T. M. Newcomb, & E. L. Hartley (Eds.), *Readings in social psychology* (3rd ed.). New York: Holt, Rinehart and Winston, 1958.

Loewi, Otto. The night prowler. In S. Rapport & H. Wright (Eds.), *Science: method and meaning.* New York: Washington Square Press, 1963.

Loftus, G. R., & Loftus, E. F. *Essentials of statistics.* Monterey, CA: Brooks/Cole, 1982.

Mann, H. B., & Whitney, D. R. On a test of whether one or two random variables is stochastically larger than the other. *Annals of Mathematical Statistics*, 1947, *18*, 50–60.

Maslach, C., Marshall, G., & Zimbardo, P. Hypnotic control of complex skin temperature. *Proceedings of the Annual Convention of the American Psychological Association*, 1971, *6*, 777–778.

Maugh, T. H. Guayule and jojoba: Agriculture in semiarid regions. *Science*, 1977, *196*, 1189–1190.

Mayo, E. *The human problems of an industrial civilization.* Boston: Harvard University, 1946.

McGaugh, J. L., & Petrinovich, L. F. Effects of drugs on learning and memory. *International review of neurobiology* (Vol. 8). New York: Academic Press, 1965.

McMullen, L., & Pearson, E. S. William Sealy Gosset, 1876–1937. *Biometrika*, 1939, *30*, 205–253.

McNemar, Q. *Psychological statistics.* New York: John Wiley, 1969.

Minium, E. W. *Statistical reasoning in psychology and education* (2nd ed.). New York: Wiley, 1978.

Minium, E. W., & Clarke, R. B. *Elements of statistical reasoning.* New York: John Wiley, 1982.

Mischel, H. N. Sex bias in the evaluation of professional achievements. *Journal of Educational Psychology,* 1974, *66,* 157–166.

Natrella, Mary G. The relation between confidence intervals and tests of significance. *American Statistician,* 1960, *14,* 20–33.

Overall, J. E. Power of chi-square tests for 2×2 contingency tables with small expected frequencies. *Psychological Bulletin,* 1980, *87,* 132–135.

Pearson, E. S. W. S. Gosset. *Dictionary of national biography: 1931–1940.* London: Oxford University Press, 1949.

Pearson, E. S., & Hartley, H. O. (eds.). *Biometrika tables for statisticians* (2nd ed., Vol. 1). New York: Cambridge University Press, 1958.

Rokeach, M., Homant, R., & Penner, L. A value analysis of the disputed Federalist papers. *Journal of Personality & Social Psychology,* 1970, *16,* 245–250.

Scheffé, H. A. method for judging all contrasts in the analysis of variance. *Biometrika,* 1953, *40,* 87–104.

Schumacher, E. F. *Good work.* New York: Harper & Row, 1979.

Sherif, M. A study of some social factors in perception. *Archives of Psychology,* 1935, No. 187.

Sherman, M. The differentiation of emotional responses in infants: The ability of observers to judge the emotional characteristics of the crying infants, and of the voice of an adult. *Journal of Comparative Psychology* 1927, *7,* 335–351.

Siegel, S. *Nonparametric statistics for the behavioral sciences.* New York: McGraw-Hill, 1956.

Snedecor, G. W., & Cochran, W. G. *Statistical methods* (6th ed.). Ames, Iowa: Iowa State University Press, 1967.

Spearman, C. Autobiography. In C. Murchison (ed.). *History of psychology in autobiography.* New York: Russell & Russell, 1961.

Spence, J. T., Cotton, J. W., Underwood, B. J., & Duncan, C. P. *Elementary statistics* (4th ed.). Englewood Cliffs, N.J: Prentice-Hall, 1983.

Stevens, S. S. On the theory of scales of measurement. *Science,* 1946, *103,* 677–680.

Stewart, I. Gauss. *Scientific American.* 1977, *237.* July, 123–131.

Stoudt, W. H., & McFarland, R. A. Height and weight of white Americans. *Human Biology,* 1960, *32,* 331.

Taylor, Janet. A personality scale of manifest anxiety. *Journal of Abnormal and Social Psychology,* 1953, *48,* 285–290.

Ury, H. In response to Noether's letter, "Needed—a new name." *American Statistician,* 1967, *21*(4), 53.

Van Cott, H. P., & Kinkade, R. G. *Human engineering guide to equipment design* (Rev. ed.). New York: McGraw-Hill, 1972.

Walker, Helen. Degrees of freedom. *Journal of Educational Psychology,* 1940, *31,* 253–269.

Walker, H. M. *Mathematics essential for elementary statistics* (Rev. ed.). New York: Holt, Rinehart and Winston, 1951.

Walker, H. M. *Studies in the history of statistical method.* Baltimore: Williams and Wilkins, 1929.

Walker, H., & Lev, J. *Statistical inference.* New York: Holt, Rinehart and Winston, 1953.

Watson, J. B. *Psychology from the standpoint of a behaviorist* (2nd ed.). Philadelphia: Lippincott, 1924.

Weinberg, G. H., Schumaker, J. A., & Oltman, D. *Statistics: An intuitive approach* (4th ed.). Monterey, Calif.: Brooks/Cole, 1981.

Wert, J. E., Neidt, C. O., & Ahmann, J. S. *Statistical methods in educational and psychological research*. New York: Appleton-Century-Crofts, 1954.

White, E. B. *The trumpet of the swan*. New York: Harper & Row, 1970.

Wilcoxon, F. Individual comparisons by ranking methods. *Biometrics*, 1945, *1*, 80–83.

Wilcoxon, F., & Wilcox, R. A. *Some rapid approximate statistical procedures* (Rev. ed.). Pearl River, N.Y.: Lederle Laboratories, 1964.

Winer, B. F. *Statistical principles in experimental design* (2nd ed.). New York: McGraw-Hill, 1971.

Woodworth, R. S. Introduction. In Henry E. Garrett, *Statistics in psychology and education*. New York: Longmans, Green, 1926.

Work, M. S., & Rogers, H. Effects of estrogen level on food-seeking dominance among male rats. *Journal of Comparative and Physiological Psychology*, 1972, *79*, 414–418.

Youden, W. J. *Experimentation and measurement*. National Science Teachers Assn., 1962.

Zajonc, R. B. Dumber by the dozen. *Psychology Today*, 1975, *8*, No. 8, 37–43.

TABLES

Table A   Critical Values for Pearson Product-Moment Correlation Coefficients, r *315*

Table B   Random Digits   *316*

Table C   Area under the Normal Curve between μ and z and beyond z *320*

Table D   The t Distribution   *322*

Table E   Chi Square Distribution   *323*

Table F   The F Distribution   *324*

Table G   Critical Values for the Mann-Whitney U Test   *328*

Table H   Critical Values for the Wilcoxon Matched-Pairs Signed-Ranks T Test   *330*

Table J   Critical Differences for the Wilcoxon-Wilcox Multiple-Comparisons Test   *331*

Table K   Critical Values for Spearman's r_s *332*

minimal values to be called statistically signif.

Table A   Critical Values for Pearson Product-Moment Correlation Coefficients, r*

of freedom (N-2)

df	α Levels (Two-Tailed Test)				
	.1	.05	.02	.01	.001
1	.98769	.99692	.999507	.999877	.9999988
2	.90000	.95000	.98000	.990000	.99900
3	.8054	.8783	.93433	.95873	.99116
4	.7293	.8114	.8822	.91720	.97406
5	.6694	.7545	.8329	.8745	.95074
6	.6215	.7067	.7887	.8343	.92493
7	.5822	.6664	.7498	.7977	.8982
8	.5494	.6319	.7155	.7646	.8721
9	.5214	.6021	.6851	.7348	.8371
10	.4973	.5760	.6581	.7079	.8233
11	.4762	.5529	.6339	.6835	.8010
12	.4575	.5324	.6120	.6614	.7800
13	.4409	.5139	.5923	.6411	.7603
14	.4259	.4973	.5742	.6226	.7420
15	.4124	.4821	.5577	.6055	.7246
16	.4000	.4683	.5425	.5897	.7084
17	.3887	.4555	.5285	.5751	.6932
18	.3783	.4438	.5155	.5614	.6787
19	.3687	.4329	.5034	.5487	.6652
20	.3598	.4227	.4921	.5368	.6524
25	.3233	.3809	.4451	.4869	.5974
30	.2960	.3494	.4093	.4487	.5541
35	.2746	.3246	.3810	.4182	.5189
40	.2573	.3044	.3578	.3932	.4896
45	.2428	.2875	.3384	.3721	.4648
50	.2306	.2732	.3218	.3541	.4433
60	.2108	.2500	.2948	.3248	.4078
70	.1954	.2319	.2737	.3017	.3799
80	.1829	.2172	.2565	.2830	.3568
90	.1726	.2050	.2422	.2673	.3375
100	.1638	.1946	.2301	.2540	.3211
	.05	.025	.01	.005	.0005

α Levels (One-Tailed Test)

*To be significant the r obtained from the data must be equal to or larger than the value shown in the table.

Table A is taken from Table VII of Fisher and Yates: *Statistical Tables for Biological, Agricultural and Medical Research*, published by Longman Group Ltd., London. (Previously published by Oliver & Boyd, Edinburgh), and by permission of the authors and publishers.

Table B Random Digits

	00–04	05–09	10–14	15–19	20–24	25–29	30–34	35–39	40–44	45–49
00	54463	22662	65905	70639	79365	67382	29085	69831	47058	08186
01	15389	85205	18850	39226	42249	90669	96325	23248	60933	26927
02	85941	40756	82414	02015	13858	78030	16269	65978	01385	15345
03	61149	69440	11286	88218	58925	03638	52862	62733	33451	77455
04	05219	81619	10651	67079	92511	59888	84502	72095	83463	75577
05	41417	98326	87719	92294	46614	50948	64886	20002	97365	30976
06	28357	94070	20652	35774	16249	75019	21145	05217	47286	76305
07	17783	00015	10806	83091	91530	36466	39981	62481	49177	75779
08	40950	84820	29881	85966	62800	70326	84740	62660	77379	90279
09	82995	64157	66164	41180	10089	41757	78258	96488	88629	37231
10	96754	17676	55659	44105	47361	34833	86679	23930	53249	27083
11	34357	88040	53364	71726	45690	66334	60332	22554	90600	71113
12	06318	37403	49927	57715	50423	67372	63116	48888	21505	80182
13	62111	52820	07243	79931	89292	84767	85693	73947	22278	11551
14	47534	09243	67879	00544	23410	12740	02540	54440	32949	13491
15	98614	75993	84460	62846	59844	14922	48730	73443	48167	34770
16	24856	03648	44898	09351	98795	18644	39765	71058	90368	44104
17	96887	12479	80621	66223	86085	78285	02432	53342	42846	94771
18	90801	21472	42815	77408	37390	76766	52615	32141	30268	18106
19	55165	77312	83666	36028	28420	70219	81369	41943	47366	41067
20	75884	12952	84318	95108	72305	64620	91318	89872	45375	85436
21	16777	37116	58550	42958	21460	43910	01175	87894	81378	10620
22	46230	43877	80207	88877	89380	32992	91380	03164	98656	59337
23	42902	66892	46134	01432	94710	23474	20423	60137	60609	13119
24	81007	00333	39693	28039	10154	95425	39220	19774	31782	49037
25	68089	01122	51111	72373	06902	74373	96199	97017	41273	21546
26	20411	67081	89950	16944	93054	87687	96693	87236	77054	33848
27	58212	13160	06468	15718	82627	76999	05999	58680	96739	63700
28	70577	42866	24969	61210	76046	67699	42054	12696	93758	03283
29	94522	74358	71659	62038	79643	79169	44741	05437	39038	13163
30	42626	86819	85651	88678	17401	03252	99547	32404	17918	62880
31	16051	33763	57194	16752	54450	19031	58580	47629	54132	60631
32	08244	27647	33851	44705	94211	46716	11738	55784	95374	72655
33	59497	04392	09419	89964	51211	04894	72882	17805	21896	83864
34	97155	13428	40293	09985	58434	01412	69124	82171	59058	82859
35	98409	66162	95763	47420	20792	61527	20441	39435	11859	41567
36	45476	84882	65109	96597	25930	66790	65706	61203	53634	22557
37	89300	69700	50741	30329	11658	23166	05400	66669	48708	03887
38	50051	95137	91631	66315	91428	12275	24816	68091	71710	33258
39	31753	85178	31310	89642	98364	02306	24617	09609	83942	22716
40	79152	53829	77250	20190	56535	18760	69942	77448	33278	48805
41	44560	38750	83635	56540	64900	42912	13953	79149	18710	68618
42	68328	83378	63369	71381	39564	05615	42451	64559	97501	65747
43	46939	38689	58625	08342	30459	85863	20781	09284	26333	91777
44	83544	86141	15707	96256	23068	13782	08467	89469	93842	55349
45	91621	00881	04900	54224	46177	55309	17852	27491	89415	23466
46	91896	67126	04151	03795	59077	11848	12630	98375	52068	60142
47	55751	62515	21108	80830	02263	29303	37204	96926	30506	09808
48	85156	87689	95493	88842	00664	55017	55539	17771	69448	87530
49	07521	56898	12236	60277	39102	62315	12239	07105	11844	01117

From *Statistical Methods* (6th ed.), by G. W. Snedecor and W. G. Cochran. Copyright © by Iowa State University Press, Ames, Iowa. Reprinted by permission.

Table B (continued)

	50–54	55–59	60–64	65–69	70–74	75–79	80–84	85–89	90–94	95–99
00	59391	58030	52098	82718	87024	82848	04190	96574	90464	29065
01	99567	76364	77204	04615	27062	96621	43918	01896	83991	51141
02	10363	97518	51400	25670	98342	61891	27101	37855	06235	33316
03	86859	19558	64432	16706	99612	59798	32803	67708	15297	28612
04	11258	24591	36863	55368	31721	94335	34936	02566	80972	08188
05	95068	88628	35911	14530	33020	80428	39936	31855	34334	64865
06	54463	47237	73800	91017	36239	71824	83671	39892	60518	37092
07	16874	62677	57412	13215	31389	62233	80827	73917	82802	84420
08	92494	63157	76593	91316	03505	72389	96363	52887	01087	66091
09	15669	56689	35682	40844	53256	81872	35213	09840	34471	74441
10	99116	75486	84989	23476	52967	67104	39495	39100	17217	74073
11	15696	10703	65178	90637	63110	17622	53988	71087	84148	11670
12	97720	15369	51269	69620	03388	13699	33423	67453	43269	56720
13	11666	13841	71681	98000	35979	39719	81899	07449	47985	46967
14	71628	73130	78783	75691	41632	09847	61547	18707	85489	69944
15	40501	51089	99943	91843	41995	88931	73631	69361	05375	15417
16	22518	55576	98215	82068	10798	86211	36584	67466	69373	40054
17	75112	30485	62173	02132	14878	92879	22281	16783	86352	00077
18	80327	02671	98191	84342	90813	49268	95441	15496	20168	09271
19	60251	45548	02146	05597	48228	81366	34598	72856	66762	17002
20	57430	82270	10421	05540	43648	75888	66049	21511	47676	33444
21	73528	39559	34434	88596	54086	71693	43132	14414	79949	85193
22	25991	65959	70769	64721	86413	33475	42740	06175	82758	66248
23	78388	16638	09134	59880	63806	48472	39318	35434	24057	74739
24	12477	09965	96657	57994	59439	76330	24596	77515	09577	91871
25	83266	32883	42451	15579	38155	29793	40914	65990	16255	17777
26	76970	80876	10237	39515	79152	74798	39357	09054	73579	92359
27	37074	65198	44785	68624	98336	84481	97610	78735	46703	98265
28	83712	06514	30101	78295	54656	85417	43189	60048	72781	72606
29	20287	56862	69727	94443	64936	08366	27227	05158	50326	59566
30	74261	32592	86538	27041	65172	85532	07571	80609	39285	65340
31	64081	49863	08478	96001	18888	14810	70545	89755	59064	07210
32	05617	75818	47750	67814	29575	10526	66192	44464	27058	40467
33	26793	74951	95466	74307	13330	42664	85515	20632	05497	33625
34	65988	72850	48737	54719	52056	01596	03845	35067	03134	70322
35	27366	42271	44300	73399	21105	03280	73457	43093	05192	48657
36	56760	10909	98147	34736	33863	95256	12731	66598	50771	83665
37	72880	43338	93643	58904	59543	23943	11231	83268	65938	81581
38	77888	38100	03062	58103	47961	83841	25878	23746	55903	44115
39	28440	07819	21580	51459	47971	29882	13990	29226	23608	15873
40	63525	94441	77033	12147	51054	49955	58312	76923	96071	05813
41	47606	93410	16359	89033	89696	47231	64498	31776	05383	39902
42	52669	45030	96279	14709	52372	87832	02735	50803	72744	88208
43	16738	60159	07425	62369	07515	82721	37875	71153	21315	00132
44	59348	11695	45751	15865	74739	05572	32688	20271	65128	14551
45	12900	71775	29845	60774	94924	21810	38636	33717	67598	82521
46	75086	23537	49939	33595	13484	97588	28617	17979	70749	35234
47	99495	51434	29181	09993	38190	42553	68922	52125	91077	40197
48	26075	31671	45386	36583	93159	48599	52022	41330	60651	91321
49	13636	93596	23377	51133	95126	61496	42474	45141	46660	42338

Table B (continued)

	00–04	05–09	10–14	15–19	20–24	25–29	30–34	35–39	40–44	45–49
50	64249	63664	39652	40646	97306	31741	07294	84149	46797	82487
51	26538	44249	04050	48174	65570	44072	40192	51153	11397	58212
52	05845	00512	78630	55328	18116	69296	91705	86224	29503	57071
53	74897	68373	67359	51014	33510	83048	17056	72506	82949	54600
54	20872	54570	35017	88132	25730	22626	86723	91691	13191	77212
55	31432	96156	89177	75541	81355	24480	77243	76690	42507	84362
56	66890	61505	01240	00660	05873	13568	76082	79172	57913	93448
57	48194	57790	79970	33106	86904	48119	52503	24130	72824	21627
58	11303	87118	81471	52936	08555	28420	49416	44448	04269	27029
59	54374	57325	16947	45356	78371	10563	97191	53798	12693	27928
60	64852	34421	61046	90849	13966	39810	42699	21753	76192	10508
61	16309	20384	09491	91588	97720	89846	30376	76970	23063	35894
62	42587	37065	24526	72602	57589	98131	37292	05967	26002	51945
63	40177	98590	97161	41682	84533	67588	62036	49967	01990	72308
64	82309	76128	93965	26743	24141	04838	40254	26065	07938	76236
65	79788	68243	59732	04257	27084	14743	17520	95401	55811	76099
66	40538	79000	89559	25026	42274	23489	34502	75508	06059	86682
67	64016	73598	18609	73150	62463	33102	45205	87440	96767	67042
68	49767	12691	17903	93871	99721	79109	09425	26904	07419	76013
69	76974	55108	29795	08404	82684	00497	51126	79935	57450	55671
70	23854	08480	85983	96025	50117	64610	99425	62291	86943	21541
71	68973	70551	25098	78033	98573	79848	31778	29555	61446	23037
72	36444	93600	65350	14971	25325	00427	52073	64280	18847	24768
73	03003	87800	07391	11594	21196	00781	32550	57158	58887	73041
74	17540	26188	36647	78386	04558	61463	57842	90382	77019	24210
75	38916	55809	47982	41968	69760	79422	80154	91486	19180	15100
76	64288	19843	69122	42502	48508	28820	59933	72998	99942	10515
77	86809	51564	38040	39418	49915	19000	58050	16899	79952	57849
78	99800	99566	14742	05028	30033	94889	53381	23656	75787	59223
79	92345	31890	95712	08279	91794	94068	49337	88674	35355	12267
80	90363	65162	32245	82279	79256	80834	06088	99462	56705	06118
81	64437	32242	48431	04835	39070	59702	31508	60935	22390	52246
82	91714	53662	28373	34333	55791	74758	51144	18827	10704	76803
83	20902	17646	31391	31459	33315	03444	55743	74701	58851	27427
84	12217	86007	70371	52281	14510	76094	96579	54863	78339	20839
85	45177	02863	42307	53571	22532	74921	17735	42201	80540	54721
86	28325	90814	08804	52746	47913	54577	47525	77705	95330	21866
87	29019	28776	56116	54791	64604	08815	46049	71186	34650	14994
88	84979	81353	56219	67062	26146	82567	33122	14124	46240	92973
89	50371	26347	48513	63915	11158	25563	91915	18431	92978	11591
90	53422	06825	69711	67950	64716	18003	49581	45378	99878	61130
91	67453	35651	89316	41620	32048	70225	47597	33137	31443	51445
92	07294	85353	74819	23445	68237	07202	99515	62282	53809	26685
93	79544	00302	45338	16015	66613	88968	14595	63836	77716	79596
94	64144	85442	82060	46471	24162	39500	87351	36637	42833	71875
95	90919	11883	58318	00042	52402	28210	34075	33272	00840	73268
96	06670	57353	86275	92276	77591	46924	60839	55437	03183	13191
97	36634	93976	52062	83678	41256	60948	18685	48992	19462	96062
98	75101	72891	85745	67106	26010	62107	60885	37503	55461	71213
99	05112	71222	72654	51583	05228	62056	57390	42746	39272	96659

Table B (continued)

	50–54	55–59	60–64	65–69	70–74	75–79	80–84	85–89	90–94	95–99
50	32847	31282	03345	89593	69214	70381	78285	20054	91018	16742
51	16916	00041	30236	55023	14253	76582	12092	86533	92426	37655
52	66176	34047	21005	27137	03191	48970	64625	22394	39622	79085
53	46299	13335	12180	16861	38043	59292	62675	63631	37020	78195
54	22847	47839	45385	23289	47526	54098	45683	55849	51575	64689
55	41851	54160	92320	69936	34803	92479	33399	71160	64777	83378
56	28444	59497	91586	95917	68553	28639	06455	34174	11130	91994
57	47520	62378	98855	83174	13088	16561	68559	26679	06238	51254
58	34978	63271	13142	82681	05271	08822	06490	44984	49307	62717
59	37404	80416	69035	92980	49486	74378	75610	74976	70056	15478
60	32400	65482	52099	53676	74648	94148	65095	69597	52771	71551
61	89262	86332	51718	70663	11623	29834	79820	73002	84886	03591
62	86866	09127	98021	03871	27789	58444	44832	36505	40672	30180
63	90814	14833	08759	74645	05046	94056	99094	65091	32663	73040
64	19192	82756	20553	58446	55376	88914	75096	26119	83898	43816
65	77585	52593	56612	95766	10019	29531	73064	20953	53523	58136
66	23757	16364	05096	03192	62386	45389	85332	18877	55710	96459
67	45989	96257	23850	26216	23309	21526	07425	50254	19455	29315
68	92970	94243	07316	41467	64837	52406	25225	51553	31220	14032
69	74346	59596	40088	98176	17896	86900	20249	77753	19099	48885
70	87646	41309	27636	45153	29988	94770	07255	70908	05340	99751
71	50099	71038	45146	06146	55211	99429	43169	66259	97786	59180
72	10127	46900	64984	75348	04115	33624	68774	60013	35515	62556
73	67995	81977	18984	64091	02785	27762	42529	97144	80407	64524
74	26304	80217	84934	82657	69291	35397	98714	35104	08187	48109
75	81994	41070	56642	64091	31229	02595	13513	45148	78722	30144
76	59537	34662	79631	89403	65212	09975	06118	86197	58208	16162
77	51228	10937	62396	81460	47331	91403	95007	06047	16846	64809
78	31089	37995	29577	07828	42272	54016	21950	86192	99046	84864
79	38207	97938	93459	75174	79460	55436	57206	87644	21296	43395
80	88666	31142	09474	89712	63153	62333	42212	06140	42594	43671
81	53365	56134	67582	92557	89520	33452	05134	70628	27612	33738
82	89807	74530	38004	90102	11693	90257	05500	79920	62700	43325
83	18682	81038	85662	90915	91631	22223	91588	80774	07716	12548
84	63571	32579	63942	25371	09234	94592	98475	76884	37635	33608
85	68927	56492	67799	95398	77642	54913	91853	08424	81450	76229
86	56401	63186	39389	88798	31356	89235	97036	32341	33292	73757
87	24333	95603	02359	72942	46287	95382	08452	62862	97869	71775
88	17025	84202	95199	62272	06366	16175	97577	99304	41587	03686
89	02804	08253	52133	20224	68034	50865	57868	22343	55111	03607
90	08298	03879	20995	19850	73090	13191	18963	82244	78479	99121
91	59883	01785	82403	96062	03785	03488	12970	64896	38336	30030
92	46982	06682	62864	91837	74021	89094	39952	64158	79614	78235
93	31121	47266	07661	02051	67599	24471	69843	83696	71402	76287
94	97867	56641	63416	17577	30161	87320	37752	73276	48969	41915
95	57364	86746	08415	14621	49430	22311	15836	72492	49372	44103
96	09559	26263	69511	28064	75999	44540	13337	10918	79846	54809
97	53873	55571	00608	42661	91332	63956	74087	59008	47493	99581
98	35531	19162	86406	05299	77511	24311	57257	22826	77555	05941
99	28229	88629	25695	94932	30721	16197	78742	34974	97528	45447

Table C   Area under the Normal Curve between μ and z and beyond z

A z	B Area between Mean and z	C Area Beyond z	A z	B Area between Mean and z	C Area Beyond z	A z	B Area between Mean and z	C Area Beyond z
0.00	.0000	.5000	0.55	.2088	.2912	1.10	.3643	.1357
0.01	.0040	.4960	0.56	.2123	.2877	1.11	.3665	.1335
0.02	.0080	.4920	0.57	.2157	.2843	1.12	.3686	.1314
0.03	.0120	.4880	0.58	.2190	.2810	1.13	.3708	.1292
0.04	.0160	.4840	0.59	.2224	.2776	1.14	.3729	.1271
0.05	.0199	.4801	0.60	.2257	.2743	1.15	.3749	.1251
0.06	.0239	.4761	0.61	.2291	.2709	1.16	.3770	.1230
0.07	.0279	.4721	0.62	.2324	.2676	1.17	.3790	.1210
0.08	.0319	.4681	0.63	.2357	.2643	1.18	.3810	.1190
0.09	.0359	.4641	0.64	.2389	.2611	1.19	.3830	.1170
0.10	.0398	.4602	0.65	.2422	.2578	1.20	.3849	.1151
0.11	.0438	.4562	0.66	.2454	.2546	1.21	.3869	.1131
0.12	.0478	.4522	0.67	.2486	.2514	1.22	.3888	.1112
0.13	.0517	.4483	0.68	.2517	.2483	1.23	.3907	.1093
0.14	.0557	.4443	0.69	.2549	.2451	1.24	.3925	.1075
0.15	.0596	.4404	0.70	.2580	.2420	1.25	.3944	.1056
0.16	.0636	.4364	0.71	.2611	.2389	1.26	.3962	.1038
0.17	.0675	.4325	0.72	.2642	.2358	1.27	.3980	.1020
0.18	.0714	.4286	0.73	.2673	.2327	1.28	.3997	.1003
0.19	.0753	.4247	0.74	.2704	.2296	1.29	.4015	.0985
0.20	.0793	.4207	0.75	.2734	.2266	1.30	.4032	.0968
0.21	.0832	.4168	0.76	.2764	.2236	1.31	.4049	.0951
0.22	.0871	.4129	0.77	.2794	.2206	1.32	.4066	.0934
0.23	.0910	.4090	0.78	.2823	.2177	1.33	.4082	.0918
0.24	.0948	.4052	0.79	.2852	.2148	1.34	.4099	.0901
0.25	.0987	.4013	0.80	.2881	.2119	1.35	.4115	.0885
0.26	.1026	.3974	0.81	.2910	.2090	1.36	.4131	.0869
0.27	.1064	.3936	0.82	.2939	.2061	1.37	.4147	.0853
0.28	.1103	.3897	0.83	.2967	.2033	1.38	.4162	.0838
0.29	.1141	.3859	0.84	.2995	.2005	1.39	.4177	.0823
0.30	.1179	.3821	0.85	.3023	.1977	1.40	.4192	.0808
0.31	.1217	.3783	0.86	.3051	.1949	1.41	.4207	.0793
0.32	.1255	.3745	0.87	.3078	.1922	1.42	.4222	.0778
0.33	.1293	.3707	0.88	.3106	.1894	1.43	.4236	.0764
0.34	.1331	.3669	0.89	.3133	.1867	1.44	.4251	.0749
0.35	.1368	.3632	0.90	.3159	.1841	1.45	.4265	.0735
0.36	.1406	.3594	0.91	.3186	.1814	1.46	.4279	.0721
0.37	.1443	.3557	0.92	.3212	.1788	1.47	.4292	.0708
0.38	.1480	.3520	0.93	.3238	.1762	1.48	.4306	.0694
0.39	.1517	.3483	0.94	.3264	.1736	1.49	.4319	.0681
0.40	.1554	.3446	0.95	.3289	.1711	1.50	.4332	.0668
0.41	.1591	.3409	0.96	.3315	.1685	1.51	.4345	.0655
0.42	.1628	.3372	0.97	.3340	.1660	1.52	.4357	.0643
0.43	.1664	.3336	0.98	.3365	.1635	1.53	.4370	.0630
0.44	.1700	.3300	0.99	.3389	.1611	1.54	.4382	.0618
0.45	.1736	.3264	1.00	.3413	.1587	1.55	.4394	.0606
0.46	.1772	.3228	1.01	.3438	.1562	1.56	.4406	.0594
0.47	.1808	.3192	1.02	.3461	.1539	1.57	.4418	.0582
0.48	.1844	.3156	1.03	.3485	.1515	1.58	.4429	.0571
0.49	.1879	.3121	1.04	.3508	.1492	1.59	.4441	.0559
0.50	.1915	.3085	1.05	.3531	.1469	1.60	.4452	.0548
0.51	.1950	.3050	1.06	.3554	.1446	1.61	.4463	.0537
0.52	.1985	.3015	1.07	.3577	.1423	1.62	.4474	.0526
0.53	.2019	.2981	1.08	.3599	.1401	1.63	.4484	.0516
0.54	.2054	.2946	1.09	.3621	.1379	1.64	.4495	.0505

[handwritten note near header: "in either direction"]

Table C (continued)

A z	B Area between Mean and z	C Area Beyond z	A z	B Area between Mean and z	C Area Beyond z	A z	B Area between Mean and z	C Area Beyond z
1.65	.4505	.0495	2.22	.4868	.0132	2.79	.4974	.0026
1.66	.4515	.0485	2.23	.4871	.0129	2.80	.4974	.0026
1.67	.4525	.0475	2.24	.4875	.0125	2.81	.4975	.0025
1.68	.4535	.0465	2.25	.4878	.0122	2.82	.4976	.0024
1.69	.4545	.0455	2.26	.4881	.0119	2.83	.4977	.0023
1.70	.4554	.0446	2.27	.4884	.0116	2.84	.4977	.0023
1.71	.4564	.0436	2.28	.4887	.0113	2.85	.4978	.0022
1.72	.4573	.0427	2.29	.4890	.0110	2.86	.4979	.0021
1.73	.4582	.0418	2.30	.4893	.0107	2.87	.4979	.0021
1.74	.4591	.0409	2.31	.4896	.0104	2.88	.4980	.0020
1.75	.4599	.0401	2.32	.4898	.0102	2.89	.4981	.0019
1.76	.4608	.0392	2.33	.4901	.0099	2.90	.4981	.0019
1.77	.4616	.0384	2.34	.4904	.0096	2.91	.4982	.0018
1.78	.4625	.0375	2.35	.4906	.0094	2.92	.4982	.0018
1.79	.4633	.0367	2.36	.4909	.0091	2.93	.4983	.0017
1.80	.4641	.0359	2.37	.4911	.0089	2.94	.4984	.0016
1.81	.4649	.0351	2.38	.4913	.0087	2.95	.4984	.0016
1.82	.4656	.0344	2.39	.4916	.0084	2.96	.4985	.0015
1.83	.4664	.0336	2.40	.4918	.0082	2.97	.4985	.0015
1.84	.4671	.0329	2.41	.4920	.0080	2.98	.4986	.0014
1.85	.4678	.0322	2.42	.4922	.0078	2.99	.4986	.0014
1.86	.4686	.0314	2.43	.4925	.0075	3.00	.4987	.0013
1.87	.4693	.0307	2.44	.4927	.0073	3.01	.4987	.0013
1.88	.4699	.0301	2.45	.4929	.0071	3.02	.4987	.0013
1.89	.4706	.0294	2.46	.4931	.0069	3.03	.4988	.0012
1.90	.4713	.0287	2.47	.4932	.0068	3.04	.4988	.0012
1.91	.4719	.0281	2.48	.4934	.0066	3.05	.4989	.0011
1.92	.4726	.0274	2.49	.4936	.0064	3.06	.4989	.0011
1.93	.4732	.0268	2.50	.4938	.0062	3.07	.4989	.0011
1.94	.4738	.0262	2.51	.4940	.0060	3.08	.4990	.0010
1.95	.4744	.0256	2.52	.4941	.0059	3.09	.4990	.0010
1.96	.4750	.0250	2.53	.4943	.0057	3.10	.4990	.0010
1.97	.4756	.0244	2.54	.4945	.0055	3.11	.4991	.0009
1.98	.4761	.0239	2.55	.4946	.0054	3.12	.4991	.0009
1.99	.4767	.0233	2.56	.4948	.0052	3.13	.4991	.0009
2.00	.4772	.0228	2.57	.4949	.0051	3.14	.4992	.0008
2.01	.4778	.0222	2.58	.4951	.0049	3.15	.4992	.0008
2.02	.4783	.0217	2.59	.4952	.0048	3.16	.4992	.0008
2.03	.4788	.0212	2.60	.4953	.0047	3.17	.4992	.0008
2.04	.4793	.0207	2.61	.4955	.0045	3.18	.4993	.0007
2.05	.4798	.0202	2.62	.4956	.0044	3.19	.4993	.0007
2.06	.4803	.0197	2.63	.4957	.0043	3.20	.4993	.0007
2.07	.4808	.0192	2.64	.4959	.0041	3.21	.4993	.0007
2.08	.4812	.0188	2.65	.4960	.0040	3.22	.4994	.0006
2.09	.4817	.0183	2.66	.4961	.0039	3.23	.4994	.0006
2.10	.4821	.0179	2.67	.4962	.0038	3.24	.4994	.0006
2.11	.4826	.0174	2.68	.4963	.0037	3.25	.4994	.0006
2.12	.4830	.0170	2.69	.4964	.0036	3.30	.4995	.0005
2.13	.4834	.0166	2.70	.4965	.0035	3.35	.4996	.0004
2.14	.4838	.0162	2.71	.4966	.0034	3.40	.4997	.0003
2.15	.4842	.0158	2.72	.4967	.0033	3.45	.4997	.0003
2.16	.4846	.0154	2.73	.4968	.0032	3.50	.4998	.0002
2.17	.4850	.0150	2.74	.4969	.0031	3.60	.4998	.0002
2.18	.4854	.0146	2.75	.4970	.0030	3.70	.4999	.0001
2.19	.4857	.0143	2.76	.4971	.0029	3.80	.4999	.0001
2.20	.4861	.0139	2.77	.4972	.0028	3.90	.49995	.00005
2.21	.4864	.0136	2.78	.4973	.0027	4.00	.49997	.00003

Table D   The t Distribution*

df	α Levels for Two-Tailed Test					
	.2	.1	.05	.02	.01	.001
1	3.078	6.314	12.706	31.821	63.657	636.619
2	1.886	2.920	4.303	6.965	9.925	31.598
3	1.638	2.353	3.182	4.541	5.841	12.924
4	1.533	2.132	2.776	3.747	4.604	8.610
5	1.476	2.015	2.571	3.365	4.032	6.869
6	1.440	1.943	2.447	3.143	3.707	5.959
7	1.415	1.895	2.365	2.998	3.499	5.408
8	1.397	1.860	2.306	2.896	3.355	5.041
9	1.383	1.833	2.262	2.821	3.250	4.781
10	1.372	1.812	2.228	2.764	3.169	4.587
11	1.363	1.796	2.201	2.718	3.106	4.437
12	1.356	1.782	2.179	2.681	3.055	4.318
13	1.350	1.771	2.160	2.650	3.012	4.221
14	1.345	1.761	2.145	2.624	2.977	4.140
15	1.341	1.753	2.131	2.602	2.947	4.073
16	1.337	1.746	2.120	2.583	2.921	4.015
17	1.333	1.740	2.110	2.567	2.898	3.965
18	1.330	1.734	2.101	2.552	2.878	3.922
19	1.328	1.729	2.093	2.539	2.861	3.883
20	1.325	1.725	2.086	2.528	2.845	3.850
21	1.323	1.721	2.080	2.518	2.831	3.819
22	1.321	1.717	2.074	2.508	2.819	3.792
23	1.319	1.714	2.069	2.500	2.807	3.767
24	1.318	1.711	2.064	2.492	2.797	3.745
25	1.316	1.708	2.060	2.485	2.787	3.725
26	1.315	1.706	2.056	2.479	2.779	3.707
27	1.314	1.703	2.052	2.473	2.771	3.690
28	1.313	1.701	2.048	2.467	2.763	3.674
29	1.311	1.699	2.045	2.462	2.756	3.659
30	1.310	1.697	2.042	2.457	2.750	3.646
40	1.303	1.684	2.021	2.423	2.704	3.551
60	1.296	1.671	2.000	2.390	2.660	3.460
120	1.289	1.658	1.980	2.358	2.617	3.373
∞	1.282	1.645	1.960	2.326	2.576	3.291
	.10	.05	.025	.01	.005	.0005
	α Levels for a One-Tailed Test					

*To be significant the t obtained from the data must be equal to or larger than the value shown in the table.

Table D is taken from Table III of Fisher and Yates: *Statistical Tables for Biological, Agricultural and Medical Research*, published by Longman Group Ltd., London. (Previously published by Oliver & Boyd, Edinburgh), and by permission of the authors and publishers.

Table E   Chi Square Distribution*

df	.10	.05	.02	.01	.001
			α levels		
1	2.71	3.84	5.41	6.64	10.38
2	4.60	5.99	7.82	9.21	13.82
3	6.25	7.82	9.84	11.34	16.27
4	7.78	9.49	11.67	13.28	18.46
5	9.24	11.07	13.39	15.09	20.52
6	10.64	12.59	15.03	16.81	22.46
7	12.02	14.07	16.62	18.48	24.32
8	13.36	15.51	18.17	20.09	26.12
9	14.68	16.92	19.68	21.67	27.88
10	15.99	18.31	21.16	23.21	29.59
11	17.28	19.68	22.62	24.72	31.26
12	18.55	21.03	24.05	26.22	32.91
13	19.81	22.36	25.47	27.69	34.53
14	21.06	23.68	26.87	29.14	36.12
15	22.31	25.00	28.26	30.58	37.70
16	23.54	26.30	29.63	32.00	39.25
17	24.77	27.59	31.00	33.41	40.79
18	25.99	28.87	32.35	34.80	42.31
19	27.20	30.14	33.69	36.19	43.82
20	28.41	31.41	35.02	37.57	45.32
21	29.62	32.67	36.34	38.93	46.80
22	30.81	33.92	37.66	40.29	48.27
23	32.01	35.17	38.97	41.64	49.73
24	33.20	36.42	40.27	42.98	51.18
25	34.38	37.65	41.57	44.31	52.62
26	35.56	38.88	42.86	45.64	54.05
27	36.74	40.11	44.14	46.96	55.48
28	37.92	41.34	45.42	48.28	56.89
29	39.09	42.56	46.69	49.59	58.30
30	40.26	43.77	47.96	50.89	59.70

*To be significant the χ^2 obtained from the data must be equal to or larger than the value shown in the table.

Table E is taken from Table IV of Fisher and Yates: *Statistical Tables for Biological, Agricultural and Medical Research*, published by Longman Group Ltd., London. (Previously published by Oliver & Boyd, Edinburgh), and by permission of the authors and publishers.

Table F The F Distribution*

.05 Level (roman) and .01 Level (bold face) α Levels for the Distribution of F

Degrees of Freedom (for the numerator)

Values shown as: .05 level / **.01 level**. Rows = Degrees of Freedom (for the denominator).

Denom. df	1	2	3	4	5	6	7	8	9	10	11	12	14	16	20	24	30	40	50	75	100	200	500	∞
1	161 / **4,052**	200 / **4,999**	216 / **5,403**	225 / **5,625**	230 / **5,764**	234 / **5,859**	237 / **5,928**	239 / **5,981**	241 / **6,022**	242 / **6,056**	243 / **6,082**	244 / **6,106**	245 / **6,142**	246 / **6,169**	248 / **6,208**	249 / **6,234**	250 / **6,258**	251 / **6,286**	252 / **6,302**	253 / **6,323**	253 / **6,334**	254 / **6,352**	254 / **6,361**	254 / **6,366**
2	18.51 / **98.49**	19.00 / **99.00**	19.16 / **99.17**	19.25 / **99.25**	19.30 / **99.30**	19.33 / **99.33**	19.36 / **99.34**	19.37 / **99.36**	19.38 / **99.38**	19.39 / **99.40**	19.40 / **99.41**	19.41 / **99.42**	19.42 / **99.43**	19.43 / **99.44**	19.44 / **99.45**	19.45 / **99.46**	19.46 / **99.47**	19.47 / **99.48**	19.47 / **99.48**	19.48 / **99.49**	19.49 / **99.49**	19.49 / **99.49**	19.50 / **99.50**	19.50 / **99.50**
3	10.13 / **34.12**	9.55 / **30.82**	9.28 / **29.46**	9.12 / **28.71**	9.01 / **28.24**	8.94 / **27.91**	8.88 / **27.67**	8.84 / **27.49**	8.81 / **27.34**	8.78 / **27.23**	8.76 / **27.13**	8.74 / **27.05**	8.71 / **26.92**	8.69 / **26.83**	8.66 / **26.69**	8.64 / **26.60**	8.62 / **26.50**	8.60 / **26.41**	8.58 / **26.35**	8.57 / **26.27**	8.56 / **26.23**	8.54 / **26.18**	8.54 / **26.14**	8.53 / **26.12**
4	7.71 / **21.20**	6.94 / **18.00**	6.59 / **16.69**	6.39 / **15.98**	6.26 / **15.52**	6.16 / **15.21**	6.09 / **14.98**	6.04 / **14.80**	6.00 / **14.66**	5.96 / **14.54**	5.93 / **14.45**	5.91 / **14.37**	5.87 / **14.24**	5.84 / **14.15**	5.80 / **14.02**	5.77 / **13.93**	5.74 / **13.83**	5.71 / **13.74**	5.70 / **13.69**	5.68 / **13.61**	5.66 / **13.57**	5.65 / **13.52**	5.64 / **13.48**	5.63 / **13.46**
5	6.61 / **16.26**	5.79 / **13.27**	5.41 / **12.06**	5.19 / **11.39**	5.05 / **10.97**	4.95 / **10.67**	4.88 / **10.45**	4.82 / **10.27**	4.78 / **10.15**	4.74 / **10.05**	4.70 / **9.96**	4.68 / **9.89**	4.64 / **9.77**	4.60 / **9.68**	4.56 / **9.55**	4.53 / **9.47**	4.50 / **9.38**	4.46 / **9.29**	4.44 / **9.24**	4.42 / **9.17**	4.40 / **9.13**	4.38 / **9.07**	4.37 / **9.04**	4.36 / **9.02**
6	5.99 / **13.74**	5.14 / **10.92**	4.76 / **9.78**	4.53 / **9.15**	4.39 / **8.75**	4.28 / **8.47**	4.21 / **8.26**	4.15 / **8.10**	4.10 / **7.98**	4.06 / **7.87**	4.03 / **7.79**	4.00 / **7.72**	3.96 / **7.60**	3.92 / **7.52**	3.87 / **7.39**	3.84 / **7.31**	3.81 / **7.23**	3.77 / **7.14**	3.75 / **7.09**	3.72 / **7.02**	3.71 / **6.99**	3.69 / **6.94**	3.68 / **6.90**	3.67 / **6.88**
7	5.59 / **12.25**	4.74 / **9.55**	4.35 / **8.45**	4.12 / **7.85**	3.97 / **7.46**	3.87 / **7.19**	3.79 / **7.00**	3.73 / **6.84**	3.68 / **6.71**	3.63 / **6.62**	3.60 / **6.54**	3.57 / **6.47**	3.52 / **6.35**	3.49 / **6.27**	3.44 / **6.15**	3.41 / **6.07**	3.38 / **5.98**	3.34 / **5.90**	3.32 / **5.85**	3.29 / **5.78**	3.28 / **5.75**	3.25 / **5.70**	3.24 / **5.67**	3.23 / **5.65**
8	5.32 / **11.26**	4.46 / **8.65**	4.07 / **7.59**	3.84 / **7.01**	3.69 / **6.63**	3.58 / **6.37**	3.50 / **6.19**	3.44 / **6.03**	3.39 / **5.91**	3.34 / **5.82**	3.31 / **5.74**	3.28 / **5.67**	3.23 / **5.56**	3.20 / **5.48**	3.15 / **5.36**	3.12 / **5.28**	3.08 / **5.20**	3.05 / **5.11**	3.03 / **5.06**	3.00 / **5.00**	2.98 / **4.96**	2.96 / **4.91**	2.94 / **4.88**	2.93 / **4.86**
9	5.12 / **10.56**	4.26 / **8.02**	3.86 / **6.99**	3.63 / **6.42**	3.48 / **6.06**	3.37 / **5.80**	3.29 / **5.62**	3.23 / **5.47**	3.18 / **5.35**	3.13 / **5.26**	3.10 / **5.18**	3.07 / **5.11**	3.02 / **5.00**	2.98 / **4.92**	2.93 / **4.80**	2.90 / **4.73**	2.86 / **4.64**	2.82 / **4.56**	2.80 / **4.51**	2.77 / **4.45**	2.76 / **4.41**	2.73 / **4.36**	2.72 / **4.33**	2.71 / **4.31**
10	4.96 / **10.04**	4.10 / **7.56**	3.71 / **6.55**	3.48 / **5.99**	3.33 / **5.64**	3.22 / **5.39**	3.14 / **5.21**	3.07 / **5.06**	3.02 / **4.95**	2.97 / **4.85**	2.94 / **4.78**	2.91 / **4.71**	2.86 / **4.60**	2.82 / **4.52**	2.77 / **4.41**	2.74 / **4.33**	2.70 / **4.25**	2.67 / **4.17**	2.64 / **4.12**	2.61 / **4.05**	2.59 / **4.01**	2.56 / **3.96**	2.55 / **3.93**	2.54 / **3.91**
11	4.84 / **9.65**	3.98 / **7.20**	3.59 / **6.22**	3.36 / **5.67**	3.20 / **5.32**	3.09 / **5.07**	3.01 / **4.88**	2.95 / **4.74**	2.90 / **4.63**	2.86 / **4.54**	2.82 / **4.46**	2.79 / **4.40**	2.74 / **4.29**	2.70 / **4.21**	2.65 / **4.10**	2.61 / **4.02**	2.57 / **3.94**	2.53 / **3.86**	2.50 / **3.80**	2.47 / **3.74**	2.45 / **3.70**	2.42 / **3.66**	2.41 / **3.62**	2.40 / **3.60**
12	4.75 / **9.33**	3.88 / **6.93**	3.49 / **5.95**	3.26 / **5.41**	3.11 / **5.06**	3.00 / **4.82**	2.92 / **4.65**	2.85 / **4.50**	2.80 / **4.39**	2.76 / **4.30**	2.72 / **4.22**	2.69 / **4.16**	2.64 / **4.05**	2.60 / **3.98**	2.54 / **3.86**	2.50 / **3.78**	2.46 / **3.70**	2.42 / **3.61**	2.40 / **3.56**	2.36 / **3.49**	2.35 / **3.46**	2.32 / **3.41**	2.31 / **3.38**	2.30 / **3.36**
13	4.67 / **9.07**	3.80 / **6.70**	3.41 / **5.74**	3.18 / **5.20**	3.02 / **4.86**	2.92 / **4.62**	2.84 / **4.44**	2.77 / **4.30**	2.72 / **4.19**	2.67 / **4.10**	2.63 / **4.02**	2.60 / **3.96**	2.55 / **3.85**	2.51 / **3.78**	2.46 / **3.67**	2.42 / **3.59**	2.38 / **3.51**	2.34 / **3.42**	2.32 / **3.37**	2.28 / **3.30**	2.26 / **3.27**	2.24 / **3.21**	2.22 / **3.18**	2.21 / **3.16**

*To be significant the F obtained from the data must be equal to or larger than the value shown in the table.
From *Statistical Methods* (6th ed.), by G. W. Snedecor and W. G. Cochran. Copyright © 1967 by Iowa State University Press, Ames, Iowa. Reprinted by permission.

Table F (continued)

Degrees of Freedom (for the numerator)

df (denom.)	1	2	3	4	5	6	7	8	9	10	11	12	14	16	20	24	30	40	50	75	100	200	500	∞
14	4.60 / **8.86**	3.74 / **6.51**	3.34 / **5.56**	3.11 / **5.03**	2.96 / **4.69**	2.85 / **4.46**	2.77 / **4.28**	2.70 / **4.14**	2.65 / **4.03**	2.60 / **3.94**	2.56 / **3.86**	2.53 / **3.80**	2.48 / **3.70**	2.44 / **3.62**	2.39 / **3.51**	2.35 / **3.43**	2.31 / **3.34**	2.27 / **3.26**	2.24 / **3.21**	2.21 / **3.14**	2.19 / **3.11**	2.16 / **3.06**	2.14 / **3.02**	2.13 / **3.00**
15	4.54 / **8.68**	3.68 / **6.36**	3.29 / **5.42**	3.06 / **4.89**	2.90 / **4.56**	2.79 / **4.32**	2.70 / **4.14**	2.64 / **4.00**	2.59 / **3.89**	2.55 / **3.80**	2.51 / **3.73**	2.48 / **3.67**	2.43 / **3.56**	2.39 / **3.48**	2.33 / **3.36**	2.29 / **3.29**	2.25 / **3.20**	2.21 / **3.12**	2.18 / **3.07**	2.15 / **3.00**	2.12 / **2.97**	2.10 / **2.92**	2.08 / **2.89**	2.07 / **2.87**
16	4.49 / **8.53**	3.63 / **6.23**	3.24 / **5.29**	3.01 / **4.77**	2.85 / **4.44**	2.74 / **4.20**	2.66 / **4.03**	2.59 / **3.89**	2.54 / **3.78**	2.49 / **3.69**	2.45 / **3.61**	2.42 / **3.55**	2.37 / **3.45**	2.33 / **3.37**	2.28 / **3.25**	2.24 / **3.18**	2.20 / **3.10**	2.16 / **3.01**	2.13 / **2.96**	2.09 / **2.89**	2.07 / **2.86**	2.04 / **2.80**	2.02 / **2.77**	2.01 / **2.75**
17	4.45 / **8.40**	3.59 / **6.11**	3.20 / **5.18**	2.96 / **4.67**	2.81 / **4.34**	2.70 / **4.10**	2.62 / **3.93**	2.55 / **3.79**	2.50 / **3.68**	2.45 / **3.59**	2.41 / **3.52**	2.38 / **3.45**	2.33 / **3.35**	2.29 / **3.27**	2.23 / **3.16**	2.19 / **3.08**	2.15 / **3.00**	2.11 / **2.92**	2.08 / **2.86**	2.04 / **2.79**	2.02 / **2.76**	1.99 / **2.70**	1.97 / **2.67**	1.96 / **2.65**
18	4.41 / **8.28**	3.55 / **6.01**	3.16 / **5.09**	2.93 / **4.58**	2.77 / **4.25**	2.66 / **4.01**	2.58 / **3.85**	2.51 / **3.71**	2.46 / **3.60**	2.41 / **3.51**	2.37 / **3.44**	2.34 / **3.37**	2.29 / **3.27**	2.25 / **3.19**	2.19 / **3.07**	2.15 / **3.00**	2.11 / **2.91**	2.07 / **2.83**	2.04 / **2.78**	2.00 / **2.71**	1.89 / **2.68**	1.95 / **2.62**	1.93 / **2.59**	1.92 / **2.57**
19	4.38 / **8.18**	3.52 / **5.93**	3.13 / **5.01**	2.90 / **4.50**	2.74 / **4.17**	2.63 / **3.94**	2.55 / **3.77**	2.48 / **3.63**	2.43 / **3.52**	2.38 / **3.43**	2.34 / **3.36**	2.31 / **3.30**	2.26 / **3.19**	2.21 / **3.12**	2.15 / **3.00**	2.11 / **2.92**	2.07 / **2.84**	2.02 / **2.76**	2.00 / **2.70**	1.96 / **2.63**	1.94 / **2.60**	1.91 / **2.54**	1.90 / **2.51**	1.88 / **2.49**
20	4.35 / **8.10**	3.49 / **5.85**	3.10 / **4.94**	2.87 / **4.43**	2.71 / **4.10**	2.60 / **3.87**	2.52 / **3.71**	2.45 / **3.56**	2.40 / **3.45**	2.35 / **3.37**	2.31 / **3.30**	2.28 / **3.23**	2.23 / **3.13**	2.18 / **3.05**	2.12 / **2.94**	2.08 / **2.86**	2.04 / **2.77**	1.99 / **2.69**	1.96 / **2.63**	1.92 / **2.56**	1.90 / **2.53**	1.87 / **2.47**	1.85 / **2.44**	1.84 / **2.42**
21	4.32 / **8.02**	3.47 / **5.78**	3.07 / **4.87**	2.84 / **4.37**	2.68 / **4.04**	2.57 / **3.81**	2.49 / **3.65**	2.42 / **3.51**	2.37 / **3.40**	2.32 / **3.31**	2.28 / **3.24**	2.25 / **3.17**	2.20 / **3.07**	2.15 / **2.99**	2.09 / **2.88**	2.05 / **2.80**	2.00 / **2.72**	1.96 / **2.63**	1.93 / **2.58**	1.89 / **2.51**	1.87 / **2.47**	1.84 / **2.42**	1.82 / **2.38**	1.81 / **2.36**
22	4.30 / **7.94**	3.44 / **5.72**	3.05 / **4.82**	2.82 / **4.31**	2.66 / **3.99**	2.55 / **3.76**	2.47 / **3.59**	2.40 / **3.45**	2.35 / **3.35**	2.30 / **3.26**	2.26 / **3.18**	2.23 / **3.12**	2.18 / **3.02**	2.13 / **2.94**	2.07 / **2.83**	2.03 / **2.75**	1.98 / **2.67**	1.93 / **2.58**	1.91 / **2.53**	1.87 / **2.46**	1.84 / **2.42**	1.81 / **2.37**	1.80 / **2.33**	1.78 / **2.31**
23	4.28 / **7.88**	3.42 / **5.66**	3.03 / **4.76**	2.80 / **4.26**	2.64 / **3.94**	2.53 / **3.71**	2.45 / **3.54**	2.38 / **3.41**	2.32 / **3.30**	2.28 / **3.21**	2.24 / **3.14**	2.20 / **3.07**	2.14 / **2.97**	2.10 / **2.89**	2.04 / **2.78**	2.00 / **2.70**	1.96 / **2.62**	1.91 / **2.53**	1.88 / **2.48**	1.84 / **2.41**	1.82 / **2.37**	1.79 / **2.32**	1.77 / **2.28**	1.76 / **2.26**
24	4.26 / **7.82**	3.40 / **5.61**	3.01 / **4.72**	2.78 / **4.22**	2.62 / **3.90**	2.51 / **3.67**	2.43 / **3.50**	2.36 / **3.36**	2.30 / **3.25**	2.26 / **3.17**	2.22 / **3.09**	2.18 / **3.03**	2.13 / **2.93**	2.09 / **2.85**	2.02 / **2.74**	1.98 / **2.66**	1.94 / **2.58**	1.89 / **2.49**	1.86 / **2.44**	1.82 / **2.36**	1.80 / **2.33**	1.76 / **2.27**	1.74 / **2.23**	1.73 / **2.21**
25	4.24 / **7.77**	3.38 / **5.57**	2.99 / **4.68**	2.76 / **4.18**	2.60 / **3.86**	2.49 / **3.63**	2.41 / **3.46**	2.34 / **3.32**	2.28 / **3.21**	2.24 / **3.13**	2.20 / **3.05**	2.16 / **2.99**	2.11 / **2.89**	2.06 / **2.81**	2.00 / **2.70**	1.96 / **2.62**	1.92 / **2.54**	1.87 / **2.45**	1.84 / **2.40**	1.80 / **2.32**	1.77 / **2.29**	1.74 / **2.23**	1.72 / **2.19**	1.71 / **2.17**
26	4.22 / **7.72**	3.37 / **5.53**	2.98 / **4.64**	2.74 / **4.14**	2.59 / **3.82**	2.47 / **3.59**	2.39 / **3.42**	2.32 / **3.29**	2.27 / **3.17**	2.22 / **3.09**	2.18 / **3.02**	2.15 / **2.96**	2.10 / **2.86**	2.05 / **2.77**	1.99 / **2.66**	1.95 / **2.58**	1.90 / **2.50**	1.85 / **2.41**	1.82 / **2.36**	1.78 / **2.28**	1.76 / **2.25**	1.72 / **2.19**	1.70 / **2.15**	1.69 / **2.13**

Degrees of Freedom (for the denominator)

Table F (continued)

Degrees of Freedom (for the numerator)

df	1	2	3	4	5	6	7	8	9	10	11	12	14	16	20	24	30	40	50	75	100	200	500	∞	df
27	4.21 **7.68**	3.35 **5.49**	2.96 **4.60**	2.73 **4.11**	2.57 **3.79**	2.46 **3.56**	2.37 **3.39**	2.30 **3.26**	2.25 **3.14**	2.20 **3.06**	2.16 **2.98**	2.13 **2.93**	2.08 **2.83**	2.03 **2.74**	1.97 **2.63**	1.93 **2.55**	1.88 **2.47**	1.84 **2.38**	1.80 **2.33**	1.76 **2.25**	1.74 **2.21**	1.71 **2.16**	1.68 **2.12**	1.67 **2.10**	27
28	4.20 **7.64**	3.34 **5.45**	2.95 **4.57**	2.71 **4.07**	2.56 **3.76**	2.44 **3.53**	2.36 **3.36**	2.29 **3.23**	2.24 **3.11**	2.19 **3.03**	2.15 **2.95**	2.12 **2.90**	2.06 **2.80**	2.02 **2.71**	1.96 **2.60**	1.91 **2.52**	1.87 **2.44**	1.81 **2.35**	1.78 **2.30**	1.75 **2.22**	1.72 **2.18**	1.69 **2.13**	1.67 **2.09**	1.65 **2.06**	28
29	4.18 **7.60**	3.33 **5.42**	2.93 **4.54**	2.70 **4.04**	2.54 **3.73**	2.43 **3.50**	2.35 **3.33**	2.28 **3.20**	2.22 **3.08**	2.18 **3.00**	2.14 **2.92**	2.10 **2.87**	2.05 **2.77**	2.00 **2.68**	1.94 **2.57**	1.90 **2.49**	1.85 **2.41**	1.80 **2.32**	1.77 **2.27**	1.73 **2.19**	1.71 **2.15**	1.68 **2.10**	1.65 **2.06**	1.64 **2.03**	29
30	4.17 **7.56**	3.32 **5.39**	2.92 **4.51**	2.69 **4.02**	2.53 **3.70**	2.42 **3.47**	2.34 **3.30**	2.27 **3.17**	2.21 **3.06**	2.16 **2.98**	2.12 **2.90**	2.09 **2.84**	2.04 **2.74**	1.99 **2.66**	1.93 **2.55**	1.89 **2.47**	1.84 **2.38**	1.79 **2.29**	1.76 **2.24**	1.72 **2.16**	1.69 **2.13**	1.66 **2.07**	1.64 **2.03**	1.62 **2.01**	30
32	4.15 **7.50**	3.30 **5.34**	2.90 **4.46**	2.67 **3.97**	2.51 **3.66**	2.40 **3.42**	2.32 **3.25**	2.25 **3.12**	2.19 **3.01**	2.14 **2.94**	2.10 **2.86**	2.07 **2.80**	2.02 **2.70**	1.97 **2.62**	1.91 **2.51**	1.86 **2.42**	1.82 **2.34**	1.76 **2.25**	1.74 **2.20**	1.69 **2.12**	1.67 **2.08**	1.64 **2.02**	1.61 **1.98**	1.59 **1.96**	32
34	4.13 **7.44**	3.28 **5.29**	2.88 **4.42**	2.65 **3.93**	2.49 **3.61**	2.38 **3.38**	2.30 **3.21**	2.23 **3.08**	2.17 **2.97**	2.12 **2.89**	2.08 **2.82**	2.05 **2.76**	2.00 **2.66**	1.95 **2.58**	1.89 **2.47**	1.84 **2.38**	1.80 **2.30**	1.74 **2.21**	1.71 **2.15**	1.67 **2.08**	1.64 **2.04**	1.61 **1.98**	1.59 **1.94**	1.57 **1.91**	34
36	4.11 **7.39**	3.26 **5.25**	2.86 **4.38**	2.63 **3.89**	2.48 **3.58**	2.36 **3.35**	2.28 **3.18**	2.21 **3.04**	2.15 **2.94**	2.10 **2.86**	2.06 **2.78**	2.03 **2.72**	1.98 **2.62**	1.93 **2.54**	1.87 **2.43**	1.82 **2.35**	1.78 **2.26**	1.72 **2.17**	1.69 **2.12**	1.65 **2.04**	1.62 **2.00**	1.59 **1.94**	1.56 **1.90**	1.55 **1.87**	36
38	4.10 **7.35**	3.25 **5.21**	2.85 **4.34**	2.62 **3.86**	2.46 **3.54**	2.35 **3.32**	2.26 **3.15**	2.19 **3.02**	2.14 **2.91**	2.09 **2.82**	2.05 **2.75**	2.02 **2.69**	1.96 **2.59**	1.92 **2.51**	1.85 **2.40**	1.80 **2.32**	1.76 **2.22**	1.71 **2.14**	1.67 **2.08**	1.63 **2.00**	1.60 **1.97**	1.57 **1.90**	1.54 **1.86**	1.53 **1.84**	38
40	4.08 **7.31**	3.23 **5.18**	2.84 **4.31**	2.61 **3.83**	2.45 **3.51**	2.34 **3.29**	2.25 **3.12**	2.18 **2.99**	2.12 **2.88**	2.07 **2.80**	2.04 **2.73**	2.00 **2.66**	1.95 **2.56**	1.90 **2.49**	1.84 **2.37**	1.79 **2.29**	1.74 **2.20**	1.69 **2.11**	1.66 **2.05**	1.61 **1.97**	1.59 **1.94**	1.55 **1.88**	1.53 **1.84**	1.51 **1.81**	40
42	4.07 **7.27**	3.22 **5.15**	2.83 **4.29**	2.59 **3.80**	2.44 **3.49**	2.32 **3.26**	2.24 **3.10**	2.17 **2.96**	2.11 **2.86**	2.06 **2.77**	2.02 **2.70**	1.99 **2.64**	1.94 **2.54**	1.89 **2.46**	1.82 **2.35**	1.78 **2.26**	1.73 **2.17**	1.68 **2.08**	1.64 **2.02**	1.60 **1.94**	1.57 **1.91**	1.54 **1.85**	1.51 **1.80**	1.49 **1.78**	42
44	4.06 **7.24**	3.21 **5.12**	2.82 **4.26**	2.58 **3.78**	2.43 **3.46**	2.31 **3.24**	2.23 **3.07**	2.16 **2.94**	2.10 **2.84**	2.05 **2.75**	2.01 **2.68**	1.98 **2.62**	1.92 **2.52**	1.88 **2.44**	1.81 **2.32**	1.76 **2.24**	1.72 **2.15**	1.66 **2.06**	1.63 **2.00**	1.58 **1.92**	1.56 **1.88**	1.52 **1.82**	1.50 **1.78**	1.48 **1.75**	44
46	4.05 **7.21**	3.20 **5.10**	2.81 **4.24**	2.57 **3.76**	2.42 **3.44**	2.30 **3.22**	2.22 **3.05**	2.14 **2.92**	2.09 **2.82**	2.04 **2.73**	2.00 **2.66**	1.97 **2.60**	1.91 **2.50**	1.87 **2.42**	1.80 **2.30**	1.75 **2.22**	1.71 **2.13**	1.65 **2.04**	1.62 **1.98**	1.57 **1.90**	1.54 **1.86**	1.51 **1.80**	1.48 **1.76**	1.46 **1.72**	46
48	4.04 **7.19**	3.19 **5.08**	2.80 **4.22**	2.56 **3.74**	2.41 **3.42**	2.30 **3.20**	2.21 **3.04**	2.14 **2.90**	2.08 **2.80**	2.03 **2.71**	1.99 **2.64**	1.96 **2.58**	1.90 **2.48**	1.86 **2.40**	1.79 **2.28**	1.74 **2.20**	1.70 **2.11**	1.64 **2.02**	1.61 **1.96**	1.56 **1.88**	1.53 **1.84**	1.50 **1.78**	1.47 **1.73**	1.45 **1.70**	48

Degrees of Freedom (for the denominator)

Table F (continued)

Degrees of Freedom (for the numerator)

df (denom)	1	2	3	4	5	6	7	8	9	10	11	12	14	16	20	24	30	40	50	75	100	200	500	∞
50	4.03 / **7.17**	3.18 / **5.06**	2.79 / **4.20**	2.56 / **3.72**	2.40 / **3.41**	2.29 / **3.18**	2.20 / **3.02**	2.13 / **2.88**	2.07 / **2.78**	2.02 / **2.70**	1.98 / **2.62**	1.95 / **2.56**	1.90 / **2.46**	1.85 / **2.39**	1.78 / **2.26**	1.74 / **2.18**	1.69 / **2.10**	1.63 / **2.00**	1.60 / **1.94**	1.55 / **1.86**	1.52 / **1.82**	1.48 / **1.76**	1.46 / **1.71**	1.44 / **1.68**
55	4.02 / **7.12**	3.17 / **5.01**	2.78 / **4.16**	2.54 / **3.68**	2.38 / **3.37**	2.27 / **3.15**	2.18 / **2.98**	2.11 / **2.85**	2.05 / **2.75**	2.00 / **2.66**	1.97 / **2.59**	1.93 / **2.53**	1.88 / **2.43**	1.83 / **2.35**	1.76 / **2.23**	1.72 / **2.15**	1.67 / **2.06**	1.61 / **1.96**	1.58 / **1.90**	1.52 / **1.82**	1.50 / **1.78**	1.46 / **1.71**	1.43 / **1.66**	1.41 / **1.64**
60	4.00 / **7.08**	3.15 / **4.98**	2.76 / **4.13**	2.52 / **3.65**	2.37 / **3.34**	2.25 / **3.12**	2.17 / **2.95**	2.10 / **2.82**	2.04 / **2.72**	1.99 / **2.63**	1.95 / **2.56**	1.92 / **2.50**	1.86 / **2.40**	1.81 / **2.32**	1.75 / **2.20**	1.70 / **2.12**	1.65 / **2.03**	1.59 / **1.93**	1.56 / **1.87**	1.50 / **1.79**	1.48 / **1.74**	1.44 / **1.68**	1.41 / **1.63**	1.39 / **1.60**
65	3.99 / **7.04**	3.14 / **4.95**	2.75 / **4.10**	2.51 / **3.62**	2.36 / **3.31**	2.24 / **3.09**	2.15 / **2.93**	2.08 / **2.79**	2.02 / **2.70**	1.98 / **2.61**	1.94 / **2.54**	1.90 / **2.47**	1.85 / **2.37**	1.80 / **2.30**	1.73 / **2.18**	1.68 / **2.09**	1.63 / **2.00**	1.57 / **1.90**	1.54 / **1.84**	1.49 / **1.76**	1.46 / **1.71**	1.42 / **1.64**	1.39 / **1.60**	1.37 / **1.56**
70	3.98 / **7.01**	3.13 / **4.92**	2.74 / **4.08**	2.50 / **3.60**	2.35 / **3.29**	2.23 / **3.07**	2.14 / **2.91**	2.07 / **2.77**	2.01 / **2.67**	1.97 / **2.59**	1.93 / **2.51**	1.89 / **2.45**	1.84 / **2.35**	1.79 / **2.28**	1.72 / **2.15**	1.67 / **2.07**	1.62 / **1.98**	1.56 / **1.88**	1.53 / **1.82**	1.47 / **1.74**	1.45 / **1.69**	1.40 / **1.62**	1.37 / **1.56**	1.35 / **1.53**
80	3.96 / **6.96**	3.11 / **4.88**	2.72 / **4.04**	2.48 / **3.56**	2.33 / **3.25**	2.21 / **3.04**	2.12 / **2.87**	2.05 / **2.74**	1.99 / **2.64**	1.95 / **2.55**	1.91 / **2.48**	1.88 / **2.41**	1.82 / **2.32**	1.77 / **2.24**	1.70 / **2.11**	1.65 / **2.03**	1.60 / **1.94**	1.54 / **1.84**	1.51 / **1.78**	1.45 / **1.70**	1.42 / **1.65**	1.38 / **1.57**	1.35 / **1.52**	1.32 / **1.49**
100	3.94 / **6.90**	3.09 / **4.82**	2.70 / **3.98**	2.46 / **3.51**	2.30 / **3.20**	2.19 / **2.99**	2.10 / **2.82**	2.03 / **2.69**	1.97 / **2.59**	1.92 / **2.51**	1.88 / **2.43**	1.85 / **2.36**	1.79 / **2.26**	1.75 / **2.19**	1.68 / **2.06**	1.63 / **1.98**	1.57 / **1.89**	1.51 / **1.79**	1.48 / **1.73**	1.42 / **1.64**	1.39 / **1.59**	1.34 / **1.51**	1.30 / **1.46**	1.28 / **1.43**
125	3.92 / **6.84**	3.07 / **4.78**	2.68 / **3.94**	2.44 / **3.47**	2.29 / **3.17**	2.17 / **2.95**	2.08 / **2.79**	2.01 / **2.65**	1.95 / **2.56**	1.90 / **2.47**	1.86 / **2.40**	1.83 / **2.33**	1.77 / **2.23**	1.72 / **2.15**	1.65 / **2.03**	1.60 / **1.94**	1.55 / **1.85**	1.49 / **1.75**	1.45 / **1.68**	1.39 / **1.59**	1.36 / **1.54**	1.31 / **1.46**	1.27 / **1.40**	1.25 / **1.37**
150	3.91 / **6.81**	3.06 / **4.75**	2.67 / **3.91**	2.43 / **3.44**	2.27 / **3.14**	2.16 / **2.92**	2.07 / **2.76**	2.00 / **2.62**	1.94 / **2.53**	1.89 / **2.44**	1.85 / **2.37**	1.82 / **2.30**	1.76 / **2.20**	1.71 / **2.12**	1.64 / **2.00**	1.59 / **1.91**	1.54 / **1.83**	1.47 / **1.72**	1.44 / **1.66**	1.37 / **1.56**	1.34 / **1.51**	1.29 / **1.43**	1.25 / **1.37**	1.22 / **1.33**
200	3.89 / **6.76**	3.04 / **4.71**	2.65 / **3.88**	2.41 / **3.41**	2.26 / **3.11**	2.14 / **2.90**	2.05 / **2.73**	1.98 / **2.60**	1.92 / **2.50**	1.87 / **2.41**	1.83 / **2.34**	1.80 / **2.28**	1.74 / **2.17**	1.69 / **2.09**	1.62 / **1.97**	1.57 / **1.88**	1.52 / **1.79**	1.45 / **1.69**	1.42 / **1.62**	1.35 / **1.53**	1.32 / **1.48**	1.26 / **1.39**	1.22 / **1.33**	1.19 / **1.28**
400	3.86 / **6.70**	3.02 / **4.66**	2.62 / **3.83**	2.39 / **3.36**	2.23 / **3.06**	2.12 / **2.85**	2.03 / **2.69**	1.96 / **2.55**	1.90 / **2.46**	1.85 / **2.37**	1.81 / **2.29**	1.78 / **2.23**	1.72 / **2.12**	1.67 / **2.04**	1.60 / **1.92**	1.54 / **1.84**	1.49 / **1.74**	1.42 / **1.64**	1.38 / **1.57**	1.32 / **1.47**	1.28 / **1.42**	1.22 / **1.32**	1.16 / **1.24**	1.13 / **1.19**
1000	3.85 / **6.66**	3.00 / **4.62**	2.61 / **3.80**	2.38 / **3.34**	2.22 / **3.04**	2.10 / **2.82**	2.02 / **2.66**	1.95 / **2.53**	1.89 / **2.43**	1.84 / **2.34**	1.80 / **2.26**	1.76 / **2.20**	1.70 / **2.09**	1.65 / **2.01**	1.58 / **1.89**	1.53 / **1.81**	1.47 / **1.71**	1.41 / **1.61**	1.36 / **1.54**	1.30 / **1.44**	1.26 / **1.38**	1.19 / **1.28**	1.13 / **1.19**	1.08 / **1.11**
∞	3.84 / **6.64**	2.99 / **4.60**	2.60 / **3.78**	2.37 / **3.32**	2.21 / **3.02**	2.09 / **2.80**	2.01 / **2.64**	1.94 / **2.51**	1.88 / **2.41**	1.83 / **2.32**	1.79 / **2.24**	1.75 / **2.18**	1.69 / **2.07**	1.64 / **1.99**	1.57 / **1.87**	1.52 / **1.79**	1.46 / **1.69**	1.40 / **1.59**	1.35 / **1.52**	1.28 / **1.41**	1.24 / **1.36**	1.17 / **1.25**	1.11 / **1.15**	1.00 / **1.00**

Degrees of Freedom (for the denominator)

Table G   Critical Values for the Mann-Whitney *U* Test*

	One-Tailed Test											*Two-Tailed Test*								
	$\alpha = .01$ (roman)											$\alpha = .02$ (roman)								
	$\alpha = .005$ (**boldface**)											$\alpha = .01$ (**boldface**)								

N_2 \ N_1	1	2	3	4	5	6	7	8	9	10	11	12	13	14	15	16	17	18	19	20
1	—	—	—	—	—	—	—	—	—	—	—	—	—	—	—	—	—	—	—	—
2	—	—	—	—	—	—	—	—	—	—	—	—	0	0	0	0	0	0	1	1
													—	—	—	—	—	—	**0**	**0**
3	—	—	—	—	—	—	—	0	0	1	1	1	2	2	2	3	3	4	4	4	5
								—	—	**0**	**0**	**0**	**1**	**1**	**1**	**2**	**2**	**2**	**2**	**3**	**3**
4	—	—	—	—	0	1	1	2	3	3	4	5	5	6	7	7	8	9	9	10
				—	—	**0**	**0**	**1**	**1**	**2**	**2**	**3**	**3**	**4**	**5**	**5**	**6**	**6**	**7**	**8**
5	—	—	—	0	1	2	3	4	5	6	7	8	9	10	11	12	13	14	15	16
				—	**0**	**1**	**1**	**2**	**3**	**4**	**5**	**6**	**7**	**7**	**8**	**9**	**10**	**11**	**12**	**13**
6	—	—	—	1	2	3	4	6	7	8	9	11	12	13	15	16	18	19	20	22
				0	**1**	**2**	**3**	**4**	**5**	**6**	**7**	**9**	**10**	**11**	**12**	**13**	**15**	**16**	**17**	**18**
7	—	—	0	1	3	4	6	7	9	11	12	14	16	17	19	21	23	24	26	28
			—	**0**	**1**	**3**	**4**	**6**	**7**	**9**	**10**	**12**	**13**	**15**	**16**	**18**	**19**	**21**	**22**	**24**
8	—	—	0	2	4	6	7	9	11	13	15	17	20	22	24	26	28	30	32	34
			—	**1**	**2**	**4**	**6**	**7**	**9**	**11**	**13**	**15**	**17**	**18**	**20**	**22**	**24**	**26**	**28**	**30**
9	—	—	1	3	5	7	9	11	14	16	18	21	23	26	28	31	33	36	38	40
		0	**1**	**3**	**5**	**7**	**9**	**11**	**13**	**16**	**18**	**20**	**22**	**24**	**27**	**29**	**31**	**33**	**36**
10	—	—	1	3	6	8	11	13	16	19	22	24	27	30	33	36	38	41	44	47
		0	**2**	**4**	**6**	**9**	**11**	**13**	**16**	**18**	**21**	**24**	**26**	**29**	**31**	**34**	**37**	**39**	**42**
11	—	—	1	4	7	9	12	15	18	22	25	28	31	34	37	41	44	47	50	53
		0	**2**	**5**	**7**	**10**	**13**	**16**	**18**	**21**	**24**	**27**	**30**	**33**	**36**	**39**	**42**	**45**	**48**
12	—	—	2	5	8	11	14	17	21	24	28	31	35	38	42	46	49	53	56	60
		1	**3**	**6**	**9**	**12**	**15**	**18**	**21**	**24**	**27**	**31**	**34**	**37**	**41**	**44**	**47**	**51**	**54**
13	—	0	2	5	9	12	16	20	23	27	31	35	39	43	47	51	55	59	63	67
	—	**1**	**3**	**7**	**10**	**13**	**17**	**20**	**24**	**27**	**31**	**34**	**38**	**42**	**45**	**49**	**53**	**56**	**60**
14	—	0	2	6	10	13	17	22	26	30	34	38	43	47	51	56	60	65	69	73
	—	**1**	**4**	**7**	**11**	**15**	**18**	**22**	**26**	**30**	**34**	**38**	**42**	**46**	**50**	**54**	**58**	**63**	**67**
15	—	0	3	7	11	15	19	24	28	33	37	42	47	51	56	61	66	70	75	80
	—	**2**	**5**	**8**	**12**	**16**	**20**	**24**	**29**	**33**	**37**	**42**	**46**	**51**	**55**	**60**	**64**	**69**	**73**
16	—	0	3	7	12	16	21	26	31	36	41	46	51	56	61	66	71	76	82	87
	—	**2**	**5**	**9**	**13**	**18**	**22**	**27**	**31**	**36**	**41**	**45**	**50**	**55**	**60**	**65**	**70**	**74**	**79**
17	—	0	4	8	13	18	23	28	33	38	44	49	55	60	66	71	77	82	88	93
	—	**2**	**6**	**10**	**15**	**19**	**24**	**29**	**34**	**39**	**44**	**49**	**54**	**60**	**65**	**70**	**75**	**81**	**86**
18	—	0	4	9	14	19	24	30	36	41	47	53	59	65	70	76	82	88	94	100
	—	**2**	**6**	**11**	**16**	**21**	**26**	**31**	**37**	**42**	**47**	**53**	**58**	**64**	**70**	**75**	**81**	**87**	**92**
19	—	1	4	9	15	20	26	32	38	44	50	56	63	69	75	82	88	94	101	107
	0	**3**	**7**	**12**	**17**	**22**	**28**	**33**	**39**	**45**	**51**	**56**	**63**	**69**	**74**	**81**	**87**	**93**	**99**
20	—	1	5	10	16	22	28	34	40	47	53	60	67	73	80	87	93	100	107	114
	0	**3**	**8**	**13**	**18**	**24**	**30**	**36**	**42**	**48**	**54**	**60**	**67**	**73**	**79**	**86**	**92**	**99**	**105**

*To be significant the *U* obtained from the data must be *equal to* or *less than* the value shown in the table. Dashes in the body of the table indicate that no decision is possible at the stated level of significance.

From *Introductory Statistics* by Roger E. Kirk. Copyright © 1978, by Wadsworth Publishing Company, Inc. Reprinted by permission of the publisher, Brooks/Cole Publishing Company, Monterey, Calif.

Table G (continued)

	One-Tailed Test $\alpha = .05$ (roman) $\alpha = .025$ (boldface)									Two-Tailed Test $\alpha = .10$ (roman) $\alpha = .05$ (boldface)										
N_2 \ N_1	1	2	3	4	5	6	7	8	9	10	11	12	13	14	15	16	17	18	19	20
1	—	—	—	—	—	—	—	—	—	—	—	—	—	—	—	—	—	—	0	0
2	—	—	—	—	0	0	0	1	1	1	1	2	2	2	3	3	3	4	4	4
	—	—	—	—	**—**	**—**	**—**	**0**	**0**	**0**	**0**	**1**	**1**	**1**	**1**	**1**	**2**	**2**	**2**	**2**
3	—	—	0	0	1	2	2	3	3	4	5	5	6	7	7	8	9	9	10	11
	—	—	—	—	**0**	**1**	**1**	**2**	**2**	**3**	**3**	**4**	**4**	**5**	**5**	**6**	**6**	**7**	**7**	**8**
4	—	—	0	1	2	3	4	5	6	7	8	9	10	11	12	14	15	16	17	18
	—	—	—	**0**	**1**	**2**	**3**	**4**	**4**	**5**	**6**	**7**	**8**	**9**	**10**	**11**	**11**	**12**	**13**	**13**
5	—	0	1	2	4	5	6	8	9	11	12	13	15	16	18	19	20	22	23	25
	—	—	**0**	**1**	**2**	**3**	**5**	**6**	**7**	**8**	**9**	**11**	**12**	**13**	**14**	**15**	**17**	**18**	**19**	**20**
6	—	0	2	3	5	7	8	10	12	14	16	17	19	21	23	25	26	28	30	32
	—	—	**1**	**2**	**3**	**5**	**6**	**8**	**10**	**11**	**13**	**14**	**16**	**17**	**19**	**21**	**22**	**24**	**25**	**27**
7	—	0	2	4	6	8	11	13	15	17	19	21	24	26	28	30	33	35	37	39
	—	—	**1**	**3**	**5**	**6**	**8**	**10**	**12**	**14**	**16**	**18**	**20**	**22**	**24**	**26**	**28**	**30**	**32**	**34**
8	—	1	3	5	8	10	13	15	18	20	23	26	28	31	33	36	39	41	44	47
	—	**0**	**2**	**4**	**6**	**8**	**10**	**13**	**15**	**17**	**19**	**22**	**24**	**26**	**29**	**31**	**34**	**36**	**38**	**41**
9	—	1	3	6	9	12	15	18	21	24	27	30	33	36	39	42	45	48	51	54
	—	**0**	**2**	**4**	**7**	**10**	**12**	**15**	**17**	**20**	**23**	**26**	**28**	**31**	**34**	**37**	**39**	**42**	**45**	**48**
10	—	1	4	7	11	14	17	20	24	27	31	34	37	41	44	48	51	55	58	62
	—	**0**	**3**	**5**	**8**	**11**	**14**	**17**	**20**	**23**	**26**	**29**	**33**	**36**	**39**	**42**	**45**	**48**	**52**	**55**
11	—	1	5	8	12	16	19	23	27	31	34	38	42	46	50	54	57	61	65	69
	—	**0**	**3**	**6**	**9**	**13**	**16**	**19**	**23**	**26**	**30**	**33**	**37**	**40**	**44**	**47**	**51**	**55**	**58**	**62**
12	—	2	5	9	13	17	21	26	30	34	38	42	47	51	55	60	64	68	72	77
	—	**1**	**4**	**7**	**11**	**14**	**18**	**22**	**26**	**29**	**33**	**37**	**41**	**45**	**49**	**53**	**57**	**61**	**65**	**69**
13	—	2	6	10	15	19	24	28	33	37	42	47	51	56	61	65	70	75	80	84
	—	**1**	**4**	**8**	**12**	**16**	**20**	**24**	**28**	**33**	**37**	**41**	**45**	**50**	**54**	**59**	**63**	**67**	**72**	**76**
14	—	2	7	11	16	21	26	31	36	41	46	51	56	61	66	71	77	82	87	92
	—	**1**	**5**	**9**	**13**	**17**	**22**	**26**	**31**	**36**	**40**	**45**	**50**	**55**	**59**	**64**	**67**	**74**	**78**	**83**
15	—	3	7	12	18	23	28	33	39	44	50	55	61	66	72	77	83	88	94	100
	—	**1**	**5**	**10**	**14**	**19**	**24**	**29**	**34**	**39**	**44**	**49**	**54**	**59**	**64**	**70**	**75**	**80**	**85**	**90**
16	—	3	8	14	19	25	30	36	42	48	54	60	65	71	77	83	89	95	101	107
	—	**1**	**6**	**11**	**15**	**21**	**26**	**31**	**37**	**42**	**47**	**53**	**59**	**64**	**70**	**75**	**81**	**86**	**92**	**98**
17	—	3	9	15	20	26	33	39	45	51	57	64	70	77	83	89	96	102	109	115
	—	**2**	**6**	**11**	**17**	**22**	**28**	**34**	**39**	**45**	**51**	**57**	**63**	**67**	**75**	**81**	**87**	**93**	**99**	**105**
18	—	4	9	16	22	28	35	41	48	55	61	68	75	82	88	95	102	109	116	123
	—	**2**	**7**	**12**	**18**	**24**	**30**	**36**	**42**	**48**	**55**	**61**	**67**	**74**	**80**	**86**	**93**	**99**	**106**	**112**
19	0	4	10	17	23	30	37	44	51	58	65	72	80	87	94	101	109	116	123	130
	—	**2**	**7**	**13**	**19**	**25**	**32**	**38**	**45**	**52**	**58**	**65**	**72**	**78**	**85**	**92**	**99**	**106**	**113**	**119**
20	0	4	11	18	25	32	39	47	54	62	69	77	84	92	100	107	115	123	130	138
	—	**2**	**8**	**13**	**20**	**27**	**34**	**41**	**48**	**55**	**62**	**69**	**76**	**83**	**90**	**98**	**105**	**112**	**119**	**127**

Table H   Critical Values for the Wilcoxon Matched-Pairs Signed-Ranks T Test*

	Level of Significance for a One-Tailed Test					Level of Significance for a One-Tailed Test			
	0.05	0.025	0.01	0.005		0.05	0.025	0.01	0.005
	Level of Significance for a Two-Tailed Test					Level of Significance for a Two-Tailed Test			
N	0.10	0.05	0.02	0.01	N	0.10	0.05	0.02	0.01
5	0	—	—	—	28	130	116	101	91
6	2	0	—	—	29	140	126	110	100
7	3	2	0	—	30	151	137	120	109
8	5	3	1	0	31	163	147	130	118
9	8	5	3	1	32	175	159	140	128
10	10	8	5	3	33	187	170	151	138
11	13	10	7	5	34	200	182	162	148
12	17	13	9	7	35	213	195	173	159
13	21	17	12	9	36	227	208	185	171
14	25	21	15	12	37	241	221	198	182
15	30	25	19	15	38	256	235	211	194
16	35	29	23	19	39	271	249	224	207
17	41	34	27	23	40	286	264	238	220
18	47	40	32	27	41	302	279	252	233
19	53	46	37	32	42	319	294	266	247
20	60	52	43	37	43	336	310	281	261
21	67	58	49	42	44	353	327	296	276
22	75	65	55	48	45	371	343	312	291
23	83	73	62	54	46	389	361	328	307
24	91	81	69	61	47	407	378	345	322
25	100	89	76	68	48	426	396	362	339
26	110	98	84	75	49	446	415	379	355
27	119	107	92	83	50	466	434	397	373

*To be significant the T obtained from the data must be *equal to or less than* the value shown in the table. From *Introductory Statistics* by Roger E. Kirk. Copyright © 1978 by Wadsworth Publishing Company, Inc. Reprinted by permission of the publisher, Brooks/Cole Publishing Company, Monterey, Calif.

Table J Critical Differences for the Wilcoxon-Wilcox Multiple-Comparisons Test*

N	$\alpha = .01$ (two-tailed)							
	$K=3$	$K=4$	$K=5$	$K=6$	$K=7$	$K=8$	$K=9$	$K=10$
1	4.1	5.7	7.3	8.9	10.5	12.2	13.9	15.6
2	10.9	15.3	19.7	24.3	28.9	33.6	38.3	43.1
3	19.5	27.5	35.7	44.0	52.5	61.1	69.8	78.6
4	29.7	41.9	54.5	67.3	80.3	93.6	107.0	120.6
5	41.2	58.2	75.8	93.6	111.9	130.4	149.1	168.1
6	53.9	76.3	99.3	122.8	146.7	171.0	195.7	220.6
7	67.6	95.8	124.8	154.4	184.6	215.2	246.3	277.7
8	82.4	116.8	152.2	188.4	225.2	262.6	300.6	339.0
9	98.1	139.2	181.4	224.5	268.5	313.1	358.4	404.2
10	114.7	162.8	212.2	262.7	314.2	366.5	419.5	473.1
11	132.1	187.6	244.6	302.9	362.2	422.6	483.7	545.6
12	150.4	213.5	278.5	344.9	412.5	481.2	551.0	621.4
13	169.4	240.6	313.8	388.7	464.9	542.4	621.0	700.5
14	189.1	268.7	350.5	434.2	519.4	606.0	693.8	782.6
15	209.6	297.8	388.5	481.3	575.8	671.9	769.3	867.7
16	230.7	327.9	427.9	530.1	634.2	740.0	847.3	955.7
17	252.5	359.0	468.4	580.3	694.4	810.2	927.8	1046.5
18	275.0	391.0	510.2	632.1	756.4	882.6	1010.6	1140.0
19	298.1	423.8	553.1	685.4	820.1	957.0	1095.8	1236.2
20	321.8	457.6	597.2	740.0	885.5	1033.3	1183.3	1334.9
21	346.1	492.2	642.4	796.0	952.6	1111.6	1273.0	1436.0
22	371.0	527.6	688.7	853.4	1021.3	1191.8	1364.8	1539.7
23	396.4	563.8	736.0	912.1	1091.5	1273.8	1458.8	1645.7
24	422.4	600.9	784.4	972.1	1163.4	1357.6	1554.8	1754.0
25	449.0	638.7	833.8	1033.3	1236.7	1443.2	1652.8	1864.6

N	$\alpha = .05$ (two-tailed)							
	$K=3$	$K=4$	$K=5$	$K=6$	$K=7$	$K=8$	$K=9$	$K=10$
1	3.3	4.7	6.1	7.5	9.0	10.5	12.0	13.5
2	8.8	12.6	16.5	20.5	24.7	28.9	33.1	37.4
3	15.7	22.7	29.9	37.3	44.8	52.5	60.3	68.2
4	23.9	34.6	45.6	57.0	68.6	80.4	92.4	104.6
5	33.1	48.1	63.5	79.3	95.5	112.0	128.8	145.8
6	43.3	62.9	83.2	104.0	125.3	147.0	169.1	191.4
7	54.4	79.1	104.6	130.8	157.6	184.9	212.8	240.9
8	66.3	96.4	127.6	159.6	192.4	225.7	259.7	294.1
9	78.9	114.8	152.0	190.2	229.3	269.1	309.6	350.6
10	92.3	134.3	177.8	222.6	268.4	315.0	362.4	410.5
11	106.3	154.8	205.0	256.6	309.4	363.2	417.9	473.3
12	120.9	176.2	233.4	292.2	352.4	413.6	476.0	539.1
13	136.2	198.5	263.0	329.3	397.1	466.2	536.5	607.7
14	152.1	221.7	293.8	367.8	443.6	520.8	599.4	679.0
15	168.6	245.7	325.7	407.8	491.9	577.4	664.6	752.8
16	185.6	270.6	358.6	449.1	541.7	635.9	732.0	829.2
17	203.1	296.2	392.6	491.7	593.1	696.3	801.5	907.9
18	221.2	322.6	427.6	535.5	646.1	758.5	873.1	989.0
19	239.8	349.7	463.6	580.6	700.5	822.4	946.7	1072.4
20	258.8	377.6	500.5	626.9	756.4	888.1	1022.3	1158.1
21	278.4	406.1	538.4	674.4	813.7	955.4	1099.8	1245.9
22	298.4	435.3	577.2	723.0	872.3	1024.3	1179.1	1335.7
23	318.9	465.2	616.9	772.7	932.4	1094.8	1260.3	1427.7
24	339.8	495.8	657.4	823.5	993.7	1166.8	1343.2	1521.7
25	361.1	527.0	698.8	875.4	1056.3	1240.4	1427.9	1617.6

*To be significant the difference obtained from the data must be equal to or larger than the tabled value. From *Some Rapid Approximate Statistical Procedures*, by F. Wilcoxon and R. Wilcox, 1964. Reprinted by permission.

Table K   Critical Values for Spearman's r_s

Size of Sample	5% Level	1% Level
4 or less	none	none
5	1.000	none
6	0.886	1.000
7	0.750	0.893
8	0.714	0.857
9	0.683	0.833
10	0.648	0.794

*To be significant the r_s obtained from the data must be equal to or larger than the value shown in the table.

Reproduced by permission of the publishers, Charles Griffin & Company Ltd. of London and High Wycombe, from M. G. Kendall, *Rank Correlation Methods*, 4th edition 1970.

GLOSSARY OF WORDS

Abscissa.   The horizontal or X axis of a graph.

Absolute Value.   The value of a number without consideration of its algebraic sign.

Additive.   Can legitimately be summed.

Alternative Hypothesis.   The hypothesis that the mean of the population treated in a certain way is not equal to the mean of the population not treated in that way; symbolized H_1.

Analysis of Variance (ANOVA).   A statistical method for determining the significance of the differences among a set of means.

Asymptotic.   A line that continually approaches, but never reaches a specified level.

Bar Graph.   A frequency graph for nominal or qualitative data. Bars are raised from each designation of a nominal variable on the X axis to the level of its frequency on the Y axis. Space is left between the bars.

Biased Sample.   A sample that does not provide all members of the population an equal probability of selection.

Bimodal Distribution.   A distribution with two modes.

Binomial Distribution.   A distribution of events that have only two possible outcomes.

Bivariate Distribution.   A joint distribution of two variables, the individual scores of which are paired in some logical way.

Cell.   The portion of an ANOVA table containing the scores of subjects that are treated alike.

Central Limit Theorem.   The theorem in mathematical statistics that the sampling distribution of the mean approaches a normal curve as N gets larger, and that the standard deviation of this sampling distribution is equal to $\sigma/\sqrt{N}$.

Central Value.   The mean, median, or mode; a statistic that describes the typical score in a distribution.

Chi Square Distribution.   A theoretical sampling distribution of chi square values. There is a chi square distribution for each number of degrees of freedom.

Class Interval.   A range of scores grouped together in a grouped frequency distribution.

Coefficient of Determination.   A squared correlation coefficient; an estimate of common variance.

Common Variance.   Variance held in common by two variables. It is assumed to be determined or caused by the same factors.

Confidence Interval.   An interval of scores within which, with specified confidence, a parameter is expected to lie.

Confidence Limits.   Two numbers that define the boundaries of a confidence interval.

Constant.   A mathematical value that remains the same within a series of operations; for example, regression coefficients a and b have the same value for all predictions from the same regression line.

Control Group.   A group in an experiment against which other groups are compared.

Correlated-Samples Design.   An experimental design in which measures from different groups are not independent of each other. Some writers call this a dependent-samples design.

Correlation.   A relationship between variables such that increases or decreases in the value of one variable tend to be accompanied by increases or decreases in the other.

Critical Region.   The area of the sampling distribution that covers the values of the test statistic that are not due to chance.

Critical Value.   The value from a sampling distribution against which a computed statistic is compared to determine whether the null hypothesis may be rejected.

Degrees of Freedom.   The number of observations minus the number of necessary relations obtaining among these observations.

Dependent Variable.   The variable that is measured and analyzed in an experiment. Its values are tested to determine whether they are dependent upon values of the independent variable.

Descriptive Statistic.   Index number that summarizes or describes a set of data.

Deviation Score.   A raw score minus the mean of the distribution from which the raw score was drawn.

Dichotomous Variable.   A variable taking two, and only two, values.

Distribution-Free Statistics.   Statistical methods that do not assume any particular population distribution.

Empirical Distribution.   An arrangement from highest to lowest of actual scores from real observations.

Error Variance.   Variance due to factors not controlled in the experiment; within-group variance.

Expected Value.   The mean value of a random variable over an infinite number of samplings. The expected value of a statistic is the mean of the sampling distribution of the statistic.

Experimental Group.   A group that receives a treatment in an experiment and whose dependent-variable scores are compared with those of a control group.

Extraneous Variable.   A variable, other than the independent variable, that may affect the dependent variable.

F Distribution.   A theoretical sampling distribution of F values. There is a different F distribution for each combination of degrees of freedom.

F Test.   A method of determining the significance of the difference among two or more means.

Factor.   Independent variable.

Factorial Design.   An experimental design using two or more levels of two or more factors and permitting an analysis of interaction effects between independent variables.

Frequency.   The number of times a score occurs in a distribution.

Frequency Polygon.   A graph with quantitative scores on the X axis and frequencies on the Y axis. Each point on the graph represents a score and the frequency of occurrence of that score. Points are connected by a line.

Goodness of Fit.   Degree to which observed data coincide with theoretical expectations.

Grand Mean.   The mean of all the scores in an experiment.

Grouped Frequency Distribution.   An arrangement of scores from highest to lowest in which scores are grouped together into equal-sized ranges called class intervals. The number of scores occurring in each class interval is placed in a column beside the appropriate class interval.

Histogram.   A graph with quantitative scores on the X axis and frequencies on the Y axis. A bar covering the range from the lower to upper limit of each score or class interval is raised to the level of that score's frequency. There is no space between the bars.

Hypothesis.   A statement about the relationship between two or more variables.

Hypothesis Testing.   The process of hypothesizing a parameter and comparing (or testing) the parameter with an empirical statistic in order to decide whether the parameter is reasonable.

Independent.   Events that have nothing to do with each other. Occurrence or variation of one does not affect the occurrence or variation of the other. Two sets of uncorrelated scores are independent of each other.

Independent-Samples Design.   An experimental design using samples whose dependent-variable scores cannot logically be paired.

Independent Variable.   The treatment variable; it is selected by the experimenter.

Inferential Statistics.   A method of deciding between two or more alternative conclusions.

Interaction.   A relationship between two factors such that the effect of one treatment on the dependent variable depends upon the level of the other treatment.

Interpolation.   A method for determining a value known to lie between two other values.

Interval Scale.   A measurement scale in which equal differences between numbers stand for equal differences in the thing measured. The zero point is arbitrarily defined.

J-curve.   A severely skewed distribution with the mode at one extreme.

Least-Squares Solution.   Method of fitting a regression line such that the sums of the squared deviations from the straight regression line will be a minimum.

Level.   A treatment chosen from an independent variable.

Level of Confidence.   The confidence $(1 - \alpha)$ that a parameter lies within a given interval.

Level of Significance.   The probability level at which the null hypothesis is rejected.

Line Graph.   A graph presenting the relationship between two variables.

Linearity.   The condition wherein the "line of best fit" through a scatterplot is a straight line.

Lower Limit.   The bottom of the range of possible values that a score on a quantitative variable can take; for example, a score of 5 has 4.5 as its lower limit.

Main Effect.   The deviation of one or more treatment means from the grand mean.

Mann-Whitney U Test.   A nonparametric method used to determine whether the two sets of ranked data based on two independent samples came from the same population.

Matched Pairs.   A correlated-samples design in which pairs of scores are matched.

Mean.   The arithmetic average; the sum of the scores divided by the number of scores.

Mean Square.   An ANOVA term for the variance; a sum of squares divided by its degrees of freedom.

Median.   The point that divides a distribution of scores into two equal halves, so that half the scores are above the median and half are below it.

Mode.   The score that occurs most frequently in a distribution.

Multiple Comparisons.   Tests of differences between treatment means or combinations of means following an ANOVA.

Multiple Correlation.   A correlation method that combines intercorrelations among more than two variables into a single statistic.

Natural Pairs.   A correlated-samples design, in which pairing occurs prior to the experiment.

Nominal Scale.   A scale of measurement in which numbers are used simply as names and have no real quantitative value.

Nonparametric Methods.   Statistical methods that do not require the estimation of parameters.

Normal Distribution.   A theoretical distribution based on frequency of occurrence of chance events.

Normality.   The condition of being distributed in the form of the normal curve.

NS.   Not statistically significant.

Null Hypothesis.   The assumption that the difference between an observed statistic and a proposed parameter is the result of chance.

Observed Frequency.   Number of observations actually occurring in a category.

One-Tailed Test.   A statistical test in which the critical region lies in one tail of the distribution.

Operational Definition.   A definition that specifies a concrete meaning for a variable. The variable is defined in terms of the operations of the experiment; for example, *hunger* may be defined as "24 hours of food deprivation."

Ordinal Scale.   A rank-ordered scale of measurement in which equal differences between numbers do not represent equal differences between the things measured.

Ordinate.   The vertical or *Y* axis of a graph.

Orthogonal.   Independent; uncorrelated.

Parameter.   Some numerical characteristic of a population.

Parameter Estimation.   Estimating one particular point to be the parameter of a population.

Partial Correlation.   Technique that allows the separation or partialing out of the effects of one variable from the correlation of two other variables.

Population.   All members of a specified group.

Proportion.   A part of a whole.

Qualitative Variable.   A variable that exists in different kinds; measured on a nominal scale.

Quantitative Variable.   A variable that exists in different amounts.

Random Sample.   A subset of a population chosen in such a way that all samples of the specified size have an equal probability of being selected.

Range.   Upper limit of highest score minus lower limit of lowest score.

Ratio Scale.   A scale that has all the characteristics of an interval scale, plus a true zero point.

Raw Score.   A score as it is obtained in an experiment.

Rectangular Distribution.   A distribution in which all scores have the same frequency.

Regression Coefficients.   The values *a* (point where the regression line intersects the *Y* axis) and *b* (slope of the regression line).

Regression Equation.   An equation used to predict particular values of *Y* for specific values of *X*.

Regression Line.   The "line of best fit" that runs through a scatterplot.

Repeated Measures.   An experimental design in which more than one dependent-variable measure is taken on each subject.

Sample.   A subset of a population.

Sampling Distribution.   A theoretical distribution of a statistic based on all possible random samples drawn from the same population; used to determine probabilities.

Sampling Error.   The tendency of sample statistics from the same population to vary from one sample to another.

Scatterplot.   The plot of points that results when a distribution of paired *X* and *Y* values are plotted on a graph.

Scheffé Test.   A method of making all possible comparisons after ANOVA.

Simple Effect.   The difference between cell means in a factorial ANOVA.

Simple Frequency Distribution.   Scores arranged from highest to lowest, with the frequency of each score placed in a column beside the score.

Skewed Distribution.   An asymmetrical distribution. The skew may be positive (more low scores than high, so that the frequency polygon is pointed toward the right) or negative (more high scores than low, so that the frequency polygon is pointed toward the left).

Spearman's r_s.   A correlation statistic for two sets of ranked data.

Standard Deviation.   The square root of the mean of the squared deviations.

Standard Error.   The standard deviation of a sampling distribution.

Standard Error of Estimate.   The standard deviation of the differences between predicted outcomes and actual outcomes.

Standard Error of the Difference.   The standard deviation of a sampling distribution of differences between means.

Standard Score.   A score expressed in standard-deviation units.

Statistic.   Some numerical characteristic of a sample.

Stratified Sample.   A sample drawn in such a way that it reflects exactly a known characteristic of the population.

Subsample.   A subset of a sample.

Sum of Squares.   The sum of the squared deviations from the mean; the numerator of the formula for the standard deviation.

t Distribution.   Theoretical distribution used to determine significance of experimental results based on small samples.

t Test.   Significance test that uses the *t* distribution.

Theoretical Distribution.   Arrangement of hypothesized scores based on mathematical formulas and logic.

Theoretical Frequency.   Number of observations expected in a category if the null hypothesis is true; expected frequency.

Treatment.   A level of an independent variable.

Truncated Range.   The range of the sample is much smaller than the range of its population.

Two-Tailed Test of Significance.   Any statistical test in which the critical region is divided into the two tails of the distribution.

Type I Error.   Rejection of the null hypothesis when it is true.

Type II Error.   Retention of the null hypothesis when it is false.

Upper Limit.   The top of the range of values a score from a quantitative variable can take; for example, the number 5 has 5.5 as its upper limit.

U Value.   Statistic used in the Mann-Whitney *U* test.

Variability.   Differences among scores in a distribution.

Variable.   Something that exists in more than one amount or in more than one form.

Variance.   The square of the standard deviation.

Wilcoxon and Wilcox Multiple Comparisons.   A nonparametric method for independent samples in which all possible pairs of treatments are compared.

Wilcoxon Rank-Sum Test.   A nonparametric test for testing the difference between two independent samples.

Wilcoxon Matched-Pairs Signed-Ranks Test.   A nonparametric method for testing the difference between two correlated samples.

Yates' Correction.   A correlation for a 2 × 2 chi square when expected frequencies are few. (Now obsolete.)

z Score.   A score expressed in standard-deviation units; used to compare the relative standing of scores in two different distributions.

GLOSSARY OF FORMULAS

Analysis of Variance

degrees of freedom in one-way
ANOVA

$$df_{bg} = K - 1.$$
$$df_{wg} = N_{tot} - K.$$
$$df_{tot} = N_{tot} - 1.$$

degrees of freedom in factorial
ANOVA

$$df_A = A - 1.$$
$$df_B = B - 1.$$
$$df_{AB} = (A - 1)(B - 1).$$
$$df_{wg} = N_{tot} - (A)(B).$$
$$df_{tot} = N_{tot} - 1.$$

F value in one-way ANOVA

$$F = \frac{MS_{bg}}{MS_{wg}}.$$

F values in factorial ANOVA

$$F_A = \frac{MS_A}{MS_{wg}}.$$

$$F_B = \frac{MS_B}{MS_{wg}}.$$

$$F_{AB} = \frac{MS_{AB}}{MS_{wg}}.$$

F'_{ob} value for Scheffé
comparison

$$F'_{ob} = \frac{(c_1 \bar{X}_1 + c_2 \bar{X}_2 + \cdots + c_k \bar{X}_k)^2}{MS_{wg} \left[\frac{(c_1)^2}{N_1} + \frac{(c_2)^2}{N_2} + \cdots + \frac{(c_k)^2}{N_k} \right]}$$

where subscripts 1, 2, and k indicate the groups from which the means and N's are taken, and c_1, c_2, and c_k indicate the orthogonal coefficients for the comparison.

critical value of F' for
Scheffé tests

$$F' = F_\alpha (K - 1),$$
where F_α is the tabled value of F at the desired α level and at the appropriate df.

mean square

$$MS = \frac{SS}{df}.$$

between-groups sum of squares

$$SS_{bg} = \sum \left[\frac{(\sum X_g)^2}{N_g} \right] - \frac{(\sum X_{tot})^2}{N_{tot}}.$$

within-groups sum of squares

$$SS_{wg} = \sum \left[\sum X_g^2 - \frac{(\sum X_g)^2}{N_g} \right].$$

total sum of squares

$$SS_{tot} = \sum X_{tot}^2 - \frac{(\sum X_{tot})^2}{N_{tot}}.$$

t value for *a priori* comparisons

$$t = \frac{c_1 \bar{X}_1 + c_2 \bar{X}_2 + \cdots + c_k \bar{X}_k}{\sqrt{MS_{wg} \left[\frac{(c_1)^2}{N_1} + \frac{(c_2)^2}{N_2} + \cdots + \frac{(c_k)}{N_k} \right]}}$$

where subscripts 1, 2, and k indicate the groups from which the means and N's are taken, and c_1, c_2, and c_k indicate the orthogonal coefficients for the comparison.

sum of squares for the interaction effect in factorial ANOVA

$$SS_{AB} = N_g \sum [(\bar{X}_{AB} - \bar{X}_A - \bar{X}_B + \bar{X}_{tot})^2].$$

Check:

$$SS_{AB} = SS_{bg} - SS_A - SS_B.$$

Sum of squares for a main effect in factorial ANOVA

$$SS = \frac{(\sum X_1)^2}{N_1} + \frac{(\sum X_2)^2}{N_2} + \cdots + \frac{(\sum X_K)^2}{N_K} - \frac{(\sum X_{tot})^2}{N_{tot}}$$

where 1 and 2 denote levels of a factor and K denotes the last level of a factor.

Chi Square

basic formula

$$\chi^2 = \sum \left[\frac{(O - E)^2}{E} \right].$$

degrees of freedom for a chi square table

$df = (R - 1)(C - 1),$
where R = the number of rows
and C = the number of columns.

shortcut formula for a 2 × 2 table

$$\chi^2 = \frac{N(AD - BC)^2}{(A + B)(C + D)(A + C)(B + D)},$$

where A, B, C, and D designate the four cells of the table, moving left to right across the top row and then across the bottom row.

Confidence Intervals

about a mean $N \geqslant 30$

$LL = \bar{X} - zs_{\bar{X}}.$
$UL = \bar{X} + zs_{\bar{X}}.$

about a mean $N < 30$

$LL = \bar{X} - ts_{\bar{X}}.$
$UL = \bar{X} + ts_{\bar{X}}.$

about a mean difference (independent samples)

$LL = (\bar{X}_1 - \bar{X}_2) - t(s_{\bar{X}_1 - \bar{X}_2}).$
$UL = (\bar{X}_1 - \bar{X}_2) + t(s_{\bar{X}_1 - \bar{X}_2}).$

about a mean difference (correlated samples)

$LL = (\bar{X} - \bar{Y}) - t(s_{\bar{D}}).$
$UL = (\bar{X} - \bar{Y}) + t(s_{\bar{D}}).$

Correlation

Pearson product-moment

definition formula

$$r = \frac{\sum (z_Y z_X)}{N}.$$

computation formulas

$$r = \frac{\frac{\sum XY}{N} - (\bar{X})(\bar{Y})}{(S_X)(S_Y)};$$

$$r = \frac{N \sum XY - (\sum X)(\sum Y)}{\sqrt{[N \sum X^2 - (\sum X)^2][N \sum Y^2 - (\sum Y)^2]}}.$$

testing significance from .00

$$t = (r)\sqrt{\frac{N-2}{1-r^2}}.$$
$$df = N - 2.$$

Spearman's r_s

computation formula

$$r_s = 1 - \frac{6 \sum D^2}{N(N^2 - 1)}.$$

Degrees of Freedom

See specific statistical tests.

Deviation Score

$$x = X - \bar{X} \quad \text{or} \quad x = X - \mu.$$

Mann-Whitney U Test

value for U

$$U_1 = (N_1)(N_2) + \frac{N_1(N_1 + 1)}{2} - \sum R_1,$$

where $\sum R_1 =$ sum of the ranks in the N_1 group,

testing significance for larger samples $N \geqslant 21$

$$z = \frac{(U_1 + c) - \mu_U}{\sigma_U},$$

$$c = .5$$

$$\mu_U = \frac{N_1 N_2}{2},$$

and

$$\sigma_U = \sqrt{\frac{(N_1)(N_2)(N_1 + N_2 + 1)}{12}}.$$

Mean

from a frequency distribution

$$\mu \text{ or } \bar{X} = \frac{\sum fX}{N}.$$

from raw data

$$\mu \text{ or } \bar{X} = \frac{\sum X}{N}.$$

of a set of means

$$\bar{\bar{X}} = \frac{\sum (N_1 \bar{X}_1 + N_2 \bar{X}_2 + \cdots + N_k \bar{X}_k)}{\sum N},$$

where N_1, N_2, and so on are the number of scores associated with their respective means.

Range

range $= X_H - X_L$,
where X_H = upper limit of highest score
X_L = lower limit of lowest score

Regression

for predicting Y from X

$$Y' = r\frac{S_Y}{S_X}(X - \bar{X}) + \bar{Y}.$$

for a straight line

$$Y' = a + bX,$$

where a = the value at the Y intercept
and b = the slope of the regression line.

The Y intercept of a regression line

$$a = \bar{Y} - b\bar{X}.$$

the slope of a regression line

$$b = r\frac{S_Y}{S_X}.$$

Standard Deviation of a Population or Sample (for description)

by the raw-score method from ungrouped data

$$\sigma \text{ or } S = \sqrt{\frac{\sum X^2 - \frac{(\sum X)^2}{N}}{N}}.$$

by the raw-score method from grouped data

$$\sigma \text{ or } S = \sqrt{\frac{\sum fX^2 - \frac{(\sum fX)^2}{N}}{N}}.$$

by the deviation method from ungrouped data

$$\sigma \text{ or } S = \sqrt{\frac{\sum x^2}{N}}.$$

by the deviation method from grouped data

$$\sigma \text{ or } S = \sqrt{\frac{\sum fx^2}{N}}.$$

Standard Deviation of a Sample (for estimation of σ)

by the raw-score method from ungrouped data

$$s = \sqrt{\frac{\sum X^2 - \frac{(\sum X)^2}{N}}{N - 1}}$$

$$= \sqrt{\frac{N\sum X^2 - (\sum X)^2}{N(N - 1)}}.$$

by the raw-score method from grouped data

$$s = \sqrt{\frac{\sum fX^2 - \frac{(\sum fX)^2}{N}}{N - 1}}$$

$$= \sqrt{\frac{N\sum fX^2 - (\sum fX)^2}{N(N - 1)}}.$$

by the deviation method from ungrouped data

$$s = \sqrt{\frac{\sum x^2}{N-1}}.$$

by the deviation method from grouped data

$$s = \sqrt{\frac{\sum fx^2}{N-1}}.$$

for correlated samples

$$s_D = \sqrt{\frac{\sum D^2 - \frac{(\sum D)^2}{N}}{N-1}}.$$

Standard Error

of the mean

where the population standard deviation is known

$$\sigma_{\bar{X}} = \frac{\sigma}{\sqrt{N}}.$$

estimated from a single sample

$$s_{\bar{X}} = \frac{s}{\sqrt{N}}.$$

of the difference between means
for large independent samples or for equal-N small independent samples

$$s_{\bar{X}_1 - \bar{X}_2} = \sqrt{s_{\bar{X}_1}{}^2 + s_{\bar{X}_2}{}^2}; \quad s_{\bar{X}_1 - \bar{X}_2}$$

$$= \sqrt{\left(\frac{s_1}{\sqrt{N}}\right)^2 + \left(\frac{s_2}{\sqrt{N}}\right)^2}.$$

for small independent samples with unequal N's

$$s_{\bar{X}_1 - \bar{X}_2} = \sqrt{\left(\frac{\sum X_1{}^2 - \frac{(\sum X_1)^2}{N_1} + \sum X_2{}^2 - \frac{(\sum X_2)^2}{N_2}}{N_1 + N_2 - 2}\right)\left(\frac{1}{N_1} + \frac{1}{N_2}\right)}$$

for correlated samples by the direct-difference method

$$s_{\bar{D}} = \frac{s_D}{\sqrt{N}}$$

$$\text{where } s_D = \sqrt{\frac{\sum D^2 - \frac{(\sum D)^2}{N}}{N-1}}.$$

for correlated samples when r is known

$$s_{\bar{D}} = \sqrt{s_{\bar{X}}{}^2 + s_{\bar{Y}}{}^2 - 2r(s_{\bar{X}})(s_{\bar{Y}})}.$$

t Test

as a test for whether a sample mean came from a population with a mean, μ

$$t = \frac{X - \mu}{s_{\bar{X}}}.$$

$$df = N - 1.$$

for correlated samples where r is known

$$t = \frac{\bar{X} - \bar{Y}}{\sqrt{s_{\bar{X}}{}^2 + s_{\bar{Y}}{}^2 - 2r_{XY}(s_{\bar{X}})(s_{\bar{Y}})}}.$$

for correlated samples where the direct-difference method is used

$$t = \frac{\bar{X} - \bar{Y}}{s_{\bar{D}}}.$$

$$df = \text{Number of pairs minus one.}$$

for independent samples

$$t = \frac{\bar{X}_1 - \bar{X}_2}{s_{\bar{X}_1 - \bar{X}_2}}.$$

$$df = N_1 + N_2 - 2.$$

for testing whether a correlation coefficient is significantly different from .00

$$t = (r)\sqrt{\frac{N - 2}{1 - r^2}}.$$

$df = $ Number of pairs minus two.

Variance

Use the formulas for the standard deviation. For s^2 and σ^2, square s and σ.

Wilcoxon Matched-Pairs Signed-Ranks Test

value for T
testing significance for
larger samples $N \geqslant 50$.

$T = $ smaller sum of the signed ranks,

$$z = \frac{(T + c) - \mu_T}{\sigma_T},$$

where:

$c = .5,$

$$\mu_T = \frac{N(N + 1)}{4},$$

$$\sigma_T = \sqrt{\frac{N(N + 1)(2N + 1)}{24}},$$

and $N = $ number of pairs.

z Score

$$z = \frac{X - \bar{X}}{S}.$$

z Tests

when μ and σ are known

$$z = \frac{X - \mu}{\sigma} = \frac{x}{\sigma}.$$

as a test for whether a sample mean came from a population with a mean μ

$$z = \frac{\bar{X} - \mu}{s_{\bar{X}}}.$$

as a test of the significance of the difference between means

$$z = \frac{\bar{X}_1 - \bar{X}_2}{s_{\bar{X}_1 - \bar{X}_2}}.$$

ANSWERS TO PROBLEMS

Chapter 1

1. **a.** Inferential. The intent is to generalize from the poll to the population.
    b. Inferential. Present enrollment figures are being used to predict future ones.
    c. Descriptive. There is no intent to generalize from this class.
    d. Descriptive. This is a statement of what happened. There is no intent to generalize.
    e. Inferential. A generalization is made from the results of the study to the more general effects of moderate anxiety.

2. Population: an arbitrarily defined set of scores that is of interest to an investigator. Sample: some subset of a population. Statistic: a numerical characteristic of a sample or subsample. Parameter: a numerical characteristic of a population.

3. **a.** 66.5 and 67.5 inches   **b.** 30.45 and 30.55 seconds
    c. 50.745 and 50.755 MPH   **d.** 1.5 and 2.5 tablespoons
    e. 2.95 and 3.05 pounds   **f.** 139.5 and 140.5
    g. 2.0125 and 2.0135 liters

4. Nominal, ordinal, interval, and ratio

5. Nominal: different numbers are assigned to different classes of things. Ordinal: nominal properties, plus the numbers carry information about greater than and less than. Interval: ordinal properties, plus the distance between units is equal. Ratio: interval properties, plus there is a true zero point.

6. **a.** ordinal   **b.** ratio   **c.** nominal   **d.** nominal   **e.** ratio   **f.** ordinal

7.

	Independent	*Dependent*
a.	amount of previous experience	problem-solving ability
b.	amount eaten	general feeling
c.	amount eaten	general feeling
d.	amount of anxiety	time to solve problems
e.	amount of sunlight	rate of growth
f.	kind of teaching	amount learned
g.	when it rains	when flowers bloom

8. I. **a.** anxiety
    b. number of words and lines reread
    c. sex
    d. female college students
    e. 25 high- (or low-) anxious female college students
    f. mean number of words and lines reread by subjects
    g. mean number of words and lines high- (or low-) anxious female college students reread
    h. sex
    i. number of words or lines reread

II. **a.** handling and petting
    b. number of trials to learn the task
    c. age
    d. rats
    e. ten rats
    f. mean number of trials to learn the task by subject
    g. mean number of trials to learn task by handled and petted rats
    h. groups
    i. number of trials

III. **a.** colors
    b. rating
    c. rating was always first item on questionnaire
    d. all ratings to a particular color
    e. 40 participants
    f. mean of 10 ratings for a particular color
    g. mean rating for a particular color from the buying public
    h. colors
    i. There are none in this study except the measure of the number of participants

IV. **a.** family relationships
    b. political attitudes
    c. used the same questionnaire
    d. there are two populations: college students and their fathers
    e. ten college students
    f. mean attitude score of ten college students
    g. mean attitude of all college students
    h. family relationship
    i. none. The attitude questionnaire scores are no better than interval scale measures—probably between ordinal and interval

Chapter 2

Answers to Pretest

1. 11.424	**2.** 27.141	**3.** 19.85	**4.** 2.065
5. 49.972	**6.** .0567	**7.** 3.02	**8.** 17.11
9. 1.375	**10.** .83	**11.** .1875	**12.** .056
13. .60	**14.** .10	**15.** 1.50	**16.** 3.11
17. 6	**18.** −22	**19.** −7	**20.** −6
21. 25	**22.** −24	**23.** 3.33	**24.** −5.25
25. 33	**26.** 32	**27.** .19	**28.** 60
29. 5	**30.** 4	**31.** 6,10	**32.** 4,22
33. 16	**34.** 6.25	**35.** .1225	**36.** 4.20
37. 4.00	**38.** 4.50	**39.** .45	**40.** 15.25
41. 8.00	**42.** 16.25	**43.** 1.02	**44.** 12.00
45. 13.00	**46.** 11.00	**47.** 2.50	**48.** −2.33

1. **a.** A sum is the answer to an addition problem.
    b. A quotient is the answer to a division problem.
    c. A product is the answer to a multiplication problem.
    d. A difference is the answer to a subtraction problem.

2.	**3.**	**4.**	**5.**
.001	14.20	1.26	143.300
10.000	−7.31	.04	16.920
3.652	6.89	.0504	2.307
2.500			8.100
16.153			170.627

6.
$$
\begin{array}{r}
76.5 \\
.04\overline{)3.06} \\
28 \\
\overline{26} \\
24 \\
\overline{20} \\
20
\end{array}
$$

7.
$$
\begin{array}{r}
2.04 \\
11.75\overline{)24.00} \\
23\,50 \\
\overline{5000} \\
4700 \\
\overline{300}
\end{array}
$$

8.
$$
\begin{array}{r}
152.12 \\
-127.40 \\
\overline{24.72}
\end{array}
$$

9.
$$
\begin{array}{r}
.07 \\
.5 \\
\overline{.035}
\end{array}
$$

10. $\dfrac{9}{10} + \dfrac{1}{2} + \dfrac{2}{5} = .90 + .50 + .40 = 1.80.$

11. $\dfrac{9}{20} \div \dfrac{19}{20} = .45 \div .95 = .47.$

12. $\left(\dfrac{1}{3}\right)\left(\dfrac{5}{6}\right) = .333 \times .833 = .28.$

13. $\dfrac{4}{5} - \dfrac{1}{6} = .80 - .167 = .63.$

14. $\dfrac{1}{3} \div \dfrac{5}{6} = .333 \div .833 = .40.$

15. $\dfrac{3}{4} \times \dfrac{5}{6} = .75 \times .833 = .62.$

16. $18 \div \dfrac{1}{3} = 18 \div .333 = 54.$

17. **a.** $(24) + (-28) = -4.$ **b.** $-23.$
c. $(-8) + (11) = 3.$ **d.** $(-15) + (8) = -7.$

18. $-40.$ **19.** $24.$
20. $48.$ **21.** $-33.$
22. $(-18) - (-9) = -9.$ **23.** $14 \div (-6) = -2.33.$
24. $12 - (-3) = 15.$ **25.** $(-6) - (-7) = 1.$
26. $(-9) \div (-3) = 3.$ **27.** $(-10) \div 5 = -2.$
28. $4 \div (-12) = -.33.$ **29.** $(-7) - 5 = -12.$
30. $6 \div 13 = .46.$ **31.** $.46 \times 100 = 46$ percent.
32. $18 \div 25 = .72.$ **33.** $85 \div 115 = .74 \times 100 = 74$ percent.
34. $|-31| = 31.$ **35.** $|21 - 25| = |-4| = 4.$
36. $12 \pm (2)(5) = 12 \pm 10 = 2,22.$ **37.** $\pm(5)(6) + 10 = \pm 30 + 10 = -20, 40.$
38. $\pm(2)(2) - 6 = \pm 4 - 6 = -10, -2.$ **39.** $(2.5)^2 = 2.5 \times 2.5 = 6.25.$
40. $9^2 = 9 \times 9 = 81.$ **41.** $26^2 = 26 \times 26 = 676.$
42. $(1/4)^2 = (.25)(.25) = .0625.$

43. $\dfrac{(4 - 2)^2 + (0 - 2)^2}{6} = \dfrac{2^2 + (-2)^2}{6} = \dfrac{4 + 4}{6} = \dfrac{8}{6} = 1.33.$

44. $\dfrac{(12 - 8)^2 + (8 - 8)^2 + (5 - 8)^2 + (7 - 8)^2}{4 - 1} = \dfrac{4^2 + 0^2 + (-3)^2 + (-1)^2}{3}$

$= \dfrac{16 + 0 + 9 + 1}{3} = \dfrac{26}{3} = 8.67.$

45. $\left(\dfrac{5 + 6}{3 + 2 - 2}\right)\left(\dfrac{1}{3} + \dfrac{1}{2}\right) = \left(\dfrac{11}{3}\right)(.33 + .500) = (3.667)(.833) = 3.05.$

46. $\left(\dfrac{13 + 18}{6 + 8 - 2}\right)\left(\dfrac{1}{6} + \dfrac{1}{8}\right) = \left(\dfrac{31}{12}\right)(.167 + .125) = (2.583)(.292) = .75.$

47. $\dfrac{8[(6 - 2)^2 - 5]}{(3)(2)(4)} = \dfrac{8[4^2 - 5]}{24} = \dfrac{8[16 - 5]}{24} = \dfrac{8[11]}{24} = \dfrac{88}{24} = 3.67.$

48. $\dfrac{[(8-2)(5-1)]^2}{5(10-7)} = \dfrac{[(6)(4)]^2}{5(3)} = \dfrac{[24]^2}{15} = \dfrac{576}{15} = 38.40.$

49. $\dfrac{6}{1/2} + \dfrac{8}{1/3} = (6 \div .5) + (8 \div .333) = 12 + 24 = 36.$

50. $\left(\dfrac{9}{2/3}\right)^2 + \left(\dfrac{8}{3/4}\right)^2 = (9 \div .667)^2 + (8 \div .75)^2$

$\quad\quad = 13.5^2 + 10.667^2 = 182.25 + 113.778 = 296.03.$

51. $\dfrac{10 - (6^2/9)}{8} = \dfrac{10 - (36/9)}{8} = \dfrac{10 - 4}{8} = \dfrac{6}{8} = .75.$

52. $\dfrac{104 - (12^2/6)}{5} = \dfrac{104 - (144/6)}{5} = \dfrac{104 - 24}{5} = \dfrac{80}{5} = 16.00.$

53. $\dfrac{x-4}{2} = 2.58,\ x - 4 = 5.16,\ x = 9.16.$

54. $\dfrac{x-21}{6.1} = 1.04,\ x - 21 = 6.344,\ x = 27.34.$

55. $x = \dfrac{14-11}{2.5} = \dfrac{3}{2.5} = 1.20.$

56. $x = \dfrac{36-41}{8.2} = \dfrac{-5}{8.2} = -.61.$

57.
a. 14
b. 126
c. 9
d. 9
e. 128
f. 13
g. 126
h. 13
i. 9

58.
a. 6.33
b. 12.97
c. .05
d. .97
e. 2.61
f. .34
g. .01
h. .02
i. .99

59.
a. 25.00
b. 15.20
c. 5.10
d. 2.50
e. 2.22
f. 129.00
g. 426
h. 70.14
i. 4.30
j. 5.14

Chapter 3

1.

X	Tally Marks	f	X	Tally Marks	f
34	/	1	18	///	3
33		0	17	//	2
32	/	1	16	///	3
31	/	1	15	//	2
30		0	14	/	1
29	//	2	13	/	1
28	/	1	12	////	4
27	/	1	11	//	2
26	/	1	10	//	2
25	//	2	9		0
24	//	2	8	/	1
23	//	2	7	//	2
22	/	1	6		0
21	///	3	5	/	1
20	///	3	4	/	1
19	////	4			

2. The order of the candidates is arbitrary. We list them here in the order in which they are given in the problem; any order is correct.

Candidate	f	Candidate	f
Granger	13	Clark	8
Bassett	11	Allen	19
		Demosthenes	5

3. a. The range is $63.5 - 4.5 = 59$. You may have tried $i = 3$ and found yourself with an extra interval. That is, $59 \div 3 = 19.67$; therefore, 20 intervals should result. However, the lowest interval must begin with 3 if it is to have as its lower limit a multiple of 3 and include the lowest score, 5. This adds 2 extra scores to the bottom of the distribution. The range then becomes $65.5 - 4 = 61$, and $61 \div 3 = 20.33$. An extra interval is necessary to handle the decimal, causing the 10–20 intervals convention to be violated. The correct i for these data is 5.

Two solutions using $i = 5$ are acceptable. One solution begins the intervals with multiples of 5. The other solution places multiples of 5 at the midpoints. Both are shown below.

Class Interval	Tally Marks	f	Class Interval	Tally Marks	f
60–64	/	1	63–67	/	1
55–59	//	2	58–62	/	1
50–54	////	4	53–57	///	3
45–49	////	4	48–52	////	4
40–44	ℳ /	6	43–47	////	4
35–39	ℳ //	7	38–42	ℳ //	7
30–34	ℳ ////	9	33–37	ℳ //	7
25–29	ℳ /	6	28–32	ℳ //	7
20–24	ℳ	5	23–27	ℳ //	7
15–19	///	3	18–22	////	4
10–14	//	2	13–17	///	3
5–9	/	1	8–12	/	1
			3–7	/	1

4.

Number of Sentences Heard Before	Tally Marks	f
16	//	2
15		0
14	//	2
13	///	3
12	///	3
11	ℳ ////	9
10	ℳ /	6
9	ℳ //	7
8	////	4
7	//	2
6	/	1

5.

Class Interval	Tally Marks	f
33–35	/	1
30–32	/	1
27–29	//	2
24–26	/	1
21–23	////	4
18–20	卌 卌	10
15–17	卌 卌	10
12–14	卌 卌 /	11
9–11	卌 卌 卌 /	16
6–8	卌 卌 卌 卌 ///	23
3–5	卌 卌 卌 ////	19
0–2	/	1

6. **a.** The number 55 is the midpoint of the interval 54–56.
   b. Eight is a frequency number representing eight students.
   c. Two. Both students are represented by the score 49, which is the midpoint of the interval 48–50.
   d. This point indicates the number of students whose scores were in the interval 24–26.

7. They should be graphed as frequency polygons.

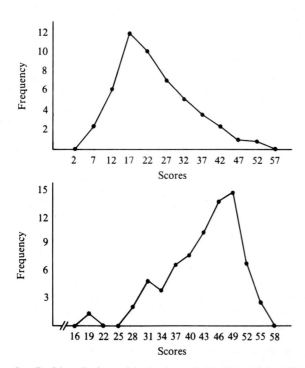

8. Problem 7a is positively skewed. Problem 7b is negatively skewed.
9. A bar graph is the proper graph for the qualitative data in Problem 2.

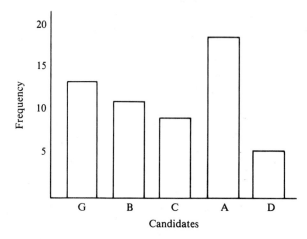

10. Check your sketches against Figures 3.6 and 3.9.
11. right.
12. In a frequency polygon the Y axis is frequency. In a line graph the Y axis is some variable other than frequency.
13. **a.** positively skewed   **b.** symmetrical   **c.** positively skewed
    d. positively skewed   **e.** negatively skewed   **f.** symmetrical
14. **a.** 5   **b.** 12.5, halfway between 12 and 13
    c. 9.5, halfway between 8 and 11.
15. Only Distribution C has two modes. They are 14 and 18.
16. Mean $= \dfrac{409}{39} = 10.49$

    Median. Since there are 39 scores, the halfway point will have 19.5 scores below and 19.5 above. Counting from the bottom of the distribution, there are 14 scores of 9 or less. Thus, you need 5.5 of the 6 scores in the interval of 9.5 to 10.5 ($19.5 - 14 = 5.5$). $5.5/6 = .92$. $9.5 + .92 = 10.42 =$ median. Mode $= 11$.

17. **a.** For these nominal data, only the mode is appropriate. The mode is Allen, which occurs more frequently than any other name.
    b. Since a precinct was covered and the student's interest was city-wide, this is a sample mode, a statistic.
    c. A simple interpretation would be, "Since there are more Allen yard signs than any other, I expect Allen to get the most votes."
18. In both cases the interest is in some larger group (5 sections of Introductory Psychology and all of Midwesternville). Thus, both sets of numbers are samples.
19. **a.** $N = 19$ and half that is 9.5. Counting from the *top*, there are 7 when you include a score of 11. You need 2.5 of the 5 scores in the interval 9.5 to 10.5. $2.5/5 = .5$, so *subtract* .5 from 10.5 to get the median, 10.0.
    b. The easiest and quickest way to deal with 20 scores is to construct a simple frequency distribution. Counting from the top again, there are 6 when you include a score of 6. You need 4 of the 5 scores that lie in the interval 4.5 to 5.5. $4/5 - .80$, so subtract .80 from 5.5 to find the median, 4.70.
20. *Mean* $\bar{X} = \sum fX / N = 1725/50 = 34.50$, or $\bar{X} = \sum fX / N = 1720/50 = 34.40$. The difference between the two means is minor and is due to the different methods of grouping. The mean computed from the ungrouped scores is 34.43, so neither solution contains much error.

Median $N/2 = 50/2 = 25$. The median will have 25 frequencies above it and 25 below it. Count frequencies from the bottom of the first distribution. There are 17 frequencies through the interval 25–29 and 26 frequencies through the interval 30–34. The median, then, must be near the top of the interval 30–34. Subtract the 17 frequencies below 30–34 from 25. $25 - 17 = 8$, so 8 more frequencies are needed. There are 9 frequencies in the interval containing the median, so you need 8/9, or .89 of the interval. The interval has 5 score points, so $.89 \times 5 = 4.44$, the number of score points needed to reach the median. The lower limit of the interval 30–34 is 29.5. Adding 4.44 to 29.5, you get 33.94 as the median.

In the second distribution, 23 frequencies are below the interval 33–37. Two more are needed to reach the median. Since there are 7 frequencies in the interval, 2/7, or .29, of the interval is also needed, or .29 of 5. Multiply $.29 \times 5$ to get 1.43. Add 1.43 points to the lower limit, 32.5, to find the median. Thus, $32.5 + 1.43 = 33.93$.

Mode In the first distribution, the mode is the midpoint of the interval 30–34, since that interval has the greatest number of frequencies. The mode is 32. In the second distribution, four class intervals in a row have seven frequencies. The mode, in this case, is the point between intervals 28–32 and 33–37, or 32.5.

21. *Mean* $\bar{X} = \sum fX/N = 1125/99 = 11.36$. *Median*: $N/2 = 99/2 = 49.5$. Working from the bottom, there are 43 frequencies below the interval 9–11 and 16 in the interval 9–11. You need $49.5 - 43 = 6.5$ of the 16 frequencies. $(6.5/16) \times 3 = 1.22$. The lower limit of the class interval, 8.5, plus 1.22 is 9.72, the median. Working from the top, there are 40 frequencies above the interval 9–11. Of the 16 frequencies in that interval, 9.5 are needed. Thus, you multiply $(9.5/16) \times 3$ to get 1.78, which is subtracted from the upper limit, 11.5, to give the median, 9.72. *Mode*: 7.

22. We hope you thought that the reasoning was statistically wrong and that the person making the statement did not know how to compute the mean of a set of means with unequal N's. The correct procedure is:

$\bar{X}$ (Semester Means)	N (Credits)	$\bar{X}N$
3.5	6	21
3.0	12	36
2.5	16	40
	$\sum = 34$	97

$$\bar{\bar{X}} \text{ (Overall G.P.A.)} = \frac{97}{34} = 2.85.$$

23. a. The mode is appropriate because the observations are made on a nominal scale.

   b. The median or mode is appropriate because the observations produce an ordinal scale.

   c. The median is appropriate for data with an open-ended category.

   d. The median is the appropriate central value to use. It is conventional to use the median for income data, because the distribution is so severely skewed. About one-half of the frequencies are in the $0–$10,000 range.

   e. The mode is appropriate because these are nominal data.

   f. The mean is appropriate because the data are not severely skewed.

   g. The median is appropriate because the distribution is severely skewed.

24. $12 \times 74 = 888.$

   $31 \times 69 = 2139.$

   $\underline{17 \times 75 = 1275.}$ Overall mean: $\dfrac{4302}{60} = 71.7.$

   $\sum 60 \qquad 4302$

25. The distribution in Problem 7a is positively skewed (mean = 23.80, median = 22.0). The distribution in Problem 7b is negatively skewed (mean = 43.19, median = 44.62).

26. This is not correct. To find his lifetime batting average he would need to add up his hits for the three years and divide by the total number of at-bats. From the description of the problem it appears that his average would be lower than .317.

Chapter 4

1. Range = 15.5 − 4.5 = 11.

X	x
15	5
13	3
12	2
10	0
8	−2
7	−3
5	−5
$\Sigma X = 70$	$\Sigma x = 0$

μ or $\bar{X}$ = 10.

2. Range 17.5 − .5 = 17.

X	x
17	11
5	−1
1	−5
1	−5
$\Sigma X = 24$	$\Sigma x = 0$

μ or $\bar{X}$ = 6.

3. Range = 3.45 − 2.55 = .90

X	x
3.4	.5
3.1	.2
2.7	−.2
2.7	−.2
2.6	−.3
Σ 14.5	0

μ or $\bar{X}$ = 2.9

4. Range = .455 − .255 = .20

X	x
.45	.10
.30	−.05
.30	−.05
Σ 1.05	0

μ or $\bar{X}$ = .35

5. σ is used to describe the variability of a population. s is used to estimate σ using a sample from the population. S is used to describe the variability of a sample when you have no desire to estimate σ.

6.

X	x	x^2
7	2	4
6	1	1
5	0	0
2	-3	9
$\sum 20$	0	14

μ or $\bar{X} = 5$.

$$\sigma \text{ or } S = \sqrt{\frac{\sum x^2}{N}} = \sqrt{\frac{14}{4}} = \sqrt{3.5} = 1.87.$$

7.

X	x	x^2
14	3.8	14.44
11	.8	.64
10	-.2	.04
8	-2.2	4.84
8	-2.2	4.84
$\sum 51$	0	24.80

μ or $\bar{X} = 10.2$.

$$\sigma \text{ or } S = \sqrt{\frac{\sum x^2}{N}} = \sqrt{\frac{24.80}{5}} = \sqrt{4.96}$$
$$= 2.23.$$

8.

X	x	x^2
107	2	4
106	1	1
105	0	0
102	-3	9
$\sum 420$	0	14

μ or $\bar{X} = 105$.

$$\sigma \text{ or } S = \sqrt{\frac{\sum x^2}{N}} = \sqrt{\frac{14}{4}} = \sqrt{3.5} = 1.87.$$

9. Although the numbers in Problem 8 are much larger than those in Problem 6, the standard deviations are the same. Thus, the size of the numbers does not give you any information about the variability of the numbers.

10. Yes. The larger the numbers, the larger the mean.

11.

City	Mean	Standard Deviation
San Francisco	56.75°	3.96°
Albuquerque	56.75°	16.24°

Although the mean temperature of the two cities is the same, Albuquerque has a wider variety of temperatures.

12. In eyeballing data for variability, use the range as a quick index.
   a. the second distribution
   b. equal variability
   c. the first distribution
   d. equal variability
   e. The second distribution is more variable; however, most of the variability is due to one extreme score, 15.

13. The second distribution (**b**) is more variable than the first.

   a.

X	X^2
6	36
5	25
4	16
3	9
2	4
$\Sigma\ 20$	20

$$S_0 = \sqrt{\dfrac{\Sigma X^2 - \dfrac{(\Sigma X)^2}{N}}{N}} = \sqrt{\dfrac{90 - \dfrac{(20)^2}{5}}{5}} = \sqrt{2.00} = 1.41.$$

   b.

X	X^2
6	36
6	36
6	36
2	4
2	4
$\Sigma\ 22$	116

$$S = \sqrt{\dfrac{\Sigma X^2 - \dfrac{(\Sigma X)^2}{N}}{N}} = \sqrt{\dfrac{116 - \dfrac{(22)^2}{5}}{5}} = \sqrt{3.84} = 1.96.$$

14. For **13a**, $5/1.41 = 3.55$. For **13b**, $5/1.96 = 2.55$. Yes, these are between 2 and 5.

15. **a.** $$\sigma = \sqrt{\dfrac{\Sigma X^2 - \dfrac{(\Sigma X)^2}{N}}{N}} = \sqrt{\dfrac{262 - \dfrac{(34)^2}{5}}{5}} = 2.48.$$

   b. $$\sigma = \sqrt{\dfrac{294 - \dfrac{(38)^2}{5}}{5}} = 1.02.$$

16. $$s = \sqrt{\dfrac{\Sigma X^2 - \dfrac{(\Sigma X)^2}{N}}{N-1}} = \sqrt{\dfrac{5064 - \dfrac{(304)^2}{21}}{20}} = \sqrt{33.16} = 5.76.$$

17. In this case, S is appropriate because there is no intention to make any inferences from this class to some larger population.

$$S = \sqrt{\dfrac{\Sigma X^2 - \dfrac{(\Sigma X)^2}{N}}{N}} = \sqrt{\dfrac{1614 - \dfrac{(134)^2}{12}}{12}} = \sqrt{9.81} = 3.13.$$

18. Before: $s = \sqrt{1.43} = 1.20$, $\bar{X} = 5.0$. After: $s = \sqrt{14.29} = 3.78$, $\bar{X} = 5.0$. It appears that, before studying poetry, students were homogeneous and neutral. After studying poetry for nine weeks, some students were turned on and some were turned off; they were no longer neutral.

19. a.

X	x	x^2	$z = \dfrac{x}{S}$
10	5	25	1.58
7	2	4	.63
4	−1	1	−.32
3	−2	4	−.63
1	−4	16	−1.26
$\sum 25$	0	50	0

$$\bar{X} = \frac{25}{5} = 5.00.$$

$$S = \sqrt{\frac{50}{5}} = 3.16.$$

b. Variance = 10.

20. $\sum z = 0$. Since $\sum x = 0$, it follows that $\sum (x/S) = 0$.

21. zero

22.

	Running Time		Errors	
Rat	X	z	X	z
1	20	1.86	22	.83
2	17	.96	23	1.07
3	16	.66	25	1.54
4	15	.36	20	.36
5	15	.36	14	−1.07
6	13	−.24	19	.12
7	12	−.54	16	−.59
8	12	−.54	17	−.36
9	10	−1.14	19	.12
10	8	−1.74	10	−2.01

$\sum X = 138.$ $\sum X = 185.$
$\sum X^2 = 2016.$ $\sum X^2 = 3601.$
$\sum x^2 = 111.60.$ $\sum x^2 = 178.50.$
$\bar{X} = 13.80.$ $\bar{X} = 18.50.$
$S = 3.34.$ $S = 4.22.$

a. Rat number 10 had the shortest running time and the fewest errors.

b. His error performance was better. He exceeded the mean of his fellows by two standard deviations ($z = -2.01$) on error commission but exceeded them in running time by only 1.74 standard deviations ($z = -1.74$).

c. Rats number 2, 3, 6, 8, and 9 all had higher z scores for errors than for speed. Since negative scores in this situation mean better performance, these rats were better at running than at staying out of blind alleys.

23.

Tobe's Apple	Zeke's Orange
$z = \dfrac{9-5}{1} = 4.00$	$z = \dfrac{10-6}{1.2} = 3.33$

Tobe's z score is larger so the answer to Hamlet must be a resounding, "To be." Notice that each fruit varies from its group mean by the same amount. It is the smaller variability of the apple weights that makes Tobe's fruit a winner.

24.

First Test	Second Test	Third Test
$z = \dfrac{79-67}{4} = 3.00.$	$z = \dfrac{125-105}{15} = 1.33.$	$z = \dfrac{51-45}{3} = 2.00.$

Milquetoast's performance was poorest on the second test.

Chapter 5

1. The statement means that variation in either variable is accompanied by predictable variation in the other. Notice that nothing is said here about direction. They may vary in the same direction (positive correlation) or opposite directions (negative correlation).

2. When correlation is positive, X and Y increase and decrease together. When correlation is negative, Y increases as X decreases, and Y decreases as X increases.

3. **a.** Yes, positive. Taller people usually weigh more than shorter people.
   b. Yes, negative. As speed increases, fewer miles can be traveled on each gallon of gasoline.
   c. No. These scores cannot be correlated, since there is no basis for pairing the scores.
   d. Yes, negative. As temperatures go up, less heat is needed, and fuel bills go down.
   e. Yes, positive. People with higher IQs score higher on reading-comprehension tests.
   f. No. There is no basis for pairing scores.

4.

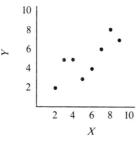

By the blanched procedure:

Twin Pair	X	Y	X²	Y²	XY
1	9	7	81	49	63
2	8	8	64	64	64
3	7	6	49	36	42
4	6	4	36	16	24
5	5	3	25	9	15
6	4	5	16	25	20
7	3	5	9	25	15
8	2	2	4	4	4
Σ	44	40	284	228	247

$\bar{X} = 5.5.$

$\bar{Y} = 5.0.$

$$S_X = \sqrt{\frac{284 - \frac{(44)^2}{8}}{8}} = \sqrt{5.25} = 2.29.$$

$$S_Y = \sqrt{\frac{228 - \frac{(40)^2}{8}}{8}} = \sqrt{3.50} = 1.87.$$

$$r = \frac{\frac{247}{8} - (5.5)(5.0)}{(2.29)(1.87)} = \frac{30.88 - 27.50}{4.28} = \frac{3.38}{4.28} = .79.$$

By the raw-score procedure:

$$r = \frac{(8)(247) - (44)(40)}{\sqrt{[(8)(284) - (44)^2][(8)(228) - (40)^2]}}$$

$$= \frac{1976 - 1760}{\sqrt{(2272 - 1936)(1824 - 1600)}} = \frac{216}{\sqrt{(336)(224)}} = \frac{216}{\sqrt{75264}}$$

$$= \frac{216}{274.34} = .79.$$

5. $\bar{X} = 46.18.$

$\bar{Y} = 30.00.$

$$S_X = \sqrt{\frac{87373 - \frac{(1755)^2}{38}}{38}} = \sqrt{166.31} = 12.90.$$

$$S_Y = \sqrt{\frac{37592 - \frac{(1140)^2}{38}}{38}} = \sqrt{89.26} = 9.45.$$

$$r = \frac{\frac{55300}{38} - (46.18)(30)}{(12.90)(9.45)} = \frac{1455.26 - 1385.40}{121.91} = \frac{69.86}{121.91} = .57.$$

6. $r = \frac{(50)(175711) - (202)(41048)}{\sqrt{[(50)(1740) - (202)^2][(50)(35451830) - (41048)^2]}} = .25.$

$$r = \frac{\frac{175711}{50} - (4.04)(820.96)}{(4.30)(187.25)} = .25.$$

7. Coefficient of determination = .20. The two tests have about 20 percent of their variances in common. Although the two tests have something in common, they are largely independent of each other.

8. Coefficient of determination = .32. The two measures of self-esteem have about 32 percent of their variance in common. Although the two measures are to some extent measuring the same traits, they are, in large part, measuring different traits.

9. **a.** $(.40)^2 = .16$, or 16 percent   **b.** $(.10)^2 = .01$, or 1 percent.

10. No. On the basis of correlational data only, cause-and-effect statements may not be made.

11. **a.** People who cannot tolerate ambiguity tend to be authoritarian.
    b. Vocational interests tend to remain similar from age 20 to age 40.
    c. Identical twins have very similar IQs.
    d. There is a slight tendency for IQ to be lower as family size increases.
    e. There is a slight tendency for taller men to have higher IQs than shorter men.
    f. The lower a person's income level is, the greater the probability that he or she will be diagnosed as psychotic.

12. **a.** There is some tendency for children with more older siblings to accept less credit or blame for their own successes and failures than children with fewer older siblings.
    b. Nothing. Correlation coefficients do not permit you to make cause-and-effect statements.
    c. The coefficient of determination is .1369 $(-.37^2)$, so we can say that about 14 percent of the variance in acceptance of responsibility is predictable from knowledge of the number of older siblings; 86 percent is not. This provides very poor ability to predict. However, if we are attempting to devise a theory about acceptance of responsibility, number of older siblings would be considered important.

13. **a.** Since $.97^2 = .94$, these two tests have 94 percent of their variance in common. They must be measuring essentially the same trait. There is a very strong tendency for persons scoring high on one test to score high on the other.
    b. No. This is a cause-and-effect statement and is not justified on the basis of correlational data.

14.

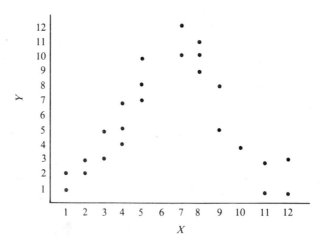

r is not appropriate; the relationship between X and Y is not linear.

15. **a.** $b = r\dfrac{S_Y}{S_X} = (.79)\dfrac{1.87}{2.29} = (.79)(.82) = .65.$

$a = \bar{Y} - b\bar{X} = 5.0 - (.65)(5.50) = 5.0 - 3.58 = 1.42,$

$Y' = a + bX = 1.42 + (.65)(9) = 1.42 + 5.85 = 7.27.$

b,c.

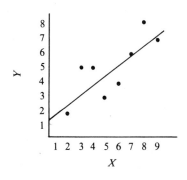

16. a. $b = r\dfrac{S_Y}{S_X} = (.57)\dfrac{9.45}{12.90} = (.57)(.73) = .42.$

$a = \bar{Y} - b\bar{X} = 30 - (.42)(46.18) = 30 - 19.40 = 10.60.$

b. $Y' = a + bX = 10.60 + (.42)(42) = 10.60 + 17.64 = 28.24.$

17. $r = \dfrac{N\sum XY - (\sum X)(\sum Y)}{\sqrt{[N\sum X^2 - (\sum X)^2][N\sum Y^2 - (\sum Y)^2]}}$

$= \dfrac{(13)(595) - (75)(125)}{\sqrt{[(13)(525) - (75)^2][(13)(1397) - (125)^2]}} = -.94.$

$b = r\dfrac{S_Y}{S_X} = (-.94)\dfrac{3.87}{2.66} = -1.37.$

$a = \bar{Y} - b\bar{X} = 9.62 - (-1.37)(5.77) = 9.62 - (-7.90) = 17.52.$

Regression equation: $Y' = a + bX = 17.52 + (-1.37X) = 17.52 - 1.37X.$

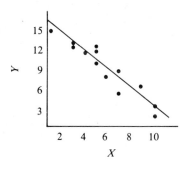

18. $Y' = (.80)\left(\dfrac{15}{16}\right)(65 - 100) + 100 = 73.75 = 74.$

19. a.

Person	Trial 1	Trial 20	X^2	Y^2	XY
1	3	21	9	441	63
2	6	25	36	625	150
3	2	20	4	400	40
4	6	33	36	1089	198
5	5	23	25	529	115
6	8	34	64	1156	272
	$\sum 30$	156	174	4240	838

$$\bar{X} = 5 \qquad \bar{Y} = 26$$

$$S_X = \sqrt{\dfrac{174 - \dfrac{(30)^2}{6}}{6}} = 2.00 \qquad S_Y = \sqrt{\dfrac{4240 - \dfrac{(156)^2}{6}}{6}} = 5.54$$

$$r = \dfrac{\dfrac{838}{6} - (5)(26)}{(2.00)(5.54)} = \dfrac{9.67}{11.08} = .87.$$

b. There is a strong tendency for those who do well initially to do well later in practice.

c.

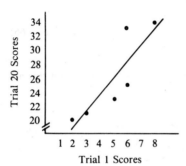

Trial 1 Scores

d. $Y' = (.87)\left(\dfrac{5.54}{2.00}\right)(9 - 5) + 26.00 = 35.64.$

Chapter 6

1. There are seven cards between the 3 and jack, each with a probability of $4/52 = .077$. So $(7)(.077) = .539$.
2. The probability of drawing a 7 is $4/52$, and there are 52 opportunities to get a 7. Thus, $(52)(4/52) = 4$.
3. There are four cards that are larger than a jack or smaller than a 3. Each has a probability of $4/52$. Thus, $(4)(4/52) = 16/52 = .308$.
4. The probability of a 5 or 6 is $4/52 + 4/52 = 8/52 = .154$. In 78 draws, $(78)(.154) = 12$ cards that are 5s or 6s.
5. .7500. Adding the probability of one head (.3750) to that of two heads (.3750), a figure of .7500 is obtained.
6. $.1250 + .1250 = .2500$
7. $16(.1250) = 2$ times
8. **a.** .0832   **b.** .4778   **c.** .2912
9. Empirical. The scores are based on observations.
10. A quick check on your answer can be made by comparing it with the proportion having IQs of 120 or greater, .0918. The proportion with IQs of 130 or greater will be less than the proportion with IQs of 120 or greater. Is your calculated proportion less?

    Following is a picture of a normal distribution in which the proportion of the population with IQs of 130 or greater is shaded. For IQ $= 130$, $z = (130 - 100)/15 = 2.00$.

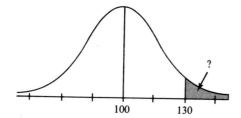

The proportion beyond $z = 2.00$ is .0228 (Column C). Thus, .0228 or 2.28 percent of the population would be expected to have IQs of 130 or greater.

11. $z = (90 - 100)/15 = -.67$. The proportion of the distribution beyond $z = -.67$ is .2514. Since an area under the curve is a probability, the probability is .2514 that a randomly selected person has an IQ of 90 or below.

12. $z = (X - \mu)/\sigma$. For IQ = 55, z score = -3.00; for IQ = 110, z score = .66; for IQ = 103, z score = .20; for IQ = 100, z score = .00.

13. $z = (110 - 100)/15 = 10/15 = .67$. The proportion beyond $z = .67$ is .2514, which is the proportion of people with IQs 110 or greater. Perhaps you got this answer without using arithmetic—by looking at your answer to Problem 11.

14. **a.** $.2514 \times 250 = 62.85$ or 63 students.   **b.** $250 - 62.85 = 187.15$ or 187 students.

   c. $1/2 \times 250 = 125$. We hope you were able to get this one immediately by thinking about the symmetrical nature of the normal distribution.

15. The z score associated with a proportion of .2500 is .67. If $z = .67$,
   $$X = 100 + (.67)(15)$$
   $$= 100 + 10.05$$
   $$= 110.05 = 110.$$
   So, an IQ score of 110 separates the top 25 percent from the lower 75 percent.

16. A z score of 1.65 separates the extreme 5 percent of the normal distribution from the rest.

   a. $X = 64.0'' + (1.65)(2.4)$
   $$= 64.0'' + 3.96''$$
   $$= 67.96'' \text{ or } 68'' \text{ or } 5'8''.$$

   b. $X = 64.0'' - (1.65)(2.4)$
   $$= 64.0'' - 3.96''$$
   $$= 60.04'' \text{ or } 60'' = 5'0''.$$

   We hope you were able to get answer **(b)** by recognizing the symmetrical nature of the normal distribution, saving yourself arithmetical steps.

17. $z = \dfrac{72 - 68.7}{2.6} = 1.27$, proportion = .1020.

18. **a.** $z = \dfrac{3.20 - 3.11}{.05} = 1.80$, proportion = .0359.

   b. The z score corresponding to a proportion of .4000 is 1.28. Thus,
   $$X = 3.11 + (1.28)(.05)$$
   $$= 3.11 + .06$$
   $$= 3.17.$$
   $$X = 3.11 - .06$$
   $$= 3.05.$$
   Thus, the middle 80 percent of the pennies weighs between 3.05 and 3.17 grams.

19.

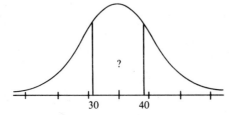

a. $z = \dfrac{30-35}{6} = -.83$, proportion $= .2967$. $z = \dfrac{40-35}{6} = .83$, proportion $= .2967$.

(2)(.2967) $= .5934 =$ the proportion of students with scores between 30 and 40.
b. The probability is also .5934.

20. No, because the distance between 30 and 40 straddles the mean, where scores that occur frequently are found. The distance between 20 and 30 is all in one tail of the curve, where scores are achieved less frequently. If you missed this problem, it was probably because you failed to draw a normal curve and write in the mean and the scores in question.

21.

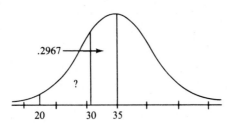

$z = \dfrac{20 - 35}{6} = -2.50$, proportion $= .4938$.

$z = \dfrac{30 - 35}{6} = -.83$, proportion $= .2967$.

.4938 $-$ .2967 $= .1971$. We have found the proportion of students with scores between 35 and 20 and subtracted from it the proportion with scores between 35 and 30. There are also other correct ways to set up this problem.

22. 800 people $\times .1971 = 157.7 = 158$ people.

23.

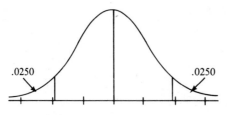

The z score corresponding to a proportion of .0250 is 1.96 (Column C). Thus,

$X = \mu + (z)(\sigma)$
$\quad = 100 + (1.96)(15)$
$\quad = 129.4 = 129$

and

$X = 100 + (-1.96)(15)$
$\quad = 70.6 = 71.$

IQ scores of 129 or greater or 71 or less are achieved by 5 percent of the population.

24. Since .05 of the curve lies outside the limits of the scores 71 and 129, the probability is .05. The probability is .025 that the randomly selected person has an IQ of 129 or higher and .025 that it is 71 or lower.

25. $z = \dfrac{.85 - .99}{.17} = \dfrac{-.14}{.17} = -.82$. The proportion associated with $z = -.82$ is .2061 (Column C). (.2061)(185,822) = 38,298 truck drivers. The usual procedure would be to report this result as "about 38,000 truck drivers," since the normal curve corresponds only approximately to the empirical curve.

26. This problem has two parts: determining the number of trees over 8 inches DBH on one acre and multiplying by 100.

$$z = \dfrac{8 - 13.68}{4.83} = -1.18, \text{proportion} = .5000 + .3810 = .8810.$$

.8810 (199) = 175.319 trees on one acre.

175.319 (100) = 17,532 trees on 100 acres. (About 17,500 trees.)

27. $\mu = \dfrac{3600}{100} = 36.0$ inches.

$$\sigma = \sqrt{\dfrac{130,000 - \dfrac{(3600)^2}{100}}{100}} = 2.00 \text{ inches.}$$

a. $z = \dfrac{32 - 36}{2} = 2.00$, proportion = .5000 + .4772 = .9772.

b. $z = \dfrac{39 - 36}{2} = 1.50$, proportion = .5000 + .4332 = .9332.

.9332 (300) = 279.96 = 280 hobbits.

c. $z = \dfrac{44 - 36}{2} = 4.00$, proportion = .00003 or, practically speaking, zero.

Chapter 7

1. A statistic is a numerical characteristic of a sample.

2. A parameter is a numerical characteristic of a population.

3. μ is the mean of a population, and $\bar{X}$ is the mean of a sample.

4. *Population*: light bulbs that were produced the previous month and shipped to stores. *Parameter*: the proportion of recently produced bulbs that would not work. *Sample*: the bulbs taken from the shelves at the 25 stores. *Statistic*: the proportion of the sample that would not work, the numerical value of which was .005.

5. Our mean was: $\bar{X} = \dfrac{\Sigma X}{N} = \dfrac{245}{7} = 35.0$.

6. In this problem it is not the self-esteem scores in your sample that are "right" or "wrong," but the method you used to get them. The proper method is to haphazardly find a starting place in the table of random numbers, record in sequence five two-digit numbers from the table (ignoring numbers over 24 and duplications), and write down the five self-esteem scores whose identification number you got from the table.

7. Proceed in the same way as for Problem 6, but obtain ten two-digit numbers from the table of random numbers.

8. If you begin every random sample at the same place, the same scores will be chosen for every sample. This violates the principle that every sample has an equal probability of being selected.

9. Again, the method you used is the important thing. You should have given each score an identifying number (01 to 38), begun at some chance starting place in Table B, and selected 12 numbers. The numbers identify 12 scores, which then constitute a random sample.

10. Yes, those with greater educational accomplishments are more likely to return the questionnaire. Such a biased sample will overestimate the accomplishments of the population.

11. The sample will be biased. Some of the general population will not even read the article and will thus have a zero probability of being selected. Of those who do read it, some will have no opinion and will not return the ballot. Of those who read it and have an opinion, many will not take the time and trouble to mark the ballot and return it. The opinions of those who do not return the ballot have a zero probability of being in the sample. In short, the results of such a sample cannot, with confidence, be generalized to a larger population.

12. A biased sample is one in which there is a systematic over- or under-representation of certain members of the population from the sample. In a representative sample, there is no such systematic bias.

13. Yes. A mean computed from a biased sample can be arithmetically "correct" but will not be representative of a population.

14. a. Biased, although the bias is not considered serious. It is biased because every person does not have an equal chance to be selected for the sample. For example, two brothers named Kotki, adjacent to each other on the list, could not both be in a sample. Thus, samples with both Kotki brothers are not possible.
  b. Biased, because freshmen, sophomores, and seniors, although they are members of the population, have no chance to selected for the sample.
  c. Random, assuming the names in the box were thoroughly mixed.
  d. Random. Although small, this sample fits the definition of a random sample.
  e. Stratified.
  f. Biased, because some members of the population have completed the English course and, therefore, have no chance to be chosen for the sample.

15. Biased. Only bulbs in the area of the factory were included in the sample. Bulbs that had to travel farther might have been found to have more need for the new packaging.

16.

| | σ | | | |
N	1	2	4	8
1	1	2	4	8
4	.50	1	2	4
16	.25	.50	1	2
64	.125	.25	.50	1

17. As N increases by a factor of 4, $\sigma_{\bar{X}}$ decreases by a factor of .50. More simply but less accurately, as N increases, $\sigma_{\bar{X}}$ decreases.

18. four times

19. when $N = 1$

20. If this standard error (whose name is the standard error of estimate) was very small, it would mean that you would be confident that the actual Y would be close to the predicted Y. A very large standard error would indicate that you

could not put much confidence in a predicted Y, that the actual Y would be subject to lots of other influences besides X.

21. **a.** "REPORTER LAMBASTED BY EDITOR FOR BIASED SAMPLING."
Children in other schools and children in that school who were not in rooms with windows next to the sidewalk were excluded from the sample.

   b. $z = \dfrac{23.3 - 25.5}{3.1/\sqrt{12}} = -2.46.$

   $p = .0069$ that the strike was having no effect on attendance.

22. $s = \sqrt{\dfrac{40464 - \dfrac{1188^2}{36}}{35}} = 6.00.$

   $s_{\bar{X}} = \dfrac{6.00}{\sqrt{36}} = 1.00.$

   $\bar{X} = \dfrac{1188}{36} = 33.00.$

   $z = \dfrac{33 - 30}{1.00} = 3.00.$

   $p = .0013.$

   The difference between the trained clients and the national norms does not seem to be due to chance. It seems more reasonable to attribute the difference to the eight-week training course.

23. $s_{\bar{X}} = \dfrac{7500}{\sqrt{38}} = 1216.66.$ Therefore, $z = \dfrac{26,900 - 28,100}{1216.66} = -.99; \; p = .1611.$ Thus, the

   junior's suspicion that the $26,900 sample mean might be just a chance fluctuation from a true mean of $28,100 has some foundation. Such a fluctuation would be expected in about 16 percent of the samples if the true campus mean were $28,100.

24. $s_{\bar{X}} = \dfrac{s}{\sqrt{N}} = \dfrac{7500}{\sqrt{400}} = 375.$ $z = \dfrac{27,725 - 28,100}{375} = -1.00; \; p = .1587.$ Thus, in this

   hypothetical case, the increased sample size, which produced a new mean, did not reduce the uncertainty associated with the decision.

25. $LL = \bar{X} - z(s_{\bar{X}}).$ $\qquad UL = \bar{X} + z(s_{\bar{X}}).$

   a. $LL = 36 - (1.96)\left(\dfrac{2}{\sqrt{100}}\right) = 35.61.$

   $UL = 36 + (1.96)\left(\dfrac{2}{\sqrt{100}}\right) = 36.39.$

   b. $LL = 36 - (1.96)\left(\dfrac{10}{\sqrt{100}}\right) = 34.04.$

   $UL = 36 + (1.96)\left(\dfrac{10}{\sqrt{100}}\right) = 37.96.$

   c. $LL = 36 - (1.96)\left(\dfrac{10}{\sqrt{1000}}\right) = 35.38.$

   $UL = 36 + (1.96)\left(\dfrac{10}{\sqrt{1000}}\right) = 36.62.$

Notice that a fivefold increase in s (**b** versus **a**) causes a fivefold increase in the size of the confidence interval. A tenfold increase in N (**c** versus **b**) causes a threefold decrease in the confidence interval.

26. $\text{LL} = 26{,}900 - (2.58)\left(\dfrac{7500}{\sqrt{38}}\right) = \$23{,}761.$

$\text{UL} = 26{,}900 + (2.58)\left(\dfrac{7500}{\sqrt{38}}\right) = \$30{,}039.$

Our two economics students can be 99 percent confident that the mean family income for their campus is between \$23,761 and \$30,039.

27. $\text{LL} = 49.5 - (1.96)\left(\dfrac{9}{\sqrt{36}}\right) = 46.56.$

$\text{UL} = 49.5 + (1.96)\left(\dfrac{9}{\sqrt{36}}\right) = 52.44.$

The investigators can be 95 percent confident that the mean standardized test score for children whose parents supplied them with answers to their homework problems is between 46.56 and 52.44.

28. The size of the standard deviation of each sample. If you happen to get a sample that produces a large standard deviation, a large standard error will result and the confidence interval will be wide.

29. **a.** 18

   b. More narrow. As you may recall from Problem 16, as the sample size goes up by 4 the standard error reduces by one-half. Thus, confidence intervals based on $N = 100$ would be one-half the size of those based on $N = 25$.

30. From a specified population, you would draw many samples of the same size, find the median of each sample, and arrange those sample medians into a frequency distribution. This frequency distribution of medians is called a sampling distribution of medians.

31. the standard error of the median.

Chapter 8

1. Your outline should include the following points.

   a. Two samples are drawn from one population.

   b. The samples are treated the same except for one thing.

   c. The samples are measured, and the difference found is attributed to chance or to the treatment difference.

2. random assignment of subjects to groups

3. Your outline should include the following points.

   a. Recognize two logical possibilities:

      i. H_0: The treatment had no effect.

      ii. H_1: The treatment had an effect.

   b. Assume H_0 to be correct.

   c. Compare the actual difference found with those in the sampling distribution of mean differences, which is based on the assumption that H_0 is true. If the difference found has a very low probability, reject H_0 and accept H_1. If the difference found has a high probability, suspend judgment, claiming that your experiment did not allow you to choose between H_0 and H_1.

4. the null hypothesis

5. The null hypothesis states that there is no difference between the populations from which the samples come. Stated another way: the samples come from the same population.

6. *Level of significance* is the arbitrary cutoff point between considering a difference "due to chance" or "not due to chance."

7. Events in the critical region have a probability less than the level of significance.

8. .05.

9. "$p < .01$" refers to the probability that the difference found was due to chance. That is, if only chance were operating, differences as large or larger than the one found would occur less than one time in a hundred.

10. 2.33.

11. **a.** Independent variable—experience with electrical switch; dependent variable—time to solve problem

    b. H_0: Experience with the light switch does not affect the time required to solve the problem.

    c. $s_{\bar{X}_1 - \bar{X}_2} = \sqrt{\left(\dfrac{2.13}{\sqrt{43}}\right)^2 + \left(\dfrac{2.31}{\sqrt{41}}\right)^2}$

    $= \sqrt{.1055 + .1301} = \sqrt{.2357} = .4854.$

    $z = \dfrac{7.40 - 5.05}{.4854} = 4.84; p < .0001.$

    d. Since the experimental group took longer, you should conclude that the previous experience with the switch *retarded* the subject's ability to recognize the solution.

12. The z-score test requires large samples (over 30). These samples do not meet that requirement. Chapters 9 and 13 describe techniques for analyzing small-sample experiments.

13. Independent variable—percent of cortex removed; dependent variable—number of errors

    $s_{0\,\text{percent}} = \sqrt{\dfrac{\Sigma X^2 - \dfrac{(\Sigma X)^2}{N}}{N - 1}} = 4.00.$

    $s_{20\,\text{percent}} = \sqrt{\dfrac{\Sigma X^2 - \dfrac{(\Sigma X)^2}{N}}{N - 1}} = 4.00.$

    $s_{\bar{X}_1 - \bar{X}_2} = \sqrt{\left(\dfrac{4.00}{\sqrt{40}}\right)^2 + \left(\dfrac{4.00}{\sqrt{40}}\right)^2} = .894.$

    $z = \dfrac{\bar{X}_0 - \bar{X}_{20}}{s_{\bar{X}_1 - \bar{X}_2}} = \dfrac{5.2 - 6.3}{.894} = -1.23.$

    Since $1.23 < 1.96$ retain the null hypothesis and conclude that a 20 percent loss of cortex does not reduce significantly a rat's memory for a simple maze.

14. $z = \dfrac{\bar{X}_1 - \bar{X}_2}{s_{\bar{X}_1 - \bar{X}_2}} = \dfrac{60 - 36}{\sqrt{\left(\dfrac{12}{\sqrt{45}}\right)^2 + \left(\dfrac{6}{\sqrt{52}}\right)^2}} = \dfrac{24}{1.97} = 12.18.$

    The meaning of very large z scores is quite clear: they indicate a very, very small probability. Thus, conclude that children whose illness begins before age 3 exhibit symptoms longer than do those whose illness begins after age 6.

15. The only good advice is to point out that the "experiment" is so poorly *designed* that a comparison of mean scores is meaningless. In the first place, there is no reason to suppose that the two groups of students were equivalent to begin with, because sampling was not random (free choice will probably lead to biased samples). In the second place, there were two professors, and one may be a more effective teacher than the other. If so, any difference in the means may be due to the professors rather than to the two methods. The professors in this problem are an example of an extraneous variable. In the third and most important place, the scores (dependent variable) for the two groups are not based on the same test. The main reason we included this tricky question was to remind you of the importance of a sound experimental design. Without it, statistics cannot produce meaningful answers.

16. A Type I error is a rejection of the null hypothesis when it is in fact true.

17. A Type II error is retaining the null hypothesis when it is in fact false.

18. No, a Type I error can be made only when the null hypothesis is true.

19. α is the probability of a Type I error. The level of significance is the cutoff point between "due to chance" and "not due to chance," and the experimenter chooses a particular α for this cutoff point.

20. The probability of a Type I error decreases from .05 to .01, and the probability of a Type II error increases.

21. First of all, you might point out that if the populations are those particular freshmen classes, no statistics are necessary; you have the population data and there is no sampling error. State U. is one-tenth of a point higher than The U.

    If the question is not about those freshmen classes but about the two schools, and the two freshmen classes can be treated as representative samples, a z-score test is called for.

    $$s_{\bar{X}_1 - \bar{X}_2} = \sqrt{\left(\frac{s_1}{\sqrt{N_1}}\right)^2 + \left(\frac{s_2}{\sqrt{N_2}}\right)^2} = .0474.$$

    $$z = \frac{\bar{X}_1 - \bar{X}_2}{s_{\bar{X}_1 - \bar{X}_2}} = \frac{21.4 - 21.5}{.0474} = -2.11, p < .05.$$

    Students at State U. have statistically significant higher admission scores than do students at The U. You may have noted how small the difference is, only one-tenth of a point. We discuss this in the following section in the text.

22. $z = 2.33$.

23. **a.** One-tailed test.
    b. Since the new Brand Z was slower than Brand Y on hand, no test is necessary.

24. Independent variable—experience with aerobics program; dependent variable—blood pressure

    $$z = \frac{125 - 116}{\sqrt{\left(\frac{15}{\sqrt{36}}\right)^2 + \left(\frac{19}{\sqrt{36}}\right)^2}} = \frac{9}{\sqrt{16.28}} = 2.23.$$

    $p = <.05$.

    Veterans of the noontime aerobics program had significantly lower blood pressure than aerobics newcomers did.

25. For us, our own words were, "Significant differences are not due to chance. Important differences are ones that change our understanding about something."

26. Our list is
    1. Sample size.
    2. Alpha level.

3. The actual difference between the population means.
4. The preciseness of measuring the dependent variable.

Chapter 9

1. **a.** normal, t **b.** t, normal   **c.** t
2. larger than
3. $df = N - 1$.   **a.** $25 - 1 = 24$.   **b.** $4 - 1 = 3$.   **c.** $42 - 1 = 41$.
4. W. S. Gosset, who wrote under the pseudonym "Student," invented the t distribution so that he could assess probabilities for small samples.
5. $t = \dfrac{\bar{X} - \mu}{s_{\bar{X}}}$ **a.** $t = 1.00$.   **b.** $t = -.33$.   **c.** $t = 4.00$.   **d.** $t = .50$.

   e. $t = -4.51$. Notice that the mean difference on **5c** is small; on **5d** it is large. The difference in the t values shows the importance of $s_{\bar{X}}$.
6. **a.** $df = 14$; $t = 1.96$; $p > .05$.   **b.** $df = 6$; $t = -2.10$; $p > .05$.
   c. $df = 23$; $t = -2.10$; $.01 < p < .05$.   **d.** $df = 20$; $t = 2.845$; $p = .01$.
   e. $df = 4$; $t = 4.90$; $p < .01$.
7. The probabilities refer to the event of getting such a sample mean with a random sample from a population with $\mu = 81.00$.
8. Correlated samples. With twins divided between the two groups, there is a logical pairing. (Natural pairs.)
9. Correlated samples. The temperature of the left forefinger is paired with the temperature of the right forefinger for each person. (Natural pairs.)
10. Correlated samples. This is a before-and-after experiment. For each group, the amount of aggression before the screen was lifted is paired with the amount after the screen was lowered. (Repeated measures.)
11. Correlated samples. Each monkey who has control over the shock is paired with another who gets the same number of shocks. (Matched pairs, yoked control design.)
12. Independent samples. The dean randomly assigned individuals to one of the two groups.
13. Correlated samples. The IQ of the firstborn is paired with the IQ of his or her sibling. (Natural pairs.)

    This paragraph is really not about statistics, and you may skip it if you wish. Were you somewhat more anxious about your decisions on Problems 10, 11, or 13 than on Problems 8 and 9? If so, and if this anxiety was based on the expectation that surely it was time for an answer to be "independent samples" and if you based your answer on this expectation rather than on the problem, you were exhibiting a *response bias*. A response bias is when a current response is made on the basis of previous responses rather than on the basis of the current stimulus. Such response biases often lead to a correct answer in textbooks, and you may learn to make some decisions (those you are not sure about) on the basis of irrelevant cues (such as what your response was on the last question). To the extent it rewards your response biases, a textbook is doing you a disservice. So, be forewarned; recognize response bias and resist it.

14.

Laboratory	Car
$N = 4$	5
$\sum X = 52$	30
$\sum X^2 = 722$	210
$\bar{X} = 13$	6

$$t = \frac{13 - 6}{\sqrt{\left(\dfrac{722 - \dfrac{52^2}{4} + 210 - \dfrac{30^2}{5}}{4 + 5 - 2}\right)\left(\dfrac{1}{4} + \dfrac{1}{5}\right)}} = \frac{7}{2.21} = 3.17 \qquad df = 7$$

Note that you must use the longer formula for $s_{\bar{X}_1 - \bar{X}_2}$ because the N's are not equal. Since the obtained $|t|$ is greater then 2.37 $[t_{.05}(7\ df) = 2.37]$, the probability of chance is less than .05 ($p < .05$), so reject the null hypothesis. Since the Car Group made fewer errors, conclude that immediate, concrete experience facilitates memory. A two-tailed test is appropriate since the interest was in "the effect of immediate, concrete experience," which could be positive or negative.

15.

	"Winners"	"Losers"
$\bar{X}$	3.46	3.52
s	.35	.30
N	16	16

$$t = \frac{\bar{X}_1 - \bar{X}_2}{\sqrt{\left(\dfrac{s_{X_1}}{\sqrt{N_1}}\right)^2 + \left(\dfrac{s_{X_2}}{\sqrt{N_2}}\right)^2}} = \frac{3.46 - 3.52}{\sqrt{\left(\dfrac{.35}{\sqrt{16}}\right)^2 + \left(\dfrac{.30}{\sqrt{16}}\right)^2}} = \frac{-.060}{.115}$$

$$= -.52$$

$t_{.05}(30\ df) = 2.04$ (two-tailed test).

There is no evidence that being in the sophomore honors course has a significant effect on grade-point average for those students who qualified for the sophomore honors course.

16.

	New Brand	Old Brand
$\sum X$	45.9	52.8
$\sum X^2$	238.73	354.08
$\bar{X}$	5.1	6.6
N	9	8
s	.76	.89

$$t = \frac{5.1 - 6.6}{\sqrt{\left(\dfrac{10.24}{15}\right)(.24)}} = \frac{-1.5}{.40} = -3.75.$$

$t_{.005}(15\ df) = 2.95$ (one-tailed test). Thus, $p < 0.005$; therefore, reject the null hypothesis and conclude that problems can be worked more quickly on the new brand of calculator than on the old. A one-tailed test is appropriate here because the only interest is whether the new calculations are better than the ones on hand. (See pp. 171–172 and Problem 23 in Chapter 8.)

17. **a.** $s_D = \sqrt{\dfrac{\sum D^2 - \dfrac{(\sum D)^2}{N}}{N - 1}}.$

s_D is the standard deviation of the distribution of differences between correlated scores.

b. $D = X - Y.$ D is the difference between two correlated scores.

c. $s_{\bar{D}} = s_D / \sqrt{N}.$ $s_{\bar{D}}$ is the standard error of the difference between means for a correlated set of scores.

d. t is the name of a theoretical probability distribution. So far in this chapter you have used it to determine the probability that two samples came from populations with the same mean. As you will see, t has other uses, too.

e. $\bar{Y} = \sum Y/N$. $\bar{Y}$ is the mean of a set of scores that is correlated with another set.

18.

	X	Y	D	D^2
	16	18	−2	4
	10	11	−1	1
	17	19	−2	4
	4	6	−2	4
	9	10	−1	1
	12	14	−2	4
$\sum$	68	78	−10	18
mean	11.3	13.0		

$$s_D = \sqrt{\frac{\sum D^2 - \frac{(\sum D)^2}{N}}{N-1}} = \sqrt{\frac{18 - \frac{(-10)^2}{6}}{5}} = \sqrt{.2667} = .516.$$

$$s_{\bar{D}} = \frac{s_D}{\sqrt{N}} = \frac{.516}{\sqrt{6}} = .211.$$

$$t = \frac{11.333 - 13.000}{.211} = \frac{-1.667}{.211} = -7.899.$$

$t_{.001}$ (5 df) = 6.86. Therefore, $p < .001$. Notice that the small difference between means (1.67) is highly significant even though the data consist of only six pairs of scores. This illustrates the power that a large correlation can have in reducing the standard error. The conclusion reached by the experimenters was that frustration (produced by seeing others treated better) leads to aggression.

19. The problem here is that you cannot use the direct-difference method, because you do not have the raw data to work with. This leaves you with the definition formula, $s_{\bar{D}} = \sqrt{s_{\bar{X}}^2 + s_{\bar{Y}}^2 - 2r_{XY}(s_{\bar{X}})(s_{\bar{Y}})}$. This, of course, requires a correlation coefficient, but you have the data necessary to calculate r. (The initial clue, for many students, is the term $\sum XY$.) Thus, $\bar{X} = 14.77$ and $\bar{Y} = 12.08$.

$$s_{\bar{X}} = \frac{3.59}{\sqrt{13}} = 1.00. \quad s_{\bar{Y}} = \frac{2.72}{\sqrt{13}} = .75. \quad r = .565.$$

$$t = \frac{\bar{X} - \bar{Y}}{\sqrt{s_{\bar{X}}^2 + s_{\bar{Y}}^2 - 2r_{XY}(s_{\bar{X}})(s_{\bar{Y}})}}$$

$$= \frac{14.77 - 12.08}{\sqrt{1.00 + .56 - (2)(.57)(1.00)(.75)}} = \frac{2.69}{\sqrt{.71}} = 3.19.$$

$t_{.05}$(12 df) = 2.18. Therefore, reject the null hypothesis and conclude that this behavior-modification program significantly reduced the anxiety-behavior scores.

20. This is a correlated-samples study.

$$\sum D = -.19. \quad \sum D^2 = .0101. \quad N = 11. \quad \bar{X} - \bar{Y} = \frac{\sum D}{N} = -.0172.$$

$$s_D = \sqrt{\dfrac{\sum D^2 - \dfrac{(\sum D)^2}{N}}{N-1}} = \sqrt{.00068} = .026.$$

$$s_{\bar{D}} = \dfrac{.026}{\sqrt{11}} = .0078.$$

$$t = \dfrac{-.0172}{.0078} = -2.21.$$

$t_{.05}$ (10 df) = 2.23. The obtained $|t|$ does not reach statistical significance. (This is a case in which a small sample led to a Type II error. On the basis of larger samples, auditory RT has been found to be faster than visual RT—approximately .14 seconds for auditory RT and .18 seconds for visual RT, using practiced subjects.)

21. We are most hopeful that you did not treat these data as correlated samples. If you did, you exhibited a response bias that led you astray. If you treated these data as the independent samples they are, you are on your (mental) toes. So much for our role as dispensers of verbal reinforcement; here is the data analysis.

Recency	Primacy
$\sum X_1 = 86.$	$\sum X_2 = 104.$
$\sum X_1^2 = 934.$	$\sum X_2^2 = 1312.$
$N_1 = 9.$	$N_2 = 9.$

$$s_{\bar{X}_1 - \bar{X}_2} = \sqrt{\dfrac{\sum X_1^2 - \dfrac{(\sum X_1)^2}{N} + \sum X_2^2 - \dfrac{(\sum X_2)^2}{N}}{N(N-1)}} = 1.75.$$

$$t = \dfrac{\bar{X}_1 - \bar{X}_2}{s_{\bar{X}_1 - \bar{X}_2}} = \dfrac{9.56 - 11.56}{1.75} = 1.14.$$

$t_{.05}$ (16 df) = 2.12. Therefore, retain the null hypothesis.

22. $LL = (\bar{X}_1 - \bar{X}_2) - t(s_{\bar{X}_1 - \bar{X}_2}) = 1.5 - (2.13)(.04) = 1.41.$
     $UL = (\bar{X}_1 - \bar{X}_2) + t(s_{\bar{X}_1 - \bar{X}_2}) = 1.5 + (2.13)(.04) = 1.59.$
23. The decision now is to buy the old brand. The difference in time per problem—between 1.41 and 1.59 minutes—is not enough to justify the higher price.
24. Buy the new machine, without a doubt (or at least with only a very tiny doubt).
25. $LL = (\bar{X} - \bar{Y}) - t(s_{\bar{D}}) = -.0172 - (3.17)(.0078) = -.0419.$
     $UL = (\bar{X} - \bar{Y}) + t(s_{\bar{D}}) = -.0172 + (3.17)(.0078) = .0075.$
     Notice that, since the interval includes .00, the null hypothesis cannot be rejected at the .01 level.
26. $\sum D = -15; \sum D^2 = 41.$

$$s_D = \sqrt{\dfrac{\sum D^2 - \dfrac{(\sum D)^2}{N}}{N-1}} = \sqrt{\dfrac{41 - \dfrac{(-15)^2}{8}}{7}} = 1.36.$$

$$s_{\bar{D}} = \dfrac{s_D}{\sqrt{N}} = \dfrac{1.36}{\sqrt{8}} = .48.$$

$t_{.10}$ (7 df) = 1.895.

$LL = (\bar{X} - \bar{Y}) - t(s_{\bar{D}}) = -1.88 - (1.895)(.48) = -2.79.$
$UL = (\bar{X} - \bar{Y}) + t(s_{\bar{D}}) = -1.88 + (1.895)(.48) = -.97.$

27. With 90 percent confidence, you can state that the real (parametric) effect of sleep is to reduce the number of CVC's forgotten from a 10-item list by .97 to 2.79 items.

28. The formulas for the limits are

$$LL = \bar{X} - t(s_{\bar{X}}) \text{ and } UL = \bar{X} + t(s_{\bar{X}}).$$

With several confidence intervals to calculate, it is convenient to concentrate on the $-t(s_{\bar{X}})$ and the $+t(s_{\bar{X}})$ parts of the formulas for the limits. In the table below, these two parts have been combined into $\pm t(s_{\bar{X}})$. In the graph below, the values for $\pm t(s_{\bar{X}})$ were simply added to the mean for the trial. For example, on Trial 1, the confidence interval extends from approximately 19 to 23, as seen on the Y axis in the figure. $t_{.05}(16 \ df) = 2.12$.

| | Trials | | | | | |
	1	2	3	4	5	6
$\bar{X}$	21	27	30	25	30	36
s	3.3	2.7	3.0	3.6	5.1	4.5
N	17	17	17	17	17	17
$s_{\bar{X}}$	.80	.65	.73	.87	1.24	1.09
$\pm(t)(s_{\bar{X}})$	±1.70	±1.39	±1.54	±1.85	±2.62	±2.31

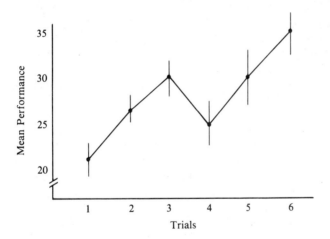

We think you will agree: the confidence intervals display the reliability of the dip in the curve.

29. Assumption 3. The scores are not independent. When one person contributes two or more scores to one sample, those scores are not independent, and Assumption 3 has been violated.

30. $t = (r)\sqrt{\dfrac{N-2}{1-r^2}}.$

    a. $t = 2.24$; $df = 8$. Since $t_{.05}(8 \ df) = 2.31$, $r = .62$ is not significantly different from .00.

    b. $t = -2.12$; $df = 120$. Since $t_{.05}(120 \ df) = 1.98$, $r = -.19$ is significantly different from .00.

    c. $t = 2.08$; $df = 13$. $t_{.05}(13 \ df) = 2.16$. Retain the null hypothesis.

    d. $t = -2.85$; $df = 62$. $t_{.05}(60 \ df) = 2.00$. Reject the null hypothesis.

31. See Table A and read the last paragraph of the chapter.

Chapter 10

1. *Example 2:* The independent variable is the amount of reinforcement; there are four levels. The dependent variable is the rate of response. The null hypothesis is that the mean rate is the same for all four populations—that amount of reward has no effect on rate of responding. *Example 3:* The independent variable is the method of teaching, and there are three such methods being tested. The dependent variable must be some measure of performance in Spanish. The null hypothesis is that all three methods have the same mean result. *Example 4:* The independent variable is the species of honeybee. There are five levels in this experiment. The dependent variable is the kilograms of honey produced. The null hypothesis is that all five species are equally productive.

2. These are correlated-samples designs, and the ANOVA technique described here is appropriate only for independent samples. ANOVA can be used on correlated-sample data using techniques described in Winer (1971, Chapter 4), and Guilford & Fruchter (1978, Chapter 13).

3. larger

4. One estimate is obtained from variability among the sample means, and another estimate is obtained by pooling the sample variances.

5. F is a ratio of the two estimates of the population variance.

6. **a.** Reject the null hypothesis and conclude that different social classes have different attitudes toward religion.

   b. Retain the null hypothesis that the rate of response is the same for 10, 20, 40, and 80 grams of reinforcement.

   c. Reject the null hypothesis. The three methods are not equally effective in teaching Spanish to fourth graders.

   d. There are no significant differences among the five species in the amount of honey produced; that is, the differences among the species can be attributed to chance.

7.

	X_1	X_2	X_3	X_4
$\sum X$	28	26	25	25
$\sum X^2$	176	162	151	147
N	5	5	5	5

$$SS_{tot} = 636 - \frac{104^2}{20} = 636 - 540.80 = 95.20.$$

$$SS_{bg} = \frac{28^2}{5} + \frac{26^2}{5} + \frac{25^2}{5} + \frac{25^2}{5} - \frac{104^2}{20}$$

$$= 156.80 + 135.20 + 125.00 + 125.00 - 540.80$$

$$= 542.00 - 540.80 = 1.20.$$

$$SS_{wg} = \left(176 - \frac{28^2}{5}\right) + \left(162 - \frac{26^2}{5}\right) + \left(151 - \frac{25^2}{5}\right) + \left(147 - \frac{25^2}{5}\right)$$

$$= 19.20 + 26.80 + 26.00 + 22.00 = 94.00.$$

Check: $95.20 = 1.20 + 94.00.$

8.

	X_1	X_2	X_3
$\sum X$	23	24	20
$\sum X^2$	181	152	118
N	3	5	4

$$SS_{tot} = 451 - \frac{67^2}{12} = 451 - 374.08 = 76.92.$$

$$SS_{bg} = \frac{23^2}{3} + \frac{24^2}{5} + \frac{20^2}{4} - \frac{67^2}{12}$$

$$= 176.33 + 115.20 + 100.00 - 374.08 = 17.45.$$

$$SS_{wg} = \left(181 - \frac{23^2}{3}\right) + \left(152 - \frac{24^2}{5}\right) + \left(118 - \frac{20^2}{4}\right)$$

$$= 4.67 + 36.80 + 18.00 = 59.47.$$

Check: $76.92 = 59.47 + 17.45$. We hope you included this check on your calculations when you worked the problem.

9.

	X_1	X_2	X_3
$\sum X$	137	88	82
$\sum X^2$	2015	836	746
$\bar{X}$	13.70	8.80	8.20
N	10	10	10

$$SS_{tot} = 3597 - \frac{307^2}{30} = 3597 - 3141.63 = 455.37.$$

$$SS_{methods} = \frac{137^2}{10} + \frac{88^2}{10} + \frac{82^2}{10} - \frac{307^2}{30}$$

$$= \frac{18769}{10} + \frac{7744}{10} + \frac{6724}{10} - 3141.63 = 182.07.$$

$$SS_{wg} = \left(2015 - \frac{137^2}{10}\right) + \left(836 - \frac{88^2}{10}\right) + \left(746 - \frac{82^2}{10}\right)$$

$$= 138.10 + 61.60 + 73.60 = 273.30.$$

Check: $455.37 = 182.07 + 273.30$.

10. Data from Problem 7.

Source	df	SS	MS	F	p
Between Groups	3	1.20	.40	.07	>.05
Within Groups	16	94.00	5.88		
Total	19	95.20			

$F_{.05}$ (3,16 df) = 3.24. Since .07 does not reach this critical value, retain the null hypothesis and conclude that these four groups could have come from the same population.

11. Data from Problem 8. For df, use

$$df_{tot} = N_{tot} - 1 = 12 - 1 = 11.$$
$$df_{bg} = K - 1 = 3 - 1 = 2.$$
$$df_{wg} = N_{tot} - K = 12 - 3 = 9.$$

Source	df	SS	MS	F	p
Between Groups	2	17.45	8.73	1.32	>.05
Within Groups	9	59.47	6.61		
Total	11	76.92			

$F_{.05}(2,9\ df) = 4.26$. Since $1.32 < 4.26$, retain the null hypothesis and conclude that there is no strong evidence that the three samples were drawn from different populations.

12. Data from Problem 9. The independent variable here is methods, which is reflected in the source column.

Source	df	SS	MS	F	p
Between Methods	2	182.07	91.04	9.00	<.01
Within Groups	27	273.30	10.12		
Total	29	455.37			

$F_{.01}\ (2,27\ df) = 5.49$. Since $9.00 > 5.49$, reject the null hypothesis at the .01 level and conclude that the teaching method had an effect on the number of errors on a comprehensive examination. By the end of this chapter, you will be able to make comparisons among the individual means to determine which method is best (or worst).

13. Six groups. The critical values of F are based on 5,70 df and are 2.35 and 3.29 at the .05 and .01 levels, respectively. If $\alpha = .01$, retain the null hypothesis; if $\alpha = .05$, reject the null hypothesis.

14. *Set B:* $(0) + (1) + (-1) = 0; (-1) + (-1) + (2) = 0$ and
$(0)(2) + (1)(-1) + (-1)(-1) = 0$.
Set C: $(0) + (1) + (-1) = 0; (2) + (-1) + (-1) = 0$ and
$(1)(-1) + (0)(2) + (-1)(-1) = 0$.

15. *Set B:*   First two comparisons: $(1)(0) + (0)(1) + (-1)(0) + (0)(-1) = 0$.
Second two comparisons: $(0)(1) + (1)(-1) + (0)(1) + (-1)(-1) = 0$.
First and third comparisons: $(1)(1) + (0)(-1) + (-1)(1) + (0)(-1) = 0$.
Set C: First two comparisons: $(3)(0) + (-1)(0) + (-1)(1) + (-1)(-1) = 0$.
Second two comparisons: $(0)(0) + (0)(2) + (1)(-1) + (-1)(-1) = 0$.
First and third comparisons: $(3)(0) + (-1)(2) + (-1)(-1) + (-1)(-1) = 0$.

16. $(1)(0) + (-1)(0) + (0)(1) + (0)(-1) = 0$.

17. $K - 1 = 3 - 1 = 2$

18. a.

	Group		
	1	2	3
1 vs. 2	1	−1	0
1 and 2 vs. 3	1	1	−2
	$(1) + (-1) + (0) = 0$		

b.   1 vs. 2 $H_0: \mu_1 - \mu_2 = 0$.

1 and 2 vs. 3 $H_0: \dfrac{\mu_1 + \mu_2}{2} - \mu_3 = 0$.

c. $t = \dfrac{c_1\bar{X}_1 + c_2\bar{X}_2}{\sqrt{MS_{wg}\left[\dfrac{(c_1)^2}{N_1} + \dfrac{(c_2)^2}{N_2}\right]}} = \dfrac{(1)(13.70) + (-1)(8.80)}{\sqrt{10.12\left[\dfrac{(1)^2}{10} + \dfrac{(-1)^2}{10}\right]}}$

$= \dfrac{4.90}{1.4227} = 3.4416$.

$$t = \frac{c_1 \bar{X}_1 + c_2 \bar{X}_2 + c_3 \bar{X}_3}{\sqrt{MS_{wg}\left[\dfrac{(c_1)^2}{N_1} + \dfrac{(c_2)^2}{N_2} + \dfrac{(c_3)^2}{N_3}\right]}}$$

$$= \frac{(1)(13.70) + (1)(8.80) + (-2)(8.20)}{\sqrt{10.12\left[\dfrac{(1)^2}{10} + \dfrac{(1)^2}{10} + \dfrac{(-2)^2}{10}\right]}}$$

$$= \frac{-6.1}{2.464} = -2.48.$$

$t_{.05}\ (df = 27) = 2.052.$

d. Both null hypotheses are rejected. Fewer errors were made by students who had the lab and lecture than by students who had only the lecture. The students who had individualized instruction made fewer errors than the average of the other two groups.

e. $$F'_{ob} = \frac{[(1)(8.80) + (-1)(8.20)]^2}{10.12\left[\dfrac{(1)^2}{10} + \dfrac{(-1)^2}{10}\right]} = \frac{.36}{2.0240} = <1.00.$$

There is no significant difference in the number of errors made by those instructed by the lecture plus lab method and those instructed by individualized instruction.

19. a. $K - 1 = 3 - 1 = 2$

b. The primary interest was in the difference between the two drugs in common use, A and B, and the new drug, C. The experimenter therefore would plan to compare Drug C with the average of the other two. The only other comparison that meets orthogonality requirements is a comparison of Drugs A and B. Since this is meaningful and important, it would be computed.

The null hypotheses for these comparisons are

$$H_0: \frac{\mu_A + \mu_B}{2} - \mu_C = 0.$$

$$H_0: \mu_A - \mu_B = 0.$$

	Drug A	Drug B	Drug C	Σ
c. 1 and 2 vs. 3	-1	-1	2	0
1 vs. 2	1	-1	0	0
Products	-1	1	0	0

d. For the hypothesis $\dfrac{\mu_A + \mu_B}{2} - \mu_C = 0.$

$$t = \frac{c_1 \bar{X}_1 + c_2 \bar{X}_2 + c_3 \bar{X}_3}{\sqrt{MS_{wg}\left[\dfrac{(c_1)^2}{N_1} + \dfrac{(c_2)^2}{N_2} + \dfrac{(c_3)^2}{N_3}\right]}} = \frac{(-1)(7.25) + (-1)(6.75) + (2)(2.25)}{\sqrt{2.69\left[\dfrac{(-1)^2}{4} + \dfrac{(-1)^2}{4} + \dfrac{(2)^2}{4}\right]}}$$

$$= \frac{-9.50}{2.009} = -4.73.$$

$df = N - K = 12 - 3 = 9.$ $t_{.01}\ (9\ df) = 3.250.$

Conclude that Drug C reduced psychotic episodes compared to the average effect of Drugs A and B.

For the hypothesis $H_0: \mu_A - \mu_B = 0$

$$t = \frac{c_1 \bar{X}_1 + c_2 \bar{X}_2}{\sqrt{MS_{wg}\left[\dfrac{(c_1)^2}{N_1} + \dfrac{(c_2)^2}{N_2}\right]}} = \frac{(1)(7.25) + (-1)(6.75)}{\sqrt{2.69\left[\dfrac{(1)^2}{4} + \dfrac{(-1)^2}{4}\right]}}$$

$$= \frac{.50}{1.1597} = <1.00.$$

Absolute values of t less than 1.00 are never significant. There is no evidence of a difference between the effectiveness of Drugs A and B.

e. $$F'_{ob} = \frac{(c_2 \bar{X}_2 + c_3 \bar{X}_3)^2}{MS_{wg}\left[\dfrac{(c_2)^2}{N_2} + \dfrac{(c_3)^2}{N_3}\right]}$$

$$= \frac{[(1)(6.75) + (-1)(2.25)]^2}{2.69\left[\dfrac{(1)^2}{4} + \dfrac{(-1)^2}{4}\right]} = \frac{20.25}{1.3450} = 15.0558.$$

$F'_{.05}(2,9) = (K-1)F_{.05}(2,9) = (2)(4.26) = 8.52.$

Since $15.06 > 8.52$, reject the hypothesis that the scores for Drug B and Drug C were drawn from the same population of scores. Drug C reduced psychotic episodes more than Drug B.

20. 1. normally distributed dependent variable   2. homogeneity of variance   3. random sampling.

21. $$F'_{ob} = \frac{(c_1 \bar{X}_1 + c_2 \bar{X}_2 + c_3 \bar{X}_3)^2}{MS_{wg}\left[\dfrac{(c_1)^2}{N_1} + \dfrac{(c_2)^2}{N_2} + \dfrac{(c_3)^2}{N_3}\right]}$$

$$= \frac{[(2)(7.6667) + (-1)(4.80) + (-1)(5)]^2}{6.61\left[\dfrac{(2)^2}{3} + \dfrac{(-1)^2}{5} + \dfrac{(-1)^2}{4}\right]}$$

$$= \frac{30.6185}{11.7876} = 2.59.$$

The critical value of $F'_{.05}$ for 2,9 df is $(2)(4.26) = 8.52$. Therefore, retain the null hypothesis.

22. a.

Source	df	SS	MS	F	p
Between Injections	2	235.64	117.82	7.98	<.05
Within Groups	23	339.40	14.76		
Total	25	575.04			

$F_{.05}(2,23\ df) = 3.42.$ Conclude that the effect of injections was seen seven days later in the number of trials necessary to relearn the avoidance task.

b. $$t = \frac{c_1 \bar{X}_1 + c_2 \bar{X}_2 + c_3 \bar{X}_3}{\sqrt{MS_{wg}\left[\dfrac{(c_1)^2}{N_1} + \dfrac{(c_2)^2}{N_2} + \dfrac{(c_3)^2}{N_3}\right]}} = \frac{(2)(12.10) + (-1)(17.25) + (-1)(19.00)}{\sqrt{14.76\left[\dfrac{(2)^2}{10} + \dfrac{(-1)^2}{8} + \dfrac{(-1)^2}{8}\right]}}$$

$$= \frac{-12.05}{3.0974} = 3.8904.$$

$df = N - K = 26 - 3 = 23.$ $t_{.05} = 2.069.$

Conclude that strychnine significantly improved memory of the first training. (If the word *improved* caught you by surprise, it was probably because you did not calculate the means for each group; or, if you did, you did not recognize that the fewer trials to criterion, the better the memory.)

23. a.

Source	df	SS	MS	F	p
Between Delays	3	160.90	53.63	3.15	<.05
Within Groups	36	613.00	17.03		
Total	39	773.90			

$F_{.05}$ (3,36 *df*) = 2.86. Therefore, reject the null hypothesis and conclude that the amount of time the strychnine is delayed makes a difference in later relearning.

b. After examining the means, we thought the difference between 30 and 60 minutes was the most interesting.

$$F'_{ob} = \frac{[(1)(13.1) + (-1)(16.8)]^2}{17.03\left[\frac{(1)^2}{10} + \frac{(-1)^2}{10}\right]} = \frac{13.69}{3.406} = 4.019.$$

The critical value of $F'_{.05}$ = (3)(2.86) = 8.58. Therefore, conclude that there is no significant difference between a delay of 30 minutes and one of 60 minutes. (Our choice as the second most interesting difference was the mean of 0 and 30 versus the mean of 60 and 120.)

24.

Source	df	SS	MS	F	p
Between Varieties	2	44.40	22.20	12.13	<.01
Within Groups	12	22.00	1.83		
Total	14	66.40			

$F_{.05}$(2,12 *df*) = 3.88. For varieties 1 and 2,

$$F'_{ob} = \frac{[(1)(7) + (-1)(11.2)]^2}{1.83\left[\frac{(1)^2}{5} + \frac{(-1)^2}{5}\right]} = \frac{17.64}{.732} = 24.10.$$

The critical value of F' for $\alpha = .05$, *df* = 2,12 is (2)(3.88) = 7.76. Therefore, conclude that Variety 2 produces significantly more soybeans on this type of soil than does Variety 1. For Varieties 2 and 3,

$$F'_{ob} = \frac{[(1)(11.2) + (-1)(9.4)]^2}{1.83\left[\frac{(1)^2}{5} + \frac{(-1)^2}{5}\right]} = \frac{3.24}{.732} = 4.426.$$

The critical value of $F'_{.05}$ is 7.76. Therefore, conclude there is no statistical difference between Varieties 2 and 3.

Chapter 11

1. a. 3×2 **b.** 4×3 **c.** 2×2 **d.** 4×3×3. This is a factorial design in which there are *three* independent variables. This text does not cover the analysis of three-way ANOVAs. If you are interested in such complex designs, see Footnote 1 in Chapter 11.

2. A main effect is the effect that the different levels of one factor have on the dependent variable.

3. An interaction occurs when the effect of one factor depends on the level of another factor.

4. main effect

5. main effect

6.   **I. a.**

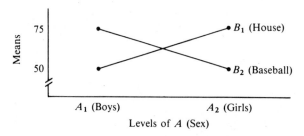

Levels of A (Sex)

  b. There appears to be an interaction.

  c. Attitude toward playing the games depends on sex. Boys would rather play baseball and girls would rather play house.

  d. There appear to be no main effects. Means for games are house = 62.5 and baseball = 62.5. Means for sex are boys = 62.5 and girls = 62.5.

  e. Later in this chapter, you will learn that when there is a significant interaction, the interpretation of main effects depends on the interaction. Though all four means are exactly the same, there is a difference in attitudes toward the two games, but this difference depends on the sex of the child.

   II. a.   -

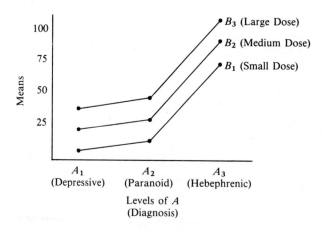

Levels of A
(Diagnosis)

  b. There appears to be no interaction.

  c. The effect of the drug dosage does not depend on the diagnosis of the patient.

  d. There appear to be main effects for both factors.

  e. For all schizophrenics, the larger the dose, the higher the "happiness score" (Factor B, main effect). Hebephrenics have higher "happiness" scores than the others at all dose levels (Factor A, main effect).

III. a.

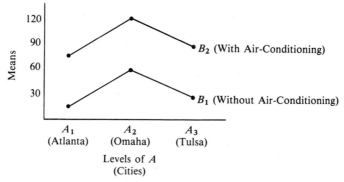

b. There appears to be no interaction.

c. Whether people buy cars with or without air conditioning does not depend upon which of the three cities they live in.

d. There appear to be main effects for both factors.

e. It appears that General Motors sells more cars in Omaha than in Tulsa or Atlanta (Factor A). More cars are sold with than without air conditioning (Factor B).

IV. a.

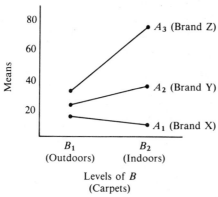

b. There appears to be an interaction.

c. The effect of placing carpet indoors or outdoors depends on the brand. Brand X seems to last a little longer outdoors than indoors, but the reverse is true for Brand Y. Brand Z lasts considerably longer indoors.

d, e. Again, when the interaction is significant, main effects must be interpreted in the light of the interaction effect. Clearly, the effect of placing the carpet indoors or outdoors causes differences in carpet life, but these differences depend on the brand. The brands of carpet differ in life, but these differences depend on whether the carpet is indoors or outdoors.

V. a. See graph on p. 382.

b. There appears to be no interaction. Though the lines in the graph are not parallel—they even cross—the departure from parallel is slight and probably should be attributed to sampling variation.

c. Sex differences in attitudes toward lowering taxes on investments do not depend upon socioeconomic status.

d. There appears to be a main effect for socioeconomic status but not for sex.

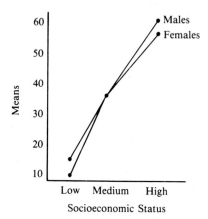

e. Males and females do not differ in their attitudes toward lowering taxes on investments (Factor A). The higher the socioeconomic status, the more positive are attitudes toward lowering taxes on investments.

7. a. The number of scores in each group (cell) must be the same.
   b. The samples in each cell must be independent.
   c. The levels of both factors must be chosen by the experimenter and not left to chance.
   d. The samples are drawn from normally distributed populations.
   e. The variances of the populations are equal.
   f. The samples are drawn randomly from the populations.

8. a. 2×3 b. diets and sex

   c. $SS_{tot} = 553 - \dfrac{95^2}{18} = 553 - 501.39 = 51.61.$

   $$SS_{bg} = \frac{(17)^2}{3} + \frac{(11)^2}{3} + \frac{(24)^2}{3} + \frac{(14)^2}{3} + \frac{(17)^2}{3} + \frac{(12)^2}{3} - \frac{(95)^2}{18}$$

   $$= 538.33 - 501.39$$

   $$= 36.94.$$

   $$SS_{diets} = \frac{(31)^2}{6} + \frac{(28)^2}{6} + \frac{(36)^2}{6} - \frac{(95)^2}{18} = 506.83 - 501.39 = 5.44.$$

   $$SS_{sex} = \frac{(52)^2}{9} + \frac{(43)^2}{9} - \frac{(95)^2}{18} = 505.89 - 501.39 = 4.50.$$

   $$\begin{aligned}
   SS_{AB} = 3[&(5.6667 - 5.1667 - 5.7778 + 5.2778)^2 \\
   &+ (4.6667 - 5.1667 - 4.7778 + 5.2778)^2 \\
   &+ (3.6667 - 4.6667 - 5.7778 + 5.2778)^2 \\
   &+ (5.6667 - 4.6667 - 4.7778 + 5.2778)^2 \\
   &+ (8.00 - 6.00 - 5.7778 + 5.2778)^2 \\
   &+ (4.00 - 6.00 - 4.7778 + 5.2778)^2] = 27.00.
   \end{aligned}$$

   Check: $SS_{AB} = 36.94 - 5.44 - 4.50 = 27.00.$

   $$SS_{wg} = \left[101 - \frac{(17)^2}{3} \right] + \left[41 - \frac{(11)^2}{3} \right] + \left[194 - \frac{(24)^2}{3} \right]$$

   $$+ \left[68 - \frac{(14)^2}{3} \right] + \left[99 - \frac{(17)^2}{3} \right] + \left[50 - \frac{(12)^2}{3} \right]$$

   $$= 4.67 + .67 + 2.00 + 2.67 + 2.67 + 2.00 = 14.68.$$

   Check: $51.61 = 36.94 + 14.67.$ (The difference between answers using the two methods is due to rounding.)

9. **a.** 2×3 **b.** teacher response and sex

   c. errors on a comprehensive examination in arithmetic

   d. $SS_{tot} = 13281 - \dfrac{845^2}{60} = 1380.58.$

$$SS_{bg} = \frac{(169)^2}{10} + \frac{(136)^2}{10} + \frac{(118)^2}{10} + \frac{(146)^2}{10} + \frac{(167)^2}{10} + \frac{(109)^2}{10} - \frac{(845)^2}{60} = 306.28.$$

$$SS_{response} = \frac{(315)^2}{20} + \frac{(303)^2}{20} + \frac{(227)^2}{20} - \frac{(845)^2}{60} = 227.73.$$

$$SS_{sex} = \frac{(423)^2}{30} + \frac{(422)^2}{30} - \frac{(845)^2}{60} = 0.02.$$

$$\begin{aligned}
SS_{AB} = 10[&(16.90 - 15.75 - 14.10 + 14.0833)^2 \\
+ &(14.60 - 15.75 - 14.0667 + 14.0833)^2 \\
+ &(13.60 - 15.15 - 14.10 + 14.0833)^2 \\
+ &(16.70 - 15.15 - 14.0667 + 14.0833)^2 \\
+ &(11.80 - 11.35 - 14.10 + 14.0833)^2 \\
+ &(10.90 - 11.35 - 14.0667 + 14.0833)^2] \\
= &\ 78.53.
\end{aligned}$$

Check: $SS_{AB} = 306.28 - 227.73 - .02 = 78.53.$

$$SS_{wg} = \left[3129 - \frac{(169)^2}{10}\right] + \left[2042 - \frac{(136)^2}{10}\right] + \left[1620 - \frac{(118)^2}{10}\right]$$
$$+ \left[2244 - \frac{(146)^2}{10}\right] + \left[2955 - \frac{(167)^2}{10}\right] + \left[1291 - \frac{(109)^2}{10}\right] = 1074.30.$$

Check: $1380.58 = 306.28 + 1074.30.$

10. $SS_{tot} = 1405 - \dfrac{145^2}{20} = 353.75.$

$$SS_{bg} = \frac{25^2}{5} + \frac{57^2}{5} + \frac{43^2}{5} + \frac{20^2}{5} - \frac{145^2}{20} = 173.35.$$

$$SS_A = \frac{68^2}{10} + \frac{77^2}{10} - \frac{145^2}{20} = 4.05.$$

$$SS_B = \frac{82^2}{10} + \frac{63^2}{10} - \frac{145^2}{20} = 18.05.$$

$$\begin{aligned}
SS_{AB} = 5[&(5.00 - 6.80 - 8.20 + 7.25)^2 + (8.60 - 6.80 - 6.30 + 7.25)^2 \\
+ &(11.40 - 7.70 - 8.20 + 7.25)^2 + (4.00 - 7.70 - 6.30 + 7.25)^2] \\
= &\ 151.25.
\end{aligned}$$

Check: $SS_{AB} = 173.35 - 4.05 - 18.05 = 151.25.$

$$SS_{wg} = \left[155 - \frac{25^2}{5}\right] + \left[750 - \frac{57^2}{5}\right] + \left[400 - \frac{43^2}{5}\right] + \left[100 - \frac{20^2}{5}\right] = 180.40.$$

Check: $353.75 = 173.35 + 180.40.$

11. There is not an equal number of scores in each cell.

12.

Source	df	SS	MS	F	p
A (diets)	2	5.44	2.72	2.23	>.05
B (sex)	1	4.50	4.50	3.69	>.05
AB	2	27.00	13.50	11.07	<.01
Within Groups	12	14.67	1.22		
Total	17	51.61			

$F_{.05}\ (2,12\ df) = 3.88.\ F_{.01}\ (2,12\ df) = 6.93.\ F_{.05}\ (1,12\ df) = 4.75.$

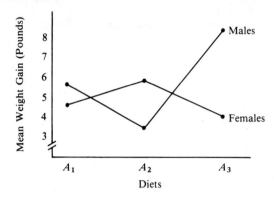

Interpretation: The null hypothesis $\mu_{A_1} = \mu_{A_2} = \mu_{A_3}$ is retained. There are no significant differences among the means of the puppies receiving different diets.

The null hypothesis $\mu_{B_1} = \mu_{B_2}$ is retained. Mean weight gain of male and female puppies did not differ significantly.

There was a significant interaction between diet and sex. Males and females were affected differently by the three diets. Whereas Diet 2 was best for female puppies, Diet 3 was best for males.

13.

Source	df	SS	MS	F	p
A (response)	2	227.73	113.87	5.72	<.01
B (sex)	1	.02	.02	<1	>.05
AB	2	78.53	39.27	1.97	>.05
Within Groups	54	1074.30	19.89		
Total	59	1380.58			

$F_{.01}$ (2,50 *df*) = 5.06. $F_{.05}$ (2,50 *df*) = 3.18.

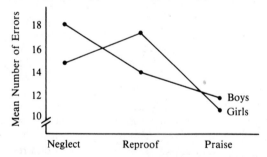

Interpretation: Only the teacher-response variable produced results that were significantly different from chance ($p < .01$). The null hypothesis $\mu_{neglect} = \mu_{reproof} = \mu_{praise}$ is rejected. The three methods of response produced different numbers of errors. Neither the sex main effect nor the interaction was statistically significant.

14.

Source	df	SS	MS	F	p
A	1	4.05	4.05	<1	>.05
B	1	18.05	18.05	1.60	>.05
AB	1	151.25	151.25	13.41	<.01
Within Groups	16	180.40	11.28		
Total	19	353.75			

$F_{.05}$ (1,16 *df*) = 4.49. $F_{.01}$ (1,16 *df*) = 8.53.

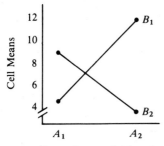

15. **a.** Independent variables: 1. treatment the subject received, 2. whether the experimenter who administered the humor test had treated the subject or not. Dependent variable: number of captions produced. $p(A) < .01$, $p(B) < .05$, $p(AB) < .05$.

The significant interaction means that the effect of the experimenter *depended* upon whether the subject had been insulted or had been treated neutrally. If the subject had been insulted, then taking the humor test from the person who had done the insulting caused a larger number of captions to be written—larger, that is, than if a different experimenter had administered the test. If the subject had not been insulted, there was little difference between the two kinds of experimenters.

The main effect of treatments was significant. On the average, more captions were produced when subjects had been insulted than when they had been treated neutrally. However, this average effect appears to be due primarily to those taking the test from the same experimenter who did the insulting (Cell $A_1 B_1$).

The main effect of the experimenters was significant, with more captions being produced when the experimenter who administered the test was the same as the one who administered the treatment. Again, this main effect seems to be heavily influenced by Cell $A_1 B_1$.

b. Independent variables: 1. name, 2. occupation of father. Dependent variable: score assigned to the theme. $p(A) > .05$, $p(B) > .05$, $p(AB) < .01$.

The significant interaction means that the effect of the name depends upon the occupation of the father. The effect of the name David is good if his father is a research chemist (Cell $A_1 B_1$) and bad if his father is unemployed (Cell $A_1 B_2$). The effect of the name Elmer is good if his father is unemployed (Cell $A_2 B_2$) and bad if his father is a research chemist (Cell $A_2 B_1$).

The main effects are not significant. The average "David" theme is about equal to the average "Elmer" theme. Similarly, the themes do not differ significantly according to whether the father was unemployed or a research chemist.

Notice that these main effects, if interpreted by themselves, would be misleading. That is, *names do have an effect*, but the effect depends upon the occupation of the father. Main effects, however, are only sensitive to average differences.

c. Independent variables: 1. sex of speaker, 2. job experience of speaker. Dependent variable: number of students looking at the speaker. $p(A) < .05$, $p(B) < .01$, $p(AB) > .05$.

Since the interaction is not significant, the interpretations of the main effects are straightforward. The sex of the speaker was significant; students paid more attention to males than to females. The experience of the speaker was significant; students paid more attention to those who had been on the job for more than two years than to those who had been on the job less than six months.

16. Neither an orthogonal comparisons test or a Scheffé test is appropriate because the interaction in that study was significant.

17.

Source	df	SS	MS	F	p
A (rewards)	3	199.31	66.44	3.90	<.05
B (classes)	1	2.25	2.25	<1	>.05
AB	3	.13	.04	<1	>.05
Within Groups	56	953.25	17.02		
Total	63	1154.94			

Up to this point the interpretation is that there is a significant difference among the rewards; that is, rewards have an effect on attitudes toward the police. Freshmen and seniors do not differ significantly and there is no significant interaction between rewards and class standing.

The suggested *a priori* test is appropriate since the interaction is not significant.

$$t = \frac{(1)(12.13) + (-1)(13.56)}{\sqrt{17.02\left[\frac{(1)^2}{16} + \frac{(-1)^2}{16}\right]}} = -.98$$

$t_{.05}(60) = 2.00$. You may conclude that there is no significant difference in attitudes between those who received 50 cents and those who received one dollar.

For the *a posteriori* test, a Scheffé test, the F value is

$$F'_{ob} = \frac{[(2)(8.94) - (+1)(12.13) + (-1)(13.56)]^2}{17.02\left[\frac{(2)^2}{16} + \frac{(-1)^2}{16} + \frac{(-1)^2}{16}\right]} = 9.56.$$

$F'_{.05}$ is $(K-1)(F[3,55\ df]) = 3(2.78) = 8.34$. Thus, the attitude toward police of those who received a ten-dollar reward is significantly lower than that of those who received a 50-cent or one-dollar reward.

The overall results of this experiment show that the less you pay a person for expressing opinions contrary to his own, the more the person will change his attitudes to fit those that he expressed.

18. $SS_{tot} = 1936.53 - \frac{(336.90)^2}{60} = 44.8365.$

$$SS_{bg} = \frac{(60.30)^2}{10} + \frac{(49.70)^2}{10} + \frac{(56.40)^2}{10} + \frac{(55.60)^2}{10}$$
$$+ \frac{(58.60)^2}{10} + \frac{(56.60)^2}{10} - \frac{(336.90)^2}{60} = 6.4015.$$

$$SS_{grade\,level} = \frac{(115.90)^2}{20} + \frac{(108)^2}{20} + \frac{(113)^2}{20} - \frac{(336.90)^2}{60} = 1.5970.$$

$$SS_{college} = \frac{(166.40)^2}{30} + \frac{(170.50)^2}{30} - \frac{(336.90)^2}{60} = .2802.$$

$SS_{AB} = 10[(6.03 - 5.7950 - 5.5467 + 5.6150)^2$
$\quad + (4.97 - 5.40 - 5.5467 + 5.6150)^2$
$\quad + (5.64 - 5.65 - 5.5467 + 5.6150)^2$
$\quad + (5.56 - 5.7950 - 5.6833 + 5.6150)^2$
$\quad + (5.83 - 5.40 - 5.6833 + 5.6150)^2$
$\quad + (5.66 - 5.65 - 5.6833 + 5.6150)^2]$
$\quad = 4.5243.$

Check: $SS_{AB} = 6.4015 - 1.5970 - .2802 = 4.5243.$

$$SS_{wg} = \left[368.69 - \frac{(60.30)^2}{10}\right] + \left[257.43 - \frac{(49.70)^2}{10}\right] + \left[323.02 - \frac{(56.40)^2}{10}\right]$$

$$+ \left[317.50 - \frac{(55.60)^2}{10}\right] + \left[344.13 - \frac{(58.30)^2}{10}\right] + \left[325.76 - \frac{(56.60)^2}{10}\right]$$

$$= 38.4350.$$

Check: $44.8365 = 6.4014 + 38.4350.$

Source	df	SS	MS	F	p
Grade Level	2	1.5970	.7985	1.12	>.05
College	1	.2802	.2802	0.39	>.05
AB	2	4.5243	2.2622	3.18	=.05
Within	54	38.4350	.7118		
Total	59	44.8365			

Only the interaction effect is significant. With $df = 2,50$, an F of 3.18 is required for significance at $\alpha = .05$, and this one is 3.18. The nonsignificant main effects, then, must be interpreted in light of the significant interaction. In a more advanced course, you would analyze for simple effects, but for our purpose, just graph the means.

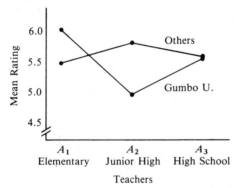

Gumbo U. is doing a supérior job of preparing elementary teachers. They are doing as well as others in their preparation of high school teachers. At the junior high level, however, they are in trouble. Their graduates are clearly rated as less satisfactory in teaching competency than are graduates of the other programs.

Chapter 12

1. $\chi^2 = \Sigma\left[\frac{(O - E)^2}{E}\right] = \frac{(316 - 300)^2}{300} + \frac{(84 - 100)^2}{100}$

    $= .8533 + 2.5600$

    $= 3.4153. \qquad df = 1.$

    $\chi^2_{.05}$ (1 df) = 3.84. Retain the null hypothesis and conclude that these data are consistent with a $3:1$ hypothesis.

2. The expected frequencies are 20 correct and 40 incorrect. $(1/3)(60) = 20$ and $(2/3)(60) = 40.$

    $$\chi^2 = \frac{(32 - 20)^2}{20} + \frac{(28 - 40)^2}{40} = 10.80, \qquad df = 1, p < .01.$$

Thus, these three emotions can be distinguished from each other. Subjects did not respond in a chance fashion.

There is an interesting sequel to this experiment. Subsequent researchers did not find the simple, clearcut results that Watson reported. One experimenter (Sherman, 1927) found that if only the infant's reactions were observed, there was a great deal of disagreement. However, if the observers also know the stimuli (dropping, stroking, and so forth) they agreed with each other. This seems to be a case in which Watson's design did not permit him to separate the effect of the infant's reaction (the independent variable) from the effect of knowing what caused the reaction (an extraneous variable).

3. The contingency table follows, with the expected frequencies in parentheses.

Evidence

		Yes	No	Σ
Species	A	13(14.70)	38(36.30)	51
	B	19(17.30)	41(42.70)	60
	Σ	32	79	111

Calculations of the expected frequencies are as follows.

$$\frac{(51)(32)}{111} = 14.70. \qquad \frac{(51)(79)}{111} = 36.30.$$

$$\frac{(60)(32)}{111} = 17.30. \qquad \frac{(60)(79)}{111} = 42.70.$$

O	E	$O - E$	$(O - E)^2$	$\dfrac{(O - E)^2}{E}$
13	14.70	−1.70	2.8900	.1966
38	36.30	1.70	2.8900	.0796
19	17.30	1.70	2.8900	.1671
41	42.70	−1.70	2.8900	.0677

$$\chi^2 = .5110.$$
$$df = 1.$$

$\chi^2_{.05}$ (1 df) = 3.84. Therefore, retain the null hypothesis and conclude that the two species seem equally subject to infestation; that is, infestation and species appear to be independent (unrelated).

$$\chi^2 = \frac{111[(13)(41) - (38)(19)]^2}{(51)(60)(32)(79)} = .513.$$

4. $$\chi^2 = \frac{100[(9)(34) - (26)(31)]^2}{(35)(65)(40)(60)} = 4.579.$$

Calculations of the expected frequencies are as follows.

$$\frac{(35)(40)}{100} = 14. \qquad \frac{(35)(60)}{100} = 21.$$

$$\frac{(65)(40)}{100} = 26. \qquad \frac{(65)(60)}{100} = 39.$$

O	E	$O - E$	$(O - E)^2$	$\dfrac{(O - E)^2}{E}$
9	14	−5	25	1.786
26	21	5	25	1.191
31	26	5	25	.962
34	39	−5	25	.641

$$\chi^2 = \overline{4.58.}$$
$$df = 1.$$

$\chi^2_{.05}$ (1 df) = 3.84. Reject the hypothesis that group size and joining are independent. Conclude that passersby are more likely to join a group of five than a group of two.

5.

	Recapture Site	
	Issaquah	East Fork
Capture Site — Issaquah	46	0
East Fork	8	19

$$\chi^2 = \frac{73[(46)(19) - (8)(0)]^2}{(46)(27)(54)(19)} = 43.76.$$

$\chi^2_{.001}$ (1 df) = 10.83.

Reject the null hypothesis (which is that the second choice is independent of the first) and conclude that choices are very consistent; salmon tend to choose the same stream each time.

6. $$\chi^2 = \frac{70[(39)(3) - (16)(12)]^2}{(51)(19)(55)(15)} = .49.$$

$\chi^2_{.05}$ (1 df) = 3.84.

Hasler's hypothesis is confirmed; fish with plugged nasal openings do not make a consistent choice of streams, but those that get olfactory cues consistently choose the same stream.

7. To get the expected frequencies, multiply the percentages given by Professor Stickler by the 340 students. Then enter these expected frequencies in the usual table.

O	E	$O - E$	$(O - E)^2$	$\dfrac{(O - E)^2}{E}$
20	23.8	−3.8	14.44	.61
74	81.6	−7.6	57.76	.71
120	129.2	−9.2	84.64	.66
88	81.6	6.4	40.96	.50
38	23.8	14.2	201.64	8.47

$$\chi^2 = \overline{10.95.}$$
$$df = 5 - 1 = 4.$$

(In this problem, the only restriction on the theoretical frequencies is that $\sum E = \sum O$.) $\chi^2_{.05}$ (4 df) = 9.49. Therefore, reject the contention of the professor that his grades conform to "the curve." By examining the data, you can also reject the contention of the colleague that the Professor is too soft. The primary reason the data do not fit the curve is that there were too many "flunks."

8. goodness of fit

9. Each of the six sides of a die is equally likely. Since the total number of throws was 1200, the expected value for any one of them is 1/6 times 1200 or 200. Thus,

O	E	$O - E$	$(O - E)^2$	$\dfrac{(O - E)^2}{E}$
195	200	−5	25	.13
200	200	0	0	.00
220	200	20	400	2.00
215	200	15	225	1.13
190	200	−10	100	.50
180	200	−20	400	2.00
$\Sigma\,1200$	1200			$\chi^2 = 5.76$

$\chi^2_{.05}(5) = 11.07.$

Thus, retain the null hypothesis. The results of the evening do not differ significantly from the "unbiased dice" model.

10. Goodness of fit.

11.

Younger Students		Older Students		Faculty	
O	E	O	E	O	E
29	23.54	6	11.64	11	10.82
31	26.10	10	12.90	10	12.00
20	23.03	13	11.38	12	10.59
7	14.33	14	7.08	7	6.59

$\chi^2 = 17.26$
$df = 3 \times 2 = 6$
$\chi^2_{.01}(6) = 16.81$

Thus, reject the null hypothesis at the .01 level. A person's opinion is not independent of group membership. By examining the separate results of each $\dfrac{(O - E)^2}{E}$, you can see that the principal contributions to the final χ^2 value were from the opinion "Intervene with Military Force." The younger and older students differed on this with the older students being pro and the younger students being con. Our political science students might conclude that age appears to be an important variable in predicting opinions.

12. Independence.

13. Expected frequencies:

$\left(\dfrac{1}{4}\right)(130) = 32.5; \left(\dfrac{3}{4}\right)(130) = 97.5.$

O	E	$O - E$	$(O - E)^2$	$\dfrac{(O - E)^2}{E}$
23	32.5	−9.5	90.25	2.777
107	97.5	9.5	90.25	.9256

$\chi^2 = 3.703.$
$df = 2 - 1 = 1.$

$\chi_{.01}{}^2$ (1 df) = 6.64. Thus, retain the null hypothesis and conclude that these data are consistent with the 1:3 ratio predicted by the theory.

14. The difference between the tests for goodness of fit and for independence is in how the expected frequencies are obtained. In the goodness-of-fit test, expected frequencies are predicted by a theory, whereas in the independence test they are obtained from the data.

15. You should be gentle with your friend but explain that the observations in his study are not independent and cannot be analyzed with χ^2.

   Explain that, since one person is making five observations, the observations are not independent; the choice of one female candidate may cause the subject to pick a male candidate next (or vice versa).

16.

	Candidates		
Signaled Turn	Hill	Dale	
Yes	57	31	88
No	11	2	13
	68	33	101

$$\chi^2 = \frac{101[(57)(2) - (31)(11)]^2}{(88)(13)(68)(33)} = 2.03$$

$\chi_{.01}{}^2$ (1 df) = 6.64. "With respect to failure to signal a left turn, there was not a statistically significant difference between cars with stickers for Hill and cars with stickers for Dale."

17. We hope you would say something like, "Gee, I've been able to analyze data like these since I learned the t test. I will need to know the standard deviations for those means, though." If you attempted to analyze these data using χ^2, you erred (which, according to Alexander Pope, is human). If you erred, forgive yourself and reread pages 262 and 276.

18.

		Houses		
		Brick	Frame	Σ
Candidates	Hill	17	59	76
	Dale	88	51	139
	Σ	105	110	215

$$\chi^2 = \frac{N(AD - BC)^2}{(A + B)(C + D)(A + C)(B + D)}$$

$$= \frac{215(867 - 5192)^2}{(76)(139)(105)(110)} = 32.96. \quad df = 1.$$

$\chi^2_{.001}$ (1 df) = 10.83. $p < .001$.

"215 houses with yard signs were classified as brick (affluent) or one-story frame (less affluent). Hill's signs were more often found at frame houses, and Dale's signs were more often found at brick houses. This difference was so statistically significant that it is not likely that it was due to chance."

19. $df = (R - 1)(C - 1)$, except for a table with only one row.
  a. 3 **b.** 12 **c.** 3 **d.** 10

Chapter 13

1. Sum of the ranks of those given estrogen = 76
   Sum of ranks of the control animals = 44

   Smaller $U = (7)(8) + \dfrac{(7)(8)}{2} - 76 = 8$

   Upon looking in Table G under $N_1 = 7$, $N_2 = 8$, for a two-tailed test with $\alpha = .05$, you see that a U value of 10 or less is required to reject the null hypothesis. Since the obtained $U = 8$, you should reject the null hypothesis. By examining the data and noting that the estrogen-injected animals were the lowest ranking ones, you can conclude that estrogen causes rats to be less dominant.

2. The sums of the ranks are 574.5 and 245.5. The smaller U is 109.5 (the larger is 274.5).

   $$U = (24)(16) + \dfrac{(24)(25)}{2} - 574.5 = 109.5.$$

   $$\mu_U = \dfrac{(24)(16)}{2} = 192.$$

   $$\sigma_U = \sqrt{\dfrac{(24)(16)(41)}{12}} = 36.22.$$

   $$z = \dfrac{(109.5 + .5) - 192}{36.22} = -2.26.$$

   With such a z value you may reject the null hypothesis that the distributions are the same. By examining the average ranks (23.9 for present-day birds and 15.4 for ten-year-ago birds) you should conclude that present-day birds have significantly fewer brain parasites. (If you gave the 0 birds a rank of 2, the average ranks are 17.1 for present-day birds and 25.6 for ten-year-ago birds. The same conclusion is reached, however.)

3. "With $\alpha = .01$, there are *no* possible results that would allow H_0 to be rejected. We must find more cars."

4. $\sum R_Y = 13, \sum R_Z = 32.$

   Smaller $U = (4)(5) + \dfrac{(5)(6)}{2} - 32 = 3$

   A U of 1 or less is required to reject H_0 (two-tailed test). Since the obtained $U = 3$, you must conclude that the quietness test did not produce evidence that Y cars are quieter than Z cars.

5. $\mu_T = \dfrac{N(N+1)}{4} = \dfrac{112(113)}{4} = 3164.$

   $$\sigma_T = \sqrt{\dfrac{N(N+1)(2N+1)}{24}} = \sqrt{\dfrac{112(113)(225)}{24}} = 344.46.$$

   $$z = \dfrac{(4077 + .5) - 3164}{344.46} = 2.65.$$

   Since 2.65 is greater than 1.96, reject H_0 and conclude that incomes were significantly different after the program. Because you do not have the actual data, you cannot tell whether the incomes were higher or lower than before.

Worker	Without Rests	With Rests	D	Signed Ranks
1	2240	2421	181	2
2	2069	2260	191	4
3	2132	2333	201	5
4	2095	2314	219	6
5	2162	2297	135	1
6	2203	2389	186	3

Check: $21 + 0 = 21$ $\sum$ positive = 21

$$\frac{(6)(7)}{2} = 21$$ $\sum$ negative = 0

$T = 0$

$N = 6$

The critical value of T at the .05 level for a two-tailed test is 1. Since $0 < 1$, reject the null hypothesis and conclude that the output with rests is greater than the output without rests. Here is the story behind this study.

From 1927 to 1932 the Western Electric Company conducted a study on a group of six workers at their Hawthorne Works near Chicago. The final conclusion reached by this study has come to be called the "Hawthorne Effect." In the study, six workers who assembled telephone relays were separated from the rest of the workers. A variety of changes in their daily routine followed, one at a time, with the following results. Five-minute rest periods morning and afternoon increased output, ten-minute rest periods increased output, company-provided snacks during rest periods increased output, and quitting 30 minutes early increased output. Going back to the original no-rest schedule increased output again, as did the reintroduction of the rest periods. Finally, the management concluded that it was the special attention paid to the workers, rather than the rest periods, that increased output.

Thus, the Hawthorne Effect is an improvement in performance that is due just to being in an experiment (getting special attention) rather than to any specific manipulation in the experiment. For a summary of this study, see Mayo (1946).

7. The appropriate test for this study is one for two independent samples. A Mann-Whitney U test will work. If high scores are given high rank (that is, 39 ranks 1):

$\sum$ (Canadians) = 208
$\sum$ (U.S.) = 257

(If low scores are given high ranks, the sums are reversed.)

$$U = (15)(15) + \frac{(15)(16)}{2} - 257 = 88.$$

By consulting the second page of Table G (bold-face type) you will find that a U value of 64 is required in order to reject the null hypothesis ($\alpha = .05$, two-tailed test). The lower of the two calculated U values for this problem is 88. (The U value using the Canadian sum is 137.) Since 88 is larger than the tabled value of 64, the null hypothesis must be retained. Thus, our political science student must conclude that he has no evidence that Canadians and people from the U.S. have different attitudes toward the regulation of business.

Once again you must be cautious in writing the conclusion when the null hypothesis is retained. You have not demonstrated that the two groups are the same; you have only shown that the groups were not significantly different, which is a rather unsatisfactory way to leave things.

8.

Student	Before	After	D	Signed Rank
1	18	4	14	13
2	14	14	0	1.5
3	20	10	10	9
4	6	9	−3	−4
5	15	10	5	6
6	17	5	12	11
7	29	16	13	12
8	5	4	1	3
9	8	8	0	−1.5
10	10	4	6	7
11	26	15	11	10
12	17	9	8	8
13	14	10	4	5
14	12	12	0	eliminated

Check: $85.5 + 5.5 = 91$

$$\frac{(13)(14)}{2} = 91$$

$\sum \text{positive} = 85.5$

$\sum \text{negative} = -5.5$

$T = 5.5$

$N = 13$

Since the tabled value for T for a two-tailed test with $\alpha = .01$ is 9, you may reject H_0 and conclude that the after distribution is from a different population than the before distribution. Since, except for one person, the number of misconceptions stayed the same or decreased, you may conclude that the course *reduced* the number of misconceptions.

We would like to remind you here of the distinction we made in Chapter 8 between statistically significant and important. There is a statistically significant drop in the number of misconceptions, but a professor might be quite dismayed at the number of misconceptions that remain.

9. $\sum \text{positive} = 81.5$ Check: $81.5 + 54.5 = 136$
     $\sum \text{negative} = -54.5$ $\dfrac{(16)(17)}{2} = 136$
          $T = 54.5$
          $N = 16$

A $T \leq 29$ is required at the .05 level. Thus, there is no significant difference in the weight ten months later. Put in the most positive language, "there is no evidence that the participants tended to gain back the weight lost during the workshop."

10.   Arrange the data into the usual summary table.

		Groups				
		1	2	3	4	5
Groups	2	85				
	3	31	54			
	4	18	67	13		
	5	103	188*	134	121	
	6	31	116	62	49	72

*$p < .05$.

For $K = 6$, $N = 8$, differences of 159.6 and 188.4 are required to reject the null hypothesis at the .05 and .01 levels, respectively. Thus, the only significant difference is between the means of Groups 2 and 5 at the .05 level. (With data such as these, the novice investigator may be tempted to report "almost significant at the .01 level." Resist that temptation.)

11.

Authoritarian		Democratic		Laissez-Faire	
X	Rank	X	Rank	X	Rank
77	5	90	10.5	50	1
86	9	92	12	62	2
90	10.5	100	16	69	3
97	13	105	21	76	4
100	16	107	22	79	6
102	19	108	23	82	7
120	27.5	110	24	84	8
121	29	118	26	99	14
128	32	125	30	100	16
130	33	131	34	101	18
135	37	132	35.5	103	20
137	38	132	35.5	114	25
141	39.5	146	41	120	27.5
147	42	156	44	126	31
153	43	161	45	141	39.5
Σ	$\overline{393.5}$		$\overline{419.5}$		$\overline{222.0}$

	Authoritarian (393.5)	Democratic (419.5)	Laissez-Faire (222)
Democratic (419.5)	26		
Laissez-Faire (222)	171.5*	197.5*	

*$p < .05$.

For $K = 3$, $N = 15$, a difference in the sum of the ranks of 168.8 is required to reject H_0 at the .05 level. Therefore, both the authoritarian leadership and the democratic leadership resulted in higher personal satisfaction scores than did laissez-faire leadership, but the authoritarian and democratic types of leadership did not significantly differ from each other.

12. The Mann-Whitney U test can be performed on ranked data from two independent samples with equal or unequal N's. The Wilcoxon matched-pairs signed-ranks test can be performed on ranked data from two correlated samples. The Wilcoxon-Wilcox multiple-comparisons test makes all possible comparisons among K independent equal-N groups of ranked data.

13.

Year	Rank in Marriages	Rank in Grain
1930	1	3
1931	2	1
1927	3	4
1925	4	2
1926	5	5
1923	6	6
1922	7	7
1919	8.5	11
1924	8.5	9
1928	10	9
1929	11	9
1918	12	12.5
1921	13	12.5
1920	14	14

This is a case of tied ranks so we will compute a Pearson r on the ranks.

$$r = \frac{N \sum XY - (\sum X)(\sum Y)}{\sqrt{[N \sum X^2 - (\sum X)^2][N \sum Y^2 - (\sum Y)^2]}}$$

$$= \frac{14(1002.5) - (105)(105)}{\sqrt{[(14)(1014.5) - (105)^2][(14)(1012.5) - (105)^2]}}$$

$$= .95$$

Consulting Table A for the significance of correlation coefficients, we find that for 12 df, a coefficient of .78 is significant at the .001 level. Our sentence of explanation would be: "There is a highly significant relationship between the number of marriages and the value of the grain crop; as one goes up the other does too."

14.

$$r_s = 1 - \frac{6 \sum D^2}{N(N^2 - 1)} = 1 - \frac{6(308)}{16(16^2 - 1)} = .55.$$

From Table A, a coefficient of .497 is required for significance. Thus, an $r_s = .55$ is significantly different from .00.

15.

Candidates	Locke	Kant	D	D^2
A	7	8	−1	1
B	10	10	0	0
C	3	5	−2	4
D	9	9	0	0
E	1	1	0	0
F	8	7	1	1
G	5	3	2	4
H	2	4	−2	4
I	6	6	0	0
J	4	2	2	4
				$\sum D^2 = \overline{18}$

$$r_s = 1 - \frac{6 \sum D^2}{N(N^2 - 1)} = 1 - \frac{6(18)}{10(99)} = 1 - .109 = .891.$$

The two professors seem to be in pretty close agreement on the selection criteria.

16.   Unfortunately, with only four pairs of scores there is no possible way to reject the hypothesis that the population correlation is .00 (see Table K). Advise your friend that more data must be obtained before any inference can be made about the population.

17.   **a.**   Wilcoxon matched-pairs signed-ranks test. This is a before-and-after study.
      b.   Wilcoxon matched-pairs signed-ranks test. Pairs are formed by family.
      c.   Spearman's r_s. The degree of relationship is desired.
      d.   Wilcoxon-Wilcox multiple-comparisons test. The experiment has three independent groups.
      e.   Wilcoxon matched-pairs signed-ranks test. Again, this is a before-and-after design.

18.

	Symbol of statistic	Appropriate for what design?	Calculated statistics must be (larger, smaller) than the tabled statistic to reject H_0
Mann-Whitney	U	two indepndent samples	smaller
Wilcoxon matched-pairs signed-ranks	T	two correlated samples	smaller
Wilcoxon-Wilcox multiple-comparisons	none	more than two independent samples, equal N's	larger

Chapter 14

Set A

1. A mean of a set of means is needed, and to find it you need to know the number of dollars invested in each division.
2. Wilcoxon matched-pairs signed-ranks test. You have good evidence that the population of dependent scores (reaction time) is not normally distributed.
3. The one-way ANOVA technique you covered in this book requires independent groups and these measures are correlated. A repeated-measures ANOVA for four groups would be appropriate.
4. A median is appropriate since the scores are skewed.
5. Assume that these measures are normally distributed and use a normal curve to find the proportion.
6. Either an independent measures t test or Mann-Whitney U test will do.
7. A 95 or 99 percent confidence interval about the mean reading achievement of those 50 sixth graders will answer the question.
8. χ^2 goodness of fit.
9. Mode.
10. Neither will do. The relationship described is nonlinear—one that first increases and then decreases. Neither r nor r_s is appropriate. A statistic, *eta*, is appropriate for this.
11. Correlated-samples t test or Wilcoxon matched-pairs signed-ranks test.
12. A regression equation will provide for the predictions and a correlation coefficient will indicate how accurate the predictions will be.
13. A 2×2 factorial ANOVA.
14. The student's wondering may be translated into a question of whether a correlation of .20 is a statistically significant one. Use a t test to find if an $r = .20$ for that class is significantly different from an $r = .00$. Or, look in Table A in the Appendix.
15. This is a χ^2 problem but it cannot be worked using the techniques in this text because these before-and-after data are correlated, not independent. Intermediate texts describe appropriate techniques for correlated χ^2 problems.
16. Finding the probability that the sample mean came from a population in which $\mu = 1000$ (a t test or a z test) will answer the question.

17. Pearson product-moment correlation coefficient.

18. One-way analysis of variance.

19. A χ^2 test of independence will determine whether a decision to report shoplifting is influenced by sex and by the dress of the shoplifter.

20. A line graph with serial position on the X axis and number of errors on the Y axis will illustrate this relationship.

21. A Mann-Whitney U test would be preferred since the dependent variable seems to be skewed. The question is whether there is a relationship between diet and cancer. (Note that if the two groups differ, two interpretations are possible. For example, if the incidence is higher among the red meat cultures, it might be because of the red meat or because of the lack of cereals.)

22. An r_s will give the degree of relationship for these two ranked variables.

23. For each species of fish, a set of z scores may be calculated. The fish with the largest z score should be declared the overall winner.

Set B

24. The critical values are 377.6 at the .05 level and 457.6 at the .01 level. You can conclude that herbicide B is significantly better than A, C, or D at the .01 level and that D is better than C at the .05 level.

25. The screening process is clearly worthwhile in making each batch profitable. Of course, the cost of screening will have to be taken into account.

26. $t_{.01}(1,40) = 3.55$.
Therefore, reject the null hypothesis and conclude that extended practice *improved* performance. Since this is just the opposite of the theory's prediction, we may conclude that the theory is not adequate to explain the serial position effect.

27. Since $\chi^2_{.05}(2\ df) = 5.99$, the null hypothesis for this goodness-of-fit test is retained. The data do fit the theory; the theory is adequate.

28. The overall F is significant so we can conclude that the differences should not be attributed to chance. The low dose is significantly better than the placebo ($F'_{.01} = 12.22$) and significantly better than the high dose (which has a smaller mean than the placebo).

29. The critical value of U for a two-tailed test with $\alpha = .01$ is 70. You may conclude that the difference between the two methods is statistically significant at the .01 level. You cannot, from the information supplied, tell which of the two methods is superior.

30. $t(16\ df) = 4.58$. $t_{.001}(16) = 4.02$.
Attitudes of college students and people in business toward the 18 groups are similar. There is little likelihood that the similarity of attitudes found in the sample is due to chance.

31. $t_{.05}(60) = 2.00$. $F_{.01}(30,40) = 2.11$.
The poetry unit appears to have no significant effect on mean attitudes toward poetry. There is a very significant effect on the variability of the attitudes of those who studied poetry. It appears that the poetry unit turned some students on and some students off, thus causing a great deal of variability in the attitude scores.

32. $F_{.05}(1,44) = 4.06$. $F_{.01}(1,44) = 7.24$.
Only the interaction is significant. A graph, as always, will help in the interpretation. See the graph on page 399. Although neither of the main effects is statistically significant, both are important. The question of whether to present one or both sides to get the most attitude change depends upon the level of education of the audience. If the members of a group have less than a high school education, present one side. If they have some college education, present both sides.

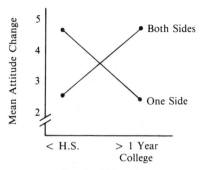

Level of Education

33. Since the tabled value of $F_{.05}(2,60) = 3.15$, the null hypothesis is retained. There is no significant difference in the verbal ability scores of third graders who are left- or right-handed or mixed.

34. Since $\chi^2_{.02}$ (2 df) = 7.82, the null hypothesis for this test of independence is rejected at the .02 level. Alcoholism and toilet training are not independent, they are related. Experiments such as this do not allow you to make cause-and-effect statements, however.

INDEX

Abscissa, 43
Absolute value, 20, 27
Algebra, simple, 21, 28–29
Alpha (α), 168–169
   affected by multiple tests, 206
   effect on rejecting H_0, 174–175
Alternative hypothesis, 156–158,
   171–172
Analysis of variance (ANOVA):
   assumptions of, 228–229
   between-means estimate, 208, 216, 247
   comparison among means, 220–228,
     254–256
   completely randomized design (*see*
     ANOVA one-way)
   degrees of freedom, 216, 247
   factorial design, 232–234
     a posteriori test, 256
     a priori test, 254–255
     compared to one-way ANOVA,
       232–234
     compared to t test, 232
     interaction, 234–238, 242
     interpretation, 252–254
     orthogonal comparisons, 254–255
     R x C notation, 233
     restrictions and limitations, 238
     Scheffe test, 256
     simple effects, 252
   grand mean, 212
   interaction, 234–238, 242
   interpretation, 206–210, 218
   mean square, 212, 216–217, 246
   one-way ANOVA:
     a posteriori test, 221, 226–228

a priori test, 221–226
   compared to factorial design,
     232–234
   null hypothesis, 207–210
   orthogonal comparison, 221–226
   Scheffé test, 221, 226–228
   rationale, 206–211
   summary table, 218, 248
   sum of squares, 211–212, 213–215,
     242–245
   within-means estimate, 208, 216, 247
a posteriori test:
   defined, 221
   factorial design, 256
   one-way ANOVA, 226–228
a priori test:
   defined, 221
   factorial design, 254–255
   one-way ANOVA, 221–226
Assumption-freer tests, 282, (*see also*
     Nonparametric tests)
Assumptions when using:
   chi square distribution, 276
   F distribution, 228–229
   Mann-Whitney U, 280–281
   normal distribution, 138, 159
   Pearson r, 85, 100–102
   t distribution, 200–201
   Wilcoxon T, 280–281
   Wilcoxon-Wilcox multiple
     comparisons, 280–281
Asymptote, 117
Autokinetic effect, 290

Bar graph, 43, 45–46
Barker, R. B., 188
Beaver, Sam, 11
Before and after design, 186
Beta (β), 169
    calculation of, 170
    defined, 169
Biased samples, 135-136
Bimodal distribution, 49–50, 140
Binominal distribution, 114–115
Biometrika, 179, 180
Birch, H. G., 166
Bivariate distribution, 85–86
Bradley, D. R., 273
Bradley, J. V., 201
Bradley, T. D., 273
Brehm, J. W., 257

Campbell, Stephen, 61
Cause and effect, 11–12, 96–97
Calhoun, J. P., 249
Camilli, G., 273
Central Limit Theorem, 138–139
Central Tendency (*see* Central value)
Central value (*see also* Mean, Median,
        Mode):
    comparisons among, 59–61
    definition, 51
Cervone, David, 173
Chi square distribution, 262–263
Chi square test:
    additive nature, 270
    combining categories, 274–276
    contingency table, 264
    distinguished from other tests, 259
    formula, 262
    goodness of fit, 262–265
        degrees of freedom, 264
        null hypothesis, 263
    of independence, 265–267
        degrees of freedom, 266
        null hypothesis, 265
    more than 1 *df,* 268–271
    restrictions, 276
    shortcut for 2 x 2 table, 267–268
    small expected frequencies, 272–276
    when appropriate, 276
    Yates' correction, 273
Clark, R. B., 176
Clark University, 13
Class intervals, conventions for
        establishing, 40–41
Closure study, 248–252

Coefficient of determination, 97–99
Cohen, A. R., 257
Columbia University, 13
Common variance, 98
Confidence interval:
    calculating limits, 147–149
    concept of, 146–147, 150
    about a mean difference, 197–198
    about a sample mean, 147–150
Confidence level, 150
Confounding variable (*see* Extraneous
        variable)
Contingency table, 264
Cornell, F. G., 263
Correlated-samples design:
    confidence interval, 195
    explained, 185–186
    matched pairs, 187
    natural pairs, 186–187
    repeated measures, 187
    t test, 192–195
    Wilcoxon matched-pairs signed-ranks
        test, 288–291
Correlation (*see also* Correlation
        coefficient)
    and causation, 96–97
    concept, 85–90, 296
    curvilinear, 102
    dichotomous variables, 102
    multiple correlation, 102
    negative, 89–90
    partial correlation, 102
    positive, 86–89
    of ranked data, 296–298
    relation to regression, 103
    scatterplot, 86
    zero, 90
Correlation coefficient (*see also*
        Correlation):
    coefficient of determination, 97–99
    other than Pearson or Spearman,
        102–103
    Pearson product moment:
        computation, 92–95
        interpretation, 95, 96–101
        linearity, 100–101
        significance test, 201–203
        z score formula, 91
    practical significance, 100
    Spearman's r_s, 102, 296–298
        calculation, 297
        compared to Pearson r, 281–282
        significance test, 297–298

spuriously low:
  nonlinearity, 100–101
  truncated range, 101–102
  Stanford-Binet and WAIS, 108
Critical region, 162, 164
Critical value, 164
Cutcomb, S. D., 273

Dallenbach, K. M., 199
Darwin, Charles, 85
Darwin, Erasmus, 85
Decimals, 20, 22–23
Degrees of freedom, 180–183
  analysis of variance:
    factorial design, 247
    one-way classification, 216
    orthogonal comparisons, 226
    Scheffé test, 227
  chi square, 264–266
  defined, 184–185
  t distribution, 180, 185
  testing Pearson r, 201
  t test:
    correlated samples, 193
    independent samples, 190
Dembo, T., 188
Dependent-samples design (*see*
  Correlated-samples design)
Dependent variable, 11–13, 16, 154–
  155, 206
Deviation scores, 69–70, 79
Distribution:
  bimodal, 49–50, 140
  binominal, 114–115
  bivariate, 85–86
  empirical (*see* Empirical distribution)
  F, 207–211
  rectangular, 112–113
  skewed, 48–49, 60–61
  theoretical (*see* Theoretical
    distributions)
  theoretical and empirical compared,
    115–116
Distribution-free statistics, 282, (*see also*
  Nonparametric tests)
Downie, N. M., 103, 114

Edwards, A. L., 98, 224, 233
Ellis, W. D., 3
Empirical distribution, 63
  characteristics, 112
  compared to theoretical distributions,
    115–116, 127–128

Error, 139, 168
Estimating answers, 32
Eta (η), 102
Expected frequencies:
  goodness-of-fit test, 263
  test of independence, 265–266
Expected value, 139
Experimental design, 154–156, 271, 368
Experiments:
  how to design, 154–156
  logic of, 154
Exponents, 21, 27
Extraneous variables, 12, 16, 155
Eyeballing, 32, 59

Factor, 232
Factorial design (*see under* Analysis of
  variance)
Family income, college students, 130
F distribution, 207–211, 218
  relation to t, 211
Federalist papers, 4
Ferguson, G. A., 103, 202, 218, 237, 238
Fisher, R. A., 204, 207, 209, 210
Forbs, R., 119
Fractions, 20, 23–24
F ratio, 207, 217, 247
  to compare means, 216, 247
  to compare variances, 218
  to test interaction, 247
Frequency distribution:
  bell-shaped, 48
  class intervals, 39–41
  empirical, 63, 112
  form, 47–50
  grouped, 39
    mean, 57–58
    median, 58–59
    mode, 59
  kinds, 43–46
  open-ended categories, 61
  rules for constructing, 38, 41–42
  simple, 37–38
    mean, 54–55
    median, 55–56
    mode, 56
  skewed, 48–49
Frequency polygon, 43, 44
Frozen pork bellies, 148–149
F test, 217, 218, 232, 247 (*see also*
  Analysis of variance)
Fruchter, B., 101, 170, 175

Galton, Francis, 85, 87, 88, 179–180, 183, 296
Gardner, P. L., 282
Garfield, S., 174
Gauss, C. Friedrich, 116
Gosset, W. S., 179–180, 195, 207, 209
Grackles, 286
Graduating with honors, 61–63
Graphs, 43–50:
  bar graph, 45–46
  describing, 47–50
  frequency polygon, 44–45
  histogram, 45
  line graph, 46–47
  pleasing appearance, 44
  skewed distributions, 48–49, 60–61
Greene, J. E., 207
Grouped frequency distributions (*see under* Frequency distribution)
Guayule, 172
Guilford, J. P., 101, 165, 170, 175
Guinness Company, 179, 180

Hamilton, Alexander, 4
Hamlet, 82
Harris, R. J., 303
Harvard University, 13
Hasler, Arthur D., 268–269
Hawthorne Effect, 393
Heath, R. W., 103, 114
Heights:
  American females, 124
  American males, 124
Histogram, 43, 45
Hobbits, 128
Hopkins, K. D., 273
Hotelling's T^2, 303
Howell, David C., 114
Huff, Darrell, 44, 61
Hunters Story, 240
Hypothesis testing:
  correlation coefficient:
    Pearson r, 201
    Spearman's r_s, 297–298
  differences among means (F), 207–211, 217–218
  differences between distributions:
    correlated samples (T), 288–291
    independent samples, 282–287, 293–295
  difference between two means:
    after ANOVA, 220–228, 254–256
    correlated samples t test, 192–193

    independent samples t test, 188
    independent samples z test, 164–165
  factorial design, 232–238, 247
  frequencies:
    goodness-of-fit (χ^2), 262–264
    independence (χ^2), 265–267
  mean, 141, 180–183

Important versus significant, 162, 173–174
Imprinting, 269
Income:
  college student's family, 130
  truck driver's wages, 120, 127
Independent events, probability of, 113, 266
Independent-samples design, 185–186, 187–188
  F test, (*see* Analysis of variance)
  Mann-Whitney U test, 282–287
  t test, 189
  Wilcoxon-Wilcox multiple comparisons test, 193–195
  z test, 163
Independent variable, 11–13, 16, 155, 206, 232, 233
Inferential statistics, 110, 131, 155
  definition, 2–3
  logic of, 156–158
Interaction:
  calculation of sums of squares, 244
  component of variance, 242
  effect on interpretation of main effects, 236–237, 249–252, 254
  explained, 234–238
  graphically presented, 236, 237, 252
Intercept of a straight line, 104
Interpolation, 55, 122
Interval scale, 10, 59
IQ scores:
  definition of "average", 124
  distribution of, 119, 274
  and sex differences, 165
  Stanford-Binet and Wechsler, 120
  theoretical and empirical, 127–128

Jay, John, 4
J-curves, 49–50
Jenkins, J. G., 199
Johnson, R. A., 303
Johnston, J. O., 119, 133, 142, 160, 249
Jojoba, 172

Keppel, G., 229, 233, 238
Kirk, Roger, E., 13, 210, 224, 229, 233, 237, 238, 252, 261, 282, 285, 298
Kish, L., 135
Knowledge of results study, 218–220, 224–228
Kotki, the brothers, 364
Kruskal-Wallis ANOVA on ranks, 293

Landon, Alf, 136
Lashley, Karl, 167
Level of significance, 162
Levels of a factor, 13, 154–155, 206, 232
Lewin, K., 188, 295
Limits, lower and upper, 7
Linear interpolation, 55, 122
Linearity and correlation, 100–101
Line graph, 46–47
Literary Digest, 136
Loewi, Otto, 175
Loftus, E. F., 114, 176
Loftus, G. R., 114, 176
Lone Ranger, 158

Madison, James, 4
Maertens, N. W., 142
Main effects, 234–238, 242 (*see also* Analysis of variance
Mann-Whitney *U* test, 282–287
    compared to *t* test, 281–282
    for large samples, 285–287
    for small samples, 283–285
Marshall, G., 188
Maslach, C., 188
Matched pairs, 186–187
Maugh, T. H., 172
McGaugh, J. L., 229
McGrath, S. G., 273
McNemar, Q., 98, 101, 103
Mean:
    characteristics of, 53
    compared to median and mode, 59–61
    interpretation of, 52–53
    sampling distribution of mean, 138–139
    sampling distribution of mean differences, 159
    of a set of means, 61–63
    skewed data, 61–62
Mean of a set of means, 61–63
Mean square (*see also* Analysis of variance)

defined, 75, 212
factorial ANOVA, 247
one-way ANOVA, 207–211
    between groups, 216
    within groups, 216
Measurement:
    defined, 6
    scales of, 9–10, 59
Median:
    compared to mean and mode, 59–61
    defined, 53
    grouped frequency distribution, 58–59
    interpretation of, 54
    situations that require, 59–61
    skewed distributions, 60–61
Michigan, University of, 13
Minium, Edward W., 32, 98, 176
Minium's First Law of Statistics, 31
Mischel, H. N., 263
Mode, 54, 56, 59
    compared to mean and median, 59–61
Modus tollens, 156
Moh scale, 16
Multiple correlation, 102
Multivariate statistics, 303

Natrella, M. G., 198
Natural pairs, 186–187
Negative numbers, 20, 25–26
Neoiconophobia, 32
Nominal scale, 9, 59
Nonparametric tests (*see also* Mann-Whitney, Wilcoxon, Wilcoxon-Wilcox, and Spearman's r_s)
    compared to parametric, 281–282
    rationale, 280–281
Normal distribution:
    comparison of empirical and theoretical answers, 127–128
    description, 116–117
    empirical examples, 119–120
    historical, 122
    IQ scores, 119, 120–126
    as "law of error", 116, 182
    relation to *t* distribution, 181
Normal distribution table, how to use, 118
Null hypothesis:
    ANOVA, 206
    chi square, 262–263, 264, 265
    described, 157, 163
    how to reject, 174–176

nonparametric tests, 282
in one-and-two tailed tests, 171–172
to test r, 201
t test, 189

Observed frequency, 262
One-tailed tests, 171–172
One-way ANOVA (*see* Analysis of
    variance, one-way)
Open-ended frequency distributions, 61
Operational definition, 280
Orcs, 128
Ordinal scale, 9–10, 59
Ordinate, 43
Orthogonal coefficients, 222–225, 227,
    254–256
Orthogonal comparisons, 221–226,
    254–255 (*see also* Analysis of
    variance)
Overall, J. E., 273–274
Oxford, 179

Parameter, 6
Parameter estimation, 146, 147
Partial correlation, 102
Pearson, E. S., 180, 262
Pearson, Karl, 85, 91, 262
Pearson product-moment r (*see under*
    Correlation coefficient)
Pennies, weight, 124
Pennsylvania, University of, 13
Percents, 20, 26–27
Petrinovich, L. F., 229
Pine trees:
    distribution of diameters, 119
    mean, 127
Playing cards, 112–113
Pope, Alexander, 391
Population, 154–156
    decisions about, 131, 157
    defined, 5
Power of a test, 174–176, 282
Probability, 112
Proportions, 20, 26–27
Puberty, variability in reaching, 77
"Publius", 4

Qualitative variables, 7–8, 43
Quantitative variables, 6–7, 43
Quételet, Adolphe, 182–183

r (*see* Correlation coefficient)
Rabbit pie, 4

Rabinowitz, H. S., 166
Racial attitudes study, 160–161
Random digits:
    table, 316
    use of, 133–134
Random samples, 154–156
    how to obtain, 132–134
    how often found, 151–152
    and uncertainty, 131
Range, 67
    relation to standard deviation, 74
Rank-order correlation (*see* Correlation
    coefficient, Spearman's r_s)
Ranks:
    sum of, 280, 287
    tests of, 282–295
    ties, 284–285, 291, 298
Rank tests (*see* Nonparametric tests *or*
    specific test)
Ratio scale, 10, 59
Reaction time, 196
Reading, active, 8
Rectangular distribution, 49–50,
    112–113, 139–140
Region of rejection, 162, 164
Regression, 104–108
    coefficents, 104
    and correlation, 103
    equation, 104–105
    Galton's use of term, 87
    line, 86–87, 103, 105–107
    origin of term, 87
    predicting a Y score, 107–109
Regression line, 86–87, 103, 105–108
    intercept, 104–105
    slope, 104–105
Repeated-measures designs, 186–187,
    302 (*see also* Correlated-samples
    design)
Representative samples, 132–135
Response bias, 369
Robustness, 229, 282
Rogers, H., 287
Rokeach, M., 4
Roosevelt, Franklin, 136
Rounding numbers, 30–31
Russian novel, 34

Salmon, 268–269
Sample:
    biased, 135–136, 144
    defined, 5–6
    needed to reject H_0, 175

nonrepresentative, 132
random, 132–134
representative, 131, 132
size and standard error, 141, 143
stratified, 135
Sampling, 155
reducing uncertainty, 145–146
and uncertainty, 131, 145, 168
Sampling distribution:
chi square, 262
described, 137
of difference between means, 159–161
mean, 159
form, 159–160
F distribution, 210
importance of, 130
of the mean, 137–142
empirical, 137–138
and sample size, 143
use of, 140–141
normal curve, 138–139
other than the mean, 150–151
standard error, 139, 151
t distribution, 180–181
use of, 150–151
Scales of measurement, 9–10, 16
and central value, 59
interval scale, 10
nominal scale, 9–10
ordinal scale, 9–10
ratio scale, 10
Scatterplot, 86
Scheffé Test
compared to Wilcoxon-Wilcox, 281–282
factorial design, 256
one-way ANOVA, 221, 226–228
Schumacher, E. F., 300
Self-esteem, Coopersmith inventory, 133
Set, 166, 369
Shakespeare, 82, 356
Sherif, M., 289
Sherman, M., 388
Significance level, 150
Significance versus importance, 173–174
Simple effects, 252
Simple frequency distributions (*see under* Frequency distribution)
Skewed distributions, 48–49
and central value, 60–61
and mean-median relationship, 61
Skiing, water, 66

Snedecor, George W., 210
Snug, Miss Annie, 11
Solar colector study, 294–295
Spatz, K. C., 133, 160
Spearman, Charles, 296
Spearman's r_s (*see under* Correlation coefficient)
Standard deviation:
computation from deviation scores, 70–72
computation from raw scores, 72–74, 76–77
as descriptive index, 68–69
as estimate of σ, 68, 75–76
explanation of $N-1$, 75–76
relation to range, 74
relation to standard error, 139–140, 151
Standard error:
defined, 139, 151
of a difference:
correlated-samples t test, 193
effect on rejecting H_0, 174–176
factors affecting, 175
independent-samples t test, 189
z test, 163
of estimate, 108
of the mean, 138, 142–143
of any statistic, 150–151
Standard score, 79
Statistic, definition, 6
Statistical symbols (*see inside covers*)
Statistician:
categories of, 13–14
Schumacher as, 300
Statistics:
descriptive, 2
and experimental design, 11–13
inferential (*see* Inferential statistics)
Statistics courses, history of, 13
Statistics texts, like Russian novel, 34
Statistics, the essence of, 300
Stevens, S. S., 9
Stratified samples, 135
Strychnine, 229–230
"Student" (*see* Gosset, W. C.)
Subsample, 5
Sum of ranks, 280, 287
Sum of squares, 211–212 (*see also* Analysis of variance)
between groups, 214, 242
interaction, 244
total, 213, 242
within groups, 214, 215, 244

Swampy Acres, 60–61
Symbols, statistical (*see inside covers*)

Tails of distributions, 164, 171–172
Taylor, J., 249
Taylor Manifest Anxiety Scale, 249
t distribution:
  assumptions for using, 200–201
  confidence intervals, 197–198
  and degrees of freedom, 180–181
  explained, 180–183
  for hypothesis testing (*see t test*)
  relation to F, 211
  relation to normal distribution, 181
  uses of, 179
Tennis players, women, 298
Theorem, Central Limit, 138–139
Theoretical distributions:
  characteristics, 112–113
  compared to empirical distributions,
    63, 112, 127–128
  and probability, 113
Ties in rank, 284–285, 291, 298
Treatment, 155
Truck drivers' wages:
  distribution of, 120
  mean, 127
Truncated range, 101
t test:
  assumptions required, 200–201, 229
  compared to nonparametric tests,
    281–282
  mean difference:
    correlated samples, 193
    independent samples, 189, 206, 232
  for r, 201
  for r_s, 297
  sample mean, 181
Two-string problem, 166
Two-tailed tests, 171–172, 182
Two-way ANOVA (*see* Analysis of
  variance, factorial design)
Type I error, 175, 273
  affected by multiple tests, 206
  defined, 168–170
Type II error, 273
  defined, 168–170
  factors affecting, 174–176
  how to avoid, 273

Upper limit, 7, 40
Ury, H., 282

Variable:
  definition, 6
  qualitative, 7–8
  quantitative, 6–7
Variability, 66 (*see also* Range, Standard
  deviation, Variance)
Variance, 75, 78–79
Variance, analysis of (*see* Analysis of
  variance)

Wages, truck drivers'
  distribution of, 120
  mean, 127
Walker, Helen, 13, 21, 184, 185
Water skiing, 66
Watson, John B., 264, 388
Wells, H. G., 4
White, E. B., 11
Whitney, D. R., 283
Wichern, D. W., 303
Widgets, 175–176
Wilcox, R. A., 293
Wilcoxon, F., 282–283, 288, 293
Wilcoxon matched-pairs signed-ranks
  test, 288–293
  compared to t test, 281–282
  for large samples, 291
  for small samples, 288–290
  tied scores, 291
Wilcoxon-Wilcox multiple comparisons
  test, 292–295
Winer, B. J., 224, 229, 233, 238, 252
Woodworth, Robert S., (facing preface)
Work, M. S., 287

Yale University, 13, 257
Yates' correction, 273

Zajonc, R. B., 189
Zeigarnik, Bliuma, 3, 5, 13
Zeigarnik Effect, 3
Zimbardo, P., 188
z score:
  correlation formula, 91
  as descriptive measure, 79–81, 116, 126
  as inferential test:
    large sample T, 291
    large sample U, 285
    mean, 140–142
    mean difference, 164–166
  range of values, 80
  table of, 320–321

Glossary of Symbols

Greek Letter Symbols

α	The probability of a Type I error.
β	The probability of a Type II error.
μ	The mean of a population.
Σ	The sum; an instruction to add.
σ	Standard deviation of a population.
σ^2	Variance of a population.
$\sigma_{\overline{x}}$	Standard error of the mean for a population.
χ^2	The chi square statistic.

Mathematical and Latin Letter Symbols

∞	Infinity.
$>$	More than.
$<$	Less than.
a	Point where the regression line intersects the Y axis.
b	The slope of the regression line.
D	The difference between two correlated scores.
$\overline{D}$	The mean of a set of difference scores.
df	Degrees of freedom.
E	In chi square, the expected frequency.
$E(\overline{X})$	The expected value of the mean; the mean of a sampling distribution.
F	A ratio of the between-means estimate to the within-groups estimate of the population variance; a sampling distribution of such ratios.
F'	Critical value for Scheffé test.
F'_{ob}	Scheffé test value obtained from the data.
f	Frequency; the number of times a score occurs.
H_0	The null hypothesis; states that a difference is equal to some constant, usually zero.
H_1	Any hypothesis alternative to the null hypothesis.
i	The interval size; the number of score points in a class interval.
LL	Lower limit of a confidence interval.
MS	Mean square; ANOVA term for the variance.
N	The number of scores or observations.